CDMA Techniques for Third Generation Mobile Systems

CDMA Techniques for Third Generation Mobile Systems

Edited by
Francis Swarts
Alcatel Research Unit for Wireless Access,
University of Pretoria, South Africa

Pieter van Rooyen
Alcatel Research Unit for Wireless Access,
University of Pretoria, South Africa

Ian Oppermann
Center for Wireless Communication (CWC),
University of Oulu, Finland

Michiel P. Lötter
Alcatel Altech Telecoms, South Africa

Kluwer Academic Publishers
Boston/Dordrecht/London

Distributors for North, Central and South America:
Kluwer Academic Publishers
101 Philip Drive
Assinippi Park
Norwell, Massachusetts 02061 USA
Telephone (781) 871-6600
Fax (781) 871-6528
E-Mail <kluwer@wkap.com>

Distributors for all other countries:
Kluwer Academic Publishers Group
Distribution Centre
Post Office Box 322
3300 AH Dordrecht, THE NETHERLANDS
Telephone 31 78 6392 392
Fax 31 78 6546 474
 E-Mail <orderdept@wkap.nl>

Electronic Services <http://www.wkap.nl>

Library of Congress Cataloging-in-Publication Data

A C.I.P. Catalogue record for this book is available
from the Library of Congress.

Copyright © 1999 by Kluwer Academic Publishers.

All rights reserved. No part of this publication may be reproduced, stored in a retrieval system or transmitted in any form or by any means, mechanical, photo-copying, recording, or otherwise, without the prior written permission of the publisher, Kluwer Academic Publishers, 101 Philip Drive, Assinippi Park, Norwell, Massachusetts 02061

Printed on acid-free paper.

Printed in the United States of America

Contents

List of Figures	ix
List of Tables	xix
Preface	xxi

1
Spreading Techniques, a Far-reaching Technology — 1
P.W. Baier, T. Weber and M. Weckerle

1.1	Introduction	1
1.2	Benefits of Spreading	4
1.3	Information Transmission Systems	6
1.4	Channel Identification Systems	9
1.5	Time and Frequency Estimation Systems	11
1.6	Application Examples	12
1.7	Promising Fields of Further Research	20
1.8	Conclusions	21

References — 21

2
A Linear Model for CDMA Signals Received with Multiple Antennas over Multipath Fading Channels — 23
Lars K. Rasmussen, Paul D. Alexander and Teng J. Lim

2.1	Introduction	24
2.2	CDMA Uplink System Model	25
2.3	Discrete-Time Baseband Uplink Model	27
2.4	Multiple-Antenna	45
2.5	Discussion	49
2.6	Concluding Remarks	53

References — 55

3
Antenna Arrays for Cellular CDMA Systems — 59
P.M. Grant, J.S. Thompson and B. Mulgrew

vi CDMA TECHNIQUES FOR THIRD GENERATION MOBILE SYSTEMS

3.1	Introduction	59
3.2	Background	59
3.3	Direct-Sequence CDMA	60
3.4	Motivations for Using Antenna Arrays	62
3.5	Channel Modelling Considerations	63
3.6	Receiver Algorithms	68
3.7	Uplink Simulation Work	71
3.8	Capacity Improvement with CDMA Antenna Arrays	76
3.9	Downlink Techniques	76
3.10	Summary	78

References   79

4
Spatial Filtering and CDMA   83
Michiel P. Lötter, Pieter van Rooyen and Ryuji Kohno

4.1	Introduction	84
4.2	Smart Antenna Techniques	85
4.3	BER Performance Calculation	88
4.4	Optimum Antenna Spacing Criteria	95
4.5	Mobile Location Distribution	97
4.6	Results	100
4.7	Conclusions	107

References   109

5
Topics in CDMA Multiuser Signal Separation   111
Y. Bar-Ness

5.1	Introduction	111
5.2	Multiuser CDMA Signal Model	113
5.3	Multiuser Decorrelating Detector	117
5.4	Adaptive Multistage Receiver for Multiuser CDMA	128
5.5	Conclusion	140

References   141

6
LMMSE Receivers for DS-CDMA Systems in Frequency-Selective Fading Channels   145
Matti Latva-aho

6.1	Introduction	146
6.2	System Model	147
6.3	LMMSE Receivers in Fading Channels	148
6.4	Bit Error Probability Analysis for the Precombining LMMSE Receiver in Fading Channels	150
6.5	Adaptive Implementations of the Precombining LMMSE Receivers	152
6.6	Delay Acquisition in the Precombining LMMSE Receiver	157
6.7	Delay Tracking in the Precombining LMMSE Receivers	160
6.8	Residual Blind Interference Suppression in PIC Receivers	162
6.9	Numerical Examples	164
6.10	Summary	174

References   176

Contents vii

7
Detection Strategies and Cancellation Schemes in a MC-CDMA System   185
Frans Kleer, Shin Hara and Ramjee Prasad
- 7.1 Introduction   185
- 7.2 MC/CDMA: System Description   187
- 7.3 Multiuser Interference in CDMA Systems   190
- 7.4 Synchronisation in a MC/CDMA System   190
- 7.5 Detection Strategies for MC-CDMA Systems   195
- 7.6 Diversity Combining Techniques   199
- 7.7 Simulations   201
- 7.8 Results   205
- 7.9 Conclusions and Recommendations   207

References   209

8
Coding vs. Spreading over Block Fading Channels   217
Ezio Biglieri, Giuseppe Caire, and Giorgio Taricco
- 8.1 Introduction   217
- 8.2 System model   219
- 8.3 System capacity versus outage probability   222
- 8.4 Receiver design and mutual information   223
- 8.5 Numerical results   228
- 8.6 Conclusions   233

References   237

9
Turbo-Codes for future mobile radio applications   239
Peter Jung, Jörg Plechinger and Markus Doetsch
- 9.1 Introduction   239
- 9.2 Code concatenation   241
- 9.3 Iterative Turbo-Code decoding   244
- 9.4 Designing rate compatible punctured Turbo-Codes   246
- 9.5 Decoding complexity   248
- 9.6 Performance results   249

References   256

10
Software Radio Receivers   257
Tim Hentschel and Gerhard Fettweis
- 10.1 Introduction   257
- 10.2 Software Radio Concept   258
- 10.3 Investigation of Critical Functionalities   264
- 10.4 Summary   280

References   282

11
Blind Space-time Receivers for CDMA Communications   285
David Gesbert, Boon Chong Ng and Arogyaswami J. Paulraj
- 11.1 Introduction   285
- 11.2 Space-time Signal Models   287
- 11.3 Single-user Receivers   289

11.3	Single-user Receivers	289
11.4	Multi-user Receivers	291
11.5	MMSE Receiver Estimation	294
11.6	Summary	299

References 299

Index 303

List of Figures

1.1	Basic system structure	2
1.2	Information transmission	2
1.3	Channel identification	3
1.4	Time and frequency estimation	3
1.5	Two different modes of spreading	4
1.6	Benefits of temporal spreading	5
1.7	Benefits of spectral spreading	6
1.8	Bandwidth luxury by FEC coding	7
1.9	Bandwidth luxury by broadband carrier	8
1.10	Decrease of required normalized SNR with increasing bandwidth luxury $B/B_{\min}$	10
1.11	Classification of channel identification systems	10
1.12	Temporal spreading	12
1.13	Thermal flow metering system	13
1.14	High resolution impulse compression radar	15
1.15	Cellular interference function $f(r)$ (downlink)	17
1.16	Available E_b/N_0 with single user detection	18
1.17	Available E_b/N_0 with multiuser detection	18
1.18	Required and available E_b/N_0 with single user detection	19
1.19	Required and available E_b/N_0 with multiuser detection	20
2.1	Block diagram of the mobile transmitter for user k.	25
2.2	Block diagram of the base station receiver.	26
2.3	Block diagram of the base station front-end.	26
2.4	Block diagram of the base station despreading unit for each user.	27
2.5	Examples of binary spreading sequences. From the top we have $\mathbf{s}_1(0)$, $\mathbf{s}_2(0)$ and $\mathbf{s}_1(1)$. Chips $s_1(3)$, $s_2(4)$ and $s_1(14)$ are specifically labelled.	28

2.6	The filter impulse responses, autocorrelation functions and power spectra for a square pulse and a raised cosine pulse with a roll-off factor of 0.5..	29
2.7	Examples of zero padded and pulse shape filtered spreading sequences, $\mathbf{x}_1(1)$, $\mathbf{x}_2(1)$ and $\mathbf{x}_1(2)$.	30
2.8	Discrete-time model of the uplink communication of a synchronous CDMA system without channel distortions.	31
2.9	The fundamental structure of the matrix $\mathbf{S}$ for a synchronous system with $K = 3$, $L = 2$, $N = 4$, $Q = 2$ and $P = 4$. The entries representing $\hat{\mathbf{s}}_k(i)$ are indicated with dots, while all other entries are zero.	32
2.10	Equivalent matched filters for user k. The top diagram corresponds to the practical approach of one receiver filter while the bottom diagram depicts the approach adopted in the model.	33
2.11	Discrete-time model of the uplink communication of a synchronous CDMA system over a mobile radio channel.	34
2.12	Diagram of the tapped-delay line filtering of $\mathbf{x}_k(i)$ with $\mathbf{c}_k(i)$.	36
2.13	Discrete-time model of the uplink communication of an asynchronous CDMA system over a mobile radio channel.	37
2.14	Fundamental structure of the matrix $\mathbf{S}$ for asynchronous CDMA with $K = 3$, $L = 2$, $N = 4$, $Q = 2$, $P = 4$ and $\tau_1 = 0$, $\tau_2 = 4T_s$ and $\tau_3 = 5T_s$.	38
2.15	An example of the transmitter and receiver mismatched waveforms for a maximum timing error of $T_s/2$ caused by off-sets between the receiver clock and the transmitter clock.	39
2.16	The fundamental structure of the matrix $\mathbf{S}$ for multipath CDMA with $K = 3$, $L = 2$, $N = 4$, $Q = 2$, $P = 4$, $M = 2$ and $\hat{M} = 5T_s$. The delays for the three users are $(0, 5T_s)$, $(3T_s, 4T_s)$ and $(4T_s, 9T_s)$, respectively.	42
2.17	Example of the resolution of paths in a multipath channel using a sliding correlator energy detector based on a square pulse shape and a raised cosine pulse shape, respectively. The actual channel is included for reference.	43
2.18	Example of the severe attenuation due to a destructive combination of multipath delays, coefficients and spreading sequence. The composite signal is practically annihilated.	44
2.19	Illustration of the transmitted signal for one user arriving at B antennas at the receiver over B different multipath channels.	45
2.20	An illustration of the ULA with the direction of arrival θ indicated.	46
2.21	Example of the resulting modulating waveform for user k with two antenna elements in a ULA.	48
2.22	Block diagram of the base station multiuser detection unit. The numbers in parentheses indicate the number of signals passed between blocks in one symbol interval.	49

LIST OF FIGURES

2.23 The fundamental structure of **R** for the multipath case in Fig. 2.16 where $K = 3$ and $L = 6$. The circles on the diagonal represent the autocorrelations. The remaining dots represent non-zero cross correlations. The band-diagonal structure is obvious. ... 50

2.24 Interference power versus desired signal power as a function of the number of active users. ... 51

2.25 The fundamental structure of a powerful iterative receiver for joint detection and decoding. ... 54

3.1 (a) Direct sequence modulation of a data sequence by a PN code, (b) A typical autocorrelation function for a PN code, (c) A typical cross-correlation function for two different users' PN codes. ... 61

3.2 Typical multipath profiles, drawn from the COST-207 report [19], for (a) rural area, (b) typical urban area, (c) bad urban area and (d) hilly terrain. ... 65

3.3 The space-time receiver structure for one user, operating with a uniform linear array in a single $120°$ sector. ... 67

3.4 The beam patterns for the fixed beam, channel estimation and PC-MMSE equaliser techniques, for a two users received at an $M = 4$ element array. The bearings of the desired user's two multipath components are shown as vertical lines. ... 71

3.5 The achieved SINR gain for the fixed beam (best and worst beam sets), bearing estimation and channel estimation algorithms, plotted against the angular width Δ. The array size $M = 8$ and the algorithms estimate the channel from $N = 50$ symbols. ... 72

3.6 The output SINR vs number of users plotted for the channel estimation and PC-MMSE equaliser methods. Two different scenarios are considered: (a) $W = 8$ and background noise -20 dB and (b) $W = 64$ and and background noise -10 dB. ... 73

3.7 The convergence performance of the PC method, plotted as output SINR vs the number of averaged symbols (snapshots) N. Results are shown for (a) array size $M = 4$ and maximum SINRs of 0, 5 and 10 dB (b) maximum SINR of 5dB and array size $M = 2, 4, 8$ or 16. ... 74

3.8 The output SINR convergence performance of the PC method with an $M = 8$ element array and angular widths of $0°, 20°$ and $60°$ ... 74

3.9 The output SINR plotted against number of averaged symbols (snapshots) N for a scenario with array size $M = 4$. The noise variance σ^2 is set to achieve an SINR ignoring CDMA interference of (a) 10 dB with processing gain $W = 64$ and 10 users; (b) 20 dB with processing gain $W = 8$ and 5 users. The maximum SINR and matched filter bounds are shown as horizontal lines. ... 75

3.10	The BER performance of a base station receiver plotted against the number of users in a cell. The antenna array size was $M = 1$ or 4; the receiver uses the single user CE method or the multi-user DFD approach. The processing gain of the system $W = 31$.	75
4.1	Cellular re-use concept.	86
4.2	SDMA system implemented using adaptive antenna arrays.	87
4.3	Space Division Multiple Access (SDMA): Allowing users in the same cell to share time/frequency and code resources.	88
4.4	Basic block diagram of cellular CDMA system.	89
4.5	Optimum antenna spacing for uniformly distributed mobile users in a cell.	96
4.6	Modeling the location of mobiles in a cellular system.	97
4.7	Qualitative description of the PDF of the mobile distributions in a typical urban micro cellular scenario.	98
4.8	Example pdf of the angular distribution of mobiles.	99
4.9	Modeling of scattering elements using a Gaussian approach.	100
4.10	PDF of the AOA of a single user as seen at the base station, caused by scattering elements surrounding the mobile.	100
4.11	Basic concept of adaptive antenna system to limit interference and increase system capacity - users in antenna nulls do not contribute to overall interference levels.	101
4.12	BER performance of cellular system with reference user located at 90°, and interfering users at 90°, 150° and 180° respectively.	102
4.13	BER performance of macro, N-LOS micro cells with uniform and non-uniform user distributions	102
4.14	BER performance of a macro cellular system with $L = 1$ and $L = 3$ multipath components	103
4.15	BER performance of a N-LOS micro cellular system with $L = 1$ and $L = 3$ multipath components	104
4.16	BER performance of a macro cellular system with $M = 1, 3, 5$ and 10 elements with users clustered at 150°.	104
4.17	BER performance of a macro cellular system with $M = 1, 3, 5$ and 10 elements with users clustered at 90°.	105
4.18	BER performance of a macro cellular system with $M = 5$ in the presence of power control errors.	106
4.19	BER performance of a macro cellular system with $M = 5$ and various user location peak widths.	106
4.20	System capacity as a function of the number of antenna element in the array.	107
5.1	One-Shot matched filter, (a) timing, (b) structure.	113
5.2	Multi-shot transformation tutorial.	116
5.3	Transformation of matched filter outputs.	117
5.4	Adaptive Bootstrap Multiuser Separator.	119

5.5	2 user synchronous bootstrap detector.	120
5.6	Theoretical error probability of user 1 as a function of the energy of user 2. $E_b/N_{0_1} = 8$dB. The asymptotes show Verdú's decorrelating detector performance.	120
5.7	Simulation results which compare performance of LMMSE and bootstrap algorithm: $K = 3$, SNR$_1 = 8$dB.	123
5.8	Performance comparison of various detectors in asynchronous multi-user CDMA communication systems. The performance improvement obtained by debiasing processing is shown.	127
5.9	Bootstrap algorithm with soft limiter control.	127
5.10	Adaptive weights convergence for various threshold levels: SNR$_1$ = SNR$_i$ = 8 dB, $i = 2, 3$.	128
5.11	Multistage receiver for multiuser CDMA.	129
5.12	Performance comparison of multistage detector with hard limiter and without hard limiter.	129
5.13	Performance comparison of multistage detector with hard limiter, soft limiter with heuristic threshold and optimal threshold setting.	131
5.14	Multi-stage receiver with adaptive canceler.	132
5.15	Error probability of user 1: $K = 5$ and SNR$_1 = 8$dB.	132
5.16	Error Probability of User 1, for $K = 2$ to $K = 5$.	133
5.17	One-shot asynchronous multistage detector with DC^3 structure.	134
5.18	Decorrelator output data combiner.	135
5.19	Canceler output data combiner.	137
5.20	The effect of data combining on the 2nd user's predecision performance (2 user asynchronous CDMA system).	139
5.21	The effect of data combining following the decorrelator on the performance of user 1 at the output of DC^3 (2-user asynchronous CDMA system).	139
5.22	BER Performance of DC^3 multistage detection structure for 2-user asynchronous CDMA system, SNR$_d$ = 8 dB. (d=1,2) while other user's SNR is variable.	140
5.23	BER Performance of DC^3 multistage detection structure for 3-user asynchronous CDMA system, SNR$_d$ = 8 dB. (d=1,2,3) while other user's SNR are variable.	140
6.1	Multiuser receiver structures.	149
6.2	General block diagram of the adaptive LMMSE-RAKE receiver.	152
6.3	Block diagram of one receiver branch in the adaptive LMMSE-RAKE receiver.	154
6.4	The block diagram of the MV-DLL suitable for the blind LS receivers.	161

6.5 Bit error probabilities (BEP) as a function of the number of users for the conventional RAKE and the precombining LMMSE (LMMSE-RAKE) receivers with exact analysis and Gaussian approximation in a two-path fading channel at vehicle speeds of 40 km/h and average SNR of 20 dB. 165

6.6 Simulated and analytical bit error probabilities as a function of the average signal-to-noise ratio for the precombining LMMSE receiver in two-path fading channels at vehicle speed 40 km/h with different numbers of users. 165

6.7 Bit error probabilities (BEP) as a function of the delay error for the precombining LMMSE receiver with exact analysis and the Gaussian approximation in two path fading channels at vehicle speeds 40 km/h and average SNR of 20 dB. 166

6.8 Bit error probabilities as a function of the average SNR for the conventional RAKE receiver and the precombining LMMSE (LMMSE-RAKE) receiver with different number of users ($K = 1, 2, 4$) in a two-path Rayleigh fading channel with maximum delay spreads of 2 μs. The data modulation is BPSK at a rate of 1.024 Mbit/s (G=4). The energy of all users is the same and no channel coding is assumed. 167

6.9 Bit error probabilities as a function of the near-far ratio for the conventional RAKE and the precombining LMMSE (LMMSE-RAKE) receiver with different spreading factors (G) in a two-path Rayleigh fading channel with maximum delay spreads of 2 μs for G=4, and 7 μs for other spreading factors. The average signal-to-noise ratio is 20 dB for the user of interest. The data modulation is BPSK and the number of users is 2. Data rates vary from 2.048 Mbit/s to 128 kbit/s. No FEC coding is assumed. 168

6.10 Excess mean squared error as a function of number of iterations for different blind adaptive receivers in a two-path fading channel with vehicle speeds of 40 km/h. The number of active users $K = 20$, SNR = 20 dB, $\mu = 10^{-1}$. 170

6.11 Excess mean squared error as a function of number of iterations for different blind adaptive receivers in a two-path fading channel with vehicle speed 40 km/h, the number of active users $K = 20$, SNR = 20 dB, $\mu = 100^{-1}$. 170

6.12 Excess mean squared error as a function of number of iterations for the one-shot blind adaptive LS receiver with different forgetting factors ($1 - 2/N$; forgetting factor $\gamma = 1 - 2/N$) in a 10-user case at a SNR of 20 dB. 171

6.13 Probability of detection for different delay acquisition schemes as a function of the number of users in a two-path Rayleigh fading channel with vehicle speeds of 40 km/h, average SNR of 10 dB, and observation intervals of 200 symbols. 172

6.14	Tracking error variances as a function of the number of users for both the conventional (DLL) and the improved non-coherent DLLs with $2\Delta = 1.0$, 0.75 and 0.50 at the SNR of 20 dB and loop bandwidth $2B_L T = 0.04$.	173
6.15	BER for the HD-PIC and LMMSE-PIC receivers in two-path fading channels at SNR of 15 dB as a function of the unknown user power offset with respect to the known users, $K = 16 + 1$.	174
6.16	BER for the HD-PIC and LMMSE-PIC receivers in two-path fading channels at SNR of 15 dB as a function of the number of users with one unknown user of 10 dB higher power.	174
7.1	MC-CDMA transmitter for the j-th user	188
7.2	MC-CDMA receiver model for user j.	189
7.3	Symbol duration compared to propagation delay.	191
7.4	Signal burst format.	192
7.5	Time Division Duplex (TDD) frame timing structure.	193
7.6	The waiting time t_{wi} changes with different propagation delays t_{pi}	194
7.7	Quasi synchronous state in the uplink.	195
7.8	The single user detection method.	196
7.9	Model of a successive interference cancellation detector.	197
7.10	Model of the MC-CDMA transmitter.	201
7.11	Simulation model for the uplink.	202
7.12	Symbolic representation of the mobile radio channel.	203
7.13	Simulation model of the single user MC-CDMA detector.	204
7.14	Flow diagram of the conventional single user detector	205
7.15	Flow diagram of the (successive interference cancellation) multi-user detector.	206
7.16	Single user detection with ORC in a MC-CDMA downlink.	211
7.17	Single User Detection MRC in a MC-CDMA downlink.	211
7.18	Single user detection with MMSEC in a MC-CDMA downlink.	212
7.19	Single User Detection EGC in a MC-CDMA uplink.	212
7.20	Multi User Detection ORC in a MC-CDMA uplink.	213
7.21	Multi-user Detection with EGC in a MC-CDMA uplink.	213
7.22	Multi-user Detection with MMSEC in a MC-CDMA uplink.	214
7.23	Performance enhancement of Multi User Detection compared to Single User Detection.	214
7.24	Imperfect channel estimation breaks down the performance of a Multi User Detection scheme using MMSEC.	215
8.1	System capacity vs. β, the signal-plus-interference to noise ratio at the receiver output for SNR= 10 dB, a single-user matched-filter receiver (continuous line) and a linear MMSE receiver (dotted line). The channel is ideal AWGN.	227

8.2 $P_{\text{out}}(\rho)$ vs. ρ/L for $M = 1$, $L = 1$ (above) and $L = 16$ (below). Curves for $N = 10, 20, \ldots, 100$ users are shown. For each family of curves, the rightmost corresponds to $N = 10$ and the leftmost to $N = 100$.   230

8.3 $P_{\text{out}}(\rho)$ vs. ρ/L for $M = 1$, $L = 16$, with residual shadowing factor $\sigma_{\text{sh}} = 2$ dB (above) and $\sigma_{\text{sh}} = 8$ dB (below). Curves for $N = 10, 20, \ldots, 100$ users are shown. For each family of curves, the rightmost corresponds to $N = 10$ and the leftmost to $N = 100$.   231

8.4 $P_{\text{out}}(\rho)$ vs. ρ/L for $M = 4$, $L = 16$, with residual shadowing factor $\sigma_{\text{sh}} = 2$ dB (above) and $\sigma_{\text{sh}} = 8$ dB (below). Curves for $N = 10, 20, \ldots, 100$ users are shown. For each family of curves, the rightmost corresponds to $N = 10$ and the leftmost to $N = 100$.   232

8.5 P_{out} vs. N for fixed $\rho/L = 7.7 \cdot 10^{-3}$, for the single-user receiver (above) and for the linear MMSE multiuser receiver (below) with interleaving depths $M = 1, 2, 4$, spreading factors $L = 1, 2, 4, 8, 16$ and ideal power control ($\sigma_{\text{sh}} = 0$ dB).   234

8.6 P_{out} vs. N for fixed $\rho/L = 7.7 \cdot 10^{-3}$, for the single-user receiver (above) and for the linear MMSE multiuser receiver (below) with interleaving depths $M = 1, 2, 4$, spreading factors $L = 2, 16$ and residual shadowing factor $\sigma_{\text{sh}} = 2$ dB.   235

8.7 P_{out} vs. N for fixed $\rho/L = 7.7 \cdot 10^{-3}$, for the single-user receiver (above) and for the linear MMSE multiuser receiver (below) with interleaving depths $M = 1, 2, 4$, spreading factors $L = 2, 16$ and residual shadowing factor $\sigma_{\text{sh}} = 8$ dB.   236

9.1 Dependence of physical signals on enviroment, network and user requirements   242

9.2 Turbo-Code encoder structure [5]   243

9.3 Turbo-Code decoder [5]   245

9.4 Structure of constituent RSC encoders   247

9.5 Performance of an RCPTC in terms of bit error ratio versus E_b/N_0 for speech transmission; block size 150 bit, code rate $\approx 1/3$, AWGN channel   250

9.6 Performance of an RCPTC in terms of bit error ratio versus E_b/N_0 for narrow band ISDN at a bearer data rate of 144 kbit/s; block size 672 bit, code rate $\approx 1/2$, fully interleaved Rayleigh fading channel   251

9.7 Performance of an RCPTC in terms of bit error ratio versus E_b/N_0; block size 672 bit, ten decoding iterations, AWGN channel   252

9.8 Performance of an RCPTC in terms of frame eror ratio versus E_b/N_0; block size 672 bit, ten decoding iterations, fully interleaved Rayleigh fading channel   253

9.9	Bit eror ratio versus variance of the log likelihood ratios at the output of the second constituent decoder; RCPTC with block size 672 bit; ten decoding iterations; AWGN channel	254
9.10	Variance of the log likelihood ratios at the output of the second constituent decoder versus E_b/N_0; RCPTC with block size 600 bit; code rate approximately $5/9$, ten decoding iterations; AWGN channel	255
10.1	Software Radio supported services sketched over frequency	259
10.2	GSM Interferer Mask	260
10.3	Receiver Architectures	261
10.4	Suggested Receiver Architecture	262
10.5	Generic Receiver for Software Radio Terminals	262
10.6	Full-Band and Partial-Band Digitization Principle	265
10.7	Generic Sigma-Delta Modulator	267
10.8	2nd order Sigma-Delta Modulator employing double integration	267
10.9	Noise shaping wide-band digitization of narrow-band signals	270
10.10	Band-pass AD conversion	272
10.11	Time-discrete down-conversion	273
10.12	FDMA Receiver	275
10.13	Spread-Spectrum Receiver	275
10.14	Generalized Platform	276
10.15	Generalized Filter, Splitting Architecture	277
10.16	Generalized Filter, Overlapping Architecture	277
10.17	Implementation of Generalized Filter	278
10.18	Filtered Code	279
10.19	Suggested Receiver	281
11.1	MSE as a function of SNR, for the proposed blind MMSE and the minimum output energy (MOE) methods ($Q = 10$, $P = 31$, $M = 1$). Perfect power control, 256 ('*') and 512 ('o') symbols are used in the estimation.	299
11.2	MSE as a function of user index for imperfect power control (user no.1 has power 10dB above user no.10). The SNR refers to the strongest user and 512 symbols are used in the estimation ($Q = 10$, $P = 9, M = 4$).	300
11.3	MSE as a function of SNR for different number of symbols (512, 1024 and 2048) used in the estimation, together with the MSE obtained using the true (asymptotic) covariance matrix. Perfect power control ($Q = 10$, $P = 9, M = 4$).	301

List of Tables

2.1	Table of the quantisation loss for different pulse shapes in terms of the maximum percentage energy loss due to sampling. Rc denotes a raised cosine filter while LPF - Hamming denotes a lowpass filter designed with a Hamming window.	39
3.1	Table of typical angular widths for different environments (after [4]).	66
3.2	Maximum number of users for different array sizes M and angular widths Δ for a BER threshold of 10^{-2}.	77
4.1	Propagation parameters for micro cellular environments.	93
4.2	Propagation parameters for macro cellular environments.	93
4.3	Summary of most important simulation parameters and their vales.	101
6.1	BERs of different blind adaptive receivers at the SNR of 20 dB in a two-path Rayleigh fading channel at vehicle speeds of 40 km/h.	169
9.1	RCPTC options for $N_e = 2$ constituent RSC encoders	247
9.2	Decoding complexity in terms of operations per uncoded bit	249
10.1	Functionalities of Generic Receiver	263
10.2	Parameters of Mobile Communications Standards	268
10.3	Dynamic Range	269
10.4	ADC resolution	270
10.5	Cost Comparison	279

Preface

The decision by the members of ETSI (European Telecommunications Standards Institute) to rely on Code Division Multiple Access (CDMA) as a basic building block for the Universal Mobile Telecommunications System (UMTS), prompted the creation of this book. Whereas CDMA "might have been" the technology of choice for third generation mobile systems at the start of 1998, it has now been propelled into the spotlight of research and development activities in Europe, Japan and other Far Eastern countries. Furthermore, the ETSI proposals for the implementation of third generation mobile networks are also being submitted to the International Telecommunications Union (ITU) as candidates for the implementation of IMT-2000 (International Mobile Telecommunications by the year 2000) systems. Clearly the concept of universal mobile telecommunication services embraces many diverse technical aspects, such as future mobile services and applications, mobile/wireless network platforms and very importantly, advanced radio interface systems. It is the latter research and development problem that is the focus of this book.

Despite vast amounts of research that has been conducted into the various aspects of CDMA technology, the choice of Wideband CDMA (W-CDMA) and Time-Division CDMA (TD-CDMA) systems as enabling technologies for the implementation of third generation mobile networks has created even more questions that require answering within the next two years. Therefore, the emphasis of this book is the introduction of some of the most important areas that will require research and development in order to make UMTS a reality. Specific attention is given to research and development advances in the areas of coding, multi-user detection, smart antenna systems and advanced hardware implementation architectures through contributions by some of the world's leading experts in these areas. Furthermore, the basic concepts of spreading techniques and the analysis of CDMA based systems through the use of linear algebra techniques are also covered making this book accessible to an audience ranging from engineers involved in the development of CDMA systems such as UMTS to research scientists focusing on specific system aspects.

Starting with a general introduction to spread spectrum techniques and the advantages thereof, the reader is taken through the analysis of a CDMA system that clearly shows the influence of various system parameters such as fading and multi-user interference on the operation and capacity of CDMA based mobile systems. From this analysis, the reader will understand the need for advanced smart antennae, multi-user detection and coding techniques. Each of these aspects is then addressed through chapters on antenna arrays and spatial filtering, multi-user detection systems such as adaptive multistage receivers and LMMSE receiver, coding techniques such as Turbo Coding and advanced software radio techniques that will facilitate the implementation of the algorithms described in the book.

The book is introduced by a chapter on the basic principles of spread spectrum communications authored by P.W. Baier et.al. In this chapter, the basic principles of operation of a spread spectrum system is reviewed and the benefits of spreading techniques in various applications, ranging from information transmission to channel identification and time and frequency estimation, is discussed. Of particular importance in this chapter is an evaluation of the ETSI decision to base the Universal Mobile Telecommunications System (UMTS) on two CDMA techniques, viz. Wideband CDMA (W-CDMA) and Time Division CDMA (TD-CDMA). The results presented on the capacity of these systems form one of the key departure points of the chapters that follow that is, what techniques can be used to increase the capacity of CDMA based 3rd generation mobile systems.

The paper by Rasmussen, Lim and Alexander lays the ground work for CDMA system analysis by providing a comprehensive linearised model of a CDMA system. A linear model of the system readily lends itself to analysis using linear algebra techniques.

The paper by Grant, Thompson and Mulgrew provides a tutorial review of antenna array processing techniques for Code Division Multiple Access networks, followed by a useful description of channel models for antenna arrays. Performance results for the cellular uplink are presented and the advantages of antenna arrays in a cellular CDMA system are discussed.

In "Spatial Filtering and CDMA", Lötter et al focus on the first step in the introduction of adaptive antenna techniques into CDMA systems namely High Sensitivity Reception (HSR). This technique utilizes adaptive beamforming techniques in the uplink of a mobile system to limit the total interference power received at the base station. The authors review some additional spatial filtering techniques that can be used in cellular CDMA systems and then proceed to derive the Bit Error Rate (BER) performance of a HSR/CDMA system. Particular attention is given to the optimum spacing of the elements of the antenna array, as well as to the influence of the user location distribution and local scatterer distribution on the performance of the cellular system. In this chapter it is shown that in addition to antenna arrays, multi-user detection algorithms (MUD), diversity and coding techniques are required to increase the overall BER performance of cellular systems.

The paper by Bar-Ness examines some late research topics in CDMA MUD. In this overview paper, some recent results on multi-user signal seperation and detection in CDMA systems are presented. It presents different decorrelating detectors for the synchronous and asynchronous channel, including adaptive approaches and multistage receivers with adaptive parallel interference cancellation. Novel structures of PIC receivers suitable for asynchronous channels are also discussed. Although non-fading environments are considered, references to extension of this work that handles these scenarios, are mentioned.

The paper by Latva-aho examines some of the possibilities for improving CDMA receiver structures with a focus on future mobile systems. The approaches considered include interference cancellation, interference suppression and hybrid techniques which perhaps hold the most promise for high efficiency applications under the multi-rate restrictions supported by UMTS proposals. Although the overall performance of the interference cancellation receiver is relatively good in the single cell case, its performance is significantly degraded in multicell environments due to the presence of unknown signal components. Blind receiver concepts developed for the down-link, are integrated into the PIC receivers for inter-cell interference suppression. The resulting LMMSEPIC receiver is capable of suppressing residual interference and results in good BER performance in the presence of unknown signal components.

The paper by Kleer, Hara and Prasad discusses detection strategies and interference cancellation schemes in a MC-CDMA system. The paper focusses on different diversity combining techniques in the up- and downlink of a cellular system. Simulation results highlight the advantage of using multi-user detection strategies to increase system performance.

Biglieri et. al. consider the optimum tradeoff between coding and spreading in their paper "Coding vs. Spreading over block fading channels". They consider the performance of both the conventional single-user receiver as well the LMMSE multiuser receiver on the block fading channel for a single-cell CDMA mobile communication system which includes ideal and non-ideal power control.

Some attention is also devoted to new coding techniques and especially to Turbo Coding. P. Jung et al provide an overview of this relatively new coding technique in "Turbo-Codes for future mobile radio applications". In addition to reviewing the basic operating principles of Turbo Codes, the paper focuses on some of the issues in designing Turbo Coding schemes for future mobile radio applications, such as UMTS. In particular, the influence of parameters such as Quality of Service, block size, multiple access scheme etc. is discussed. A possible design paradigm of rate compatible punctured Turbo Codes, which generalize the classical Turbo Codes is also presented, and a variety of performance results for various channel and service offerings are presented.

In the paper "Software radio receivers" Hentschel and Fettweis discuss the idea that future mobile radio terminals need to provide support for multiple standards and then present a detailed discussion on the various issues that are involved in implementing multi-standard software radio receivers. Various radio architectures are considered and the critical functionalities identified and discussed which includes some of the digital signal processing factors involved.

An advanced MUD receiver that utilizes both spatial and temporal processing methods is presented by Gesbert et al in "Blind Space-Time Receivers for CDMA Communications". In this chapter, blind space-time receivers for CDMA communications are considered with specific emphasis on blind signal detection. In order to develop blind processing algorithms, the authors develop channel and signal models useful for blind processing and use these models to consider the important problem of MUD.

Specifically, a novel technique allowing the estimation of the minimum-mean-square error linear MUD is derived. The interesting features of this technique include (i) its simplicity, (ii) its ability to bypass the channel estimation step, and (iii) its generality, as it requires only knowledge of the spreading sequence for the user of interest in order to develop the multiuser receiver.

We as editors would like to thank all of the contributing authors for their efforts and for being willing to contribute to this project. To John Wood of Alcatel Altech Telecoms, thank you very much for the very thorough proof-reading of the entire manuscript, on very short notice. We also thank our publisher at Kluwer Academic Publishers, Alex Greene, for his willingness to undertake the project and also for his kind and patient management of the whole project.

F.S.
P.v.R.
I.O.
M.P.L.

Some of the authors, the editors and the publisher during ISSSTA '98, Sun City, South Africa

From left to right: P.W. Baier, M.P. Lötter, Y. Bar-Ness, P.M. Grant, P. Jung, A. Greene, F. Swarts, P.G.W. van Rooyen, E. Biglieri, I. Oppermann, S. Hara, R. Kohno, R. Prasad, L. Rasmussen, T.J. Lim.

1 SPREADING TECHNIQUES, A FAR-REACHING TECHNOLOGY

P.W. Baier, T.Weber and M.Weckerle

University of Kaiserslautern,
Research Group for RF Communications,
P.O. Box 3049, D-67653 Kaiserslautern,
Germany

baier@rhrk.uni-kl.de

Abstract: Spreading techniques are characterized by a deliberate temporal or spectral spreading which leads to signals having time-bandwidth-products much larger than unity. Spreading techniques offer a number of advantages in various fields of modern information technology like information transmission, channel identification or time and frequency estimation. This can be manifested by illustrative examples of applications. Recently, spreading techniques experienced a major push by the decision of the European Telecommunications Standards Institute to include Code Division Multiple Access into the European standard for third generation mobile radio systems.

1.1 INTRODUCTION

Fig. 1.1 shows a basic system structure well known in information technology. This structure consists of a transmitter, a receiver and - placed between these two units - a connecting link comprising a channel and a device for adding noise or interference. There are three main reasons to operate systems structured such as shown in Fig. 1.1, and these reasons lead to the embodiments of the basic structure shown in Figs. 1.2, 1.3 and 1.4:

- Information transmission,
- Channel identification,
- Time and frequency estimation.

In the first case, see Fig. 1.2, the output signal of the transmitter carries data which are a priori unknown at the receiver, whereas the channel properties are known; conse-

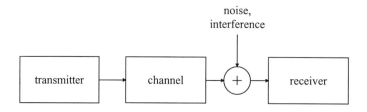

Figure 1.1   Basic system structure

quently, the task of the receiver consists in obtaining an estimate as accurate as possible of the sent data. Important parameters of information transmission systems are the rate R of the data to be transmitted and the available bandwidth W of the channel. In the second case, see Fig. 1.3, the sent signal, which is of course known at the transmitter, is also made a priori known to the receiver; however, the channel properties are unknown, and the receiver has the task to estimate these properties on the basis of the knowledge of the transmitted and the received signals. Key parameters of channel identification systems are the observation time T and the observation bandwidth W. In the third case, namely in the case of time and frequency estimation, see Fig. 1.4, the unknown time delay t_0 and frequency offset f_0 of the output bandpass signals of a transmitter have to be estimated at the receiver side based on the transmission of a test signal $\mathrm{Re}\{\underline{q}(t-t_0)\cdot\exp[\mathrm{j}2\pi(f-f_0)(t-t_0)]\}$ which is also known at the receiver except for t_0 and f_0. $\underline{q}(t)$ is the complex envelope of the test signal. The signals sent by

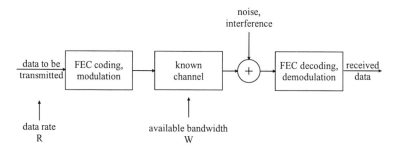

Figure 1.2   Information transmission

the transmitter in the three above mentioned applications can be characterized by their duration T and their bandwidth W. Even in the case of a continuous transmission, such a duration T can be defined. For instance, in the case of information transmission, T may be the length of the transmitted data symbols. By the quantities T and W the time-bandwidth-product TW of a system is determined, and a rough classification into systems with TW more or less close to unity on the one side and systems with TW much larger than 1 on the other side can be made; of course, the borderline between these two classes is somewhat arbitrary. Starting from a reference system with TW

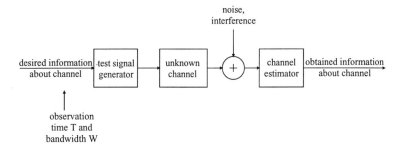

Figure 1.3   Channel identification

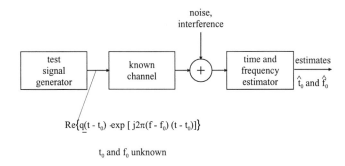

$\mathrm{Re}\{\underline{q}(t - t_0) \cdot \exp[\,j2\pi(f - f_0)(t - t_0)]\}$

t_0 and f_0 unknown

Figure 1.4   Time and frequency estimation

close to unity, TW much greater than 1 results from an extension of the signals along the time axis or an extension of the signal spectra along the frequency axis. An example for extending the spectrum is FEC (Forward Error Correction) coding in information transmission systems, see Fig. 1.2, where the bandwidth is increased by the coding algorithm when introducing redundancy.

A subclass within the class of systems with TW much larger than 1 is constituted by systems characterized by the attribute "spreading". With such systems, TW much greater than 1 is brought about in the transmitter by a very direct approach to extending the signals along the time or frequency axes, which leads to systems termed spread time or spread spectrum systems, respectively, or, more generally, to spreading systems. In such systems, the temporal or spectral extension is performed quasi in addition to the conventional algorithms of signal generation. Typically, but not necessarily, such a spreading is the last step in the signal generation chain. In contrast to the situation with spreading systems, the temporal or spectral extension observed in other systems does not result from an additional and more or less isolated step especially introduced for the purpose of spreading, but originates as a secondary effect of, for instance, FEC coding. As indicated above, the ideas behind spreading techniques can manifest themselves in a spreading along the time or frequency axes. In Fig. 1.5 these two different modes of spreading are illustrated. Temporal spreading of an unspread signal

$a(t)$ can be considered as a multiplication of its Fourier transform $\underline{A}(f)$ by a fine structured spreading function $\underline{P}(f)$. Spectral spreading of an unspread signal $a(t)$ can be interpreted as a multiplication of $a(t)$ by a broadband, i.e. fine structured spreading function $p(t)$. The spreading principle can be even applied to functions which depend neither on frequency nor time, for instance to the radiation pattern of a directive antenna, which can be spread in the angular directions of azimuth and/or elevation. Today, among the different possible modes of spreading, spectral spreading is the most important one, and, consequently, spread spectrum systems enjoy a broad variety of applications. However, interestingly, the maybe first patent on spreading techniques [1] filed in 1940 deals with the spreading of radar impulses in the time domain. Since the 1940s, spreading techniques have developed into a very mature technical field, and modern information technology is unthinkable without spreading techniques. This can also be underlined by the fact that large international conferences, see for instance [2], are dedicated to these techniques.

This paper has the aim to explain briefly the principles behind spreading techniques in a unified way, - this aim has already been accomplished to a certain extent by the above considerations - to illustrate the benefits achievable by these techniques in Section 1.2, to give some additional insights into spreading systems in Sections 1.3, 1.4 and 1.5, to present a few application examples in Section 1.6 and to address some open questions and promising fields of further research in Section 1.7.

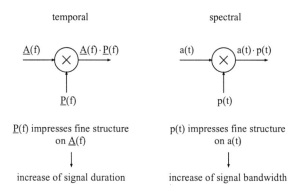

Figure 1.5   Two different modes of spreading

1.2   BENEFITS OF SPREADING

Spreading along the frequency or time axes according to the spreading principle explained in Section 1.1 can be considered as the deliberate introduction of diversity in the temporal or spectral domains, respectively. Basically, the same information content is transmitted simultaneously in different time intervals by temporal spreading or at different frequencies by spectral spreading. It may also be the case that, before being transmitted, an unspread signal is split up into temporal or spectral portions which are sent in different individual time slots or frequency bands. In the following, the benefits of the different kinds of spreading when applied in systems designed for information transmission, channel identification or time and frequency estimation, respectively, are briefly pointed out.

In Fig. 1.6 the benefits of temporal spreading are listed. Benefits experienced both in the case of information transmission systems, channel identification systems as well as time and frequency estimation systems are the following:

- Low instantaneous power of the transmitted signals. For a given amount of transmitted energy, the instantaneous transmitter power decreases with increasing temporal spreading. This is advantageous with respect to the implementation of systems with unexpensive, that is low power transmitter amplifiers, and also with respect to the achievement of a low probability of interception (LPI).

- Reduction of impulsive interference. If a transmitted desired signal is hit by an interfering impulse, the instantaneous interfering power effective in the receiver after processing the received signal can be reduced by spreading the desired signal along the time axis.

information transmission	channel identification	time and frequency estimation
time diversity		improved frequency resolution
low instantaneous transmitter power reduction of impulsive interference		

Figure 1.6 Benefits of temporal spreading

In the case of information transmission systems, the inherent time diversity of temporal spreading has the additional advantage of improving the transmission quality, if the channel is time variant. In the case of frequency estimation systems, the observation time and, consequently, the frequency resolution can be improved by temporal spreading. Of course, the time necessary for the transmission of a certain amount of data or for performing a channel measurement is increased if temporal spreading is applied.

The benefits of spectral spreading are summarized in Fig. 1.7. Benefits valid in both information transmission, channel identification as well as time and frequency estimation systems are the following:

- Low spectral power density of the transmitted signal. This feature is advantageous with respect to applications, where narrowband systems working in the same frequency band as the spread spectrum system shall not be severely disturbed by the latter, or where it is important to hide the own signals in the environmental noise in order to achieve a low LPI.

- Reduction of band limited interference. Information theory tells us that the impact of interference is not so much determined by the total power of the interfering signal, but rather by the spectral power density generated by the interferer in the frequency band occupied by the disturbed signal. For a given limited power of the interferer, the processing of the spectrally spread desired signals at the receiver has the effect, with respect to the interferer, of a simultaneous spectral spreading of the interferer power, which beneficially reduces the effective spectral power density of the interferer. Unfortunately, this mechanism of interference reduction does not work if the bandwidth of the band limited interferer is much larger than the bandwidth of the spectrally spread desired signal, or if the interferer would have a spectral power density non-vanishing over all frequencies - an example would be white noise - , which would also imply unlimited interferer power.

information transmission	channel identification	time and frequency estimation
frequency diversity	improved temporal resolution	improved time resolution

low spectral power density

reduction of band limited interference

Figure 1.7 Benefits of spectral spreading

As additional advantages of spectral spreading, one should mention the inherent frequency diversity, which helps to combat the detrimental effects of frequency selective channels in information transmission systems, and an improved temporal resolution in the case of channel identification or time estimation systems.

1.3 INFORMATION TRANSMISSION SYSTEMS

In the case of information transmission systems, temporal spreading is not a favourite spreading candidate because of the time delay going along with this technique. Usually, spreading in information transmission systems is performed in the spectral domain so that the expression spread spectrum systems is appropriate. As mentioned earlier, important parameters of an information transmission system are the source data rate R which has to be supported, and the bandwidth W covered by the transmitted signal. If $W \gg R$ holds, the system is said to operate with bandwidth luxury.

Fig. 1.8 shows the basic transmitter structure of an information transmission system working with bandwidth luxury, however, without being a spread spectrum system. The source data with rate R at the input of an FEC coder results in a data stream with the symbol rate $R_{\text{coded}} > R$ at the coder output, if binary symbols are assumed. The output signal of the coder is fed to a modulator, where it modulates a narrowband carrier of bandwidth $W_{\text{carrier}} \ll R_{\text{coded}}$. A suitable modulation scheme might be binary PSK (Phase Shift Keying). Typically, the carrier is a simple sinusoid with the nominal bandwidth W_{carrier} equal to zero. Then, the output signal of the modulator has the bandwidth

$$W \approx R_{\text{coded}} + W_{\text{carrier}} \approx R_{\text{coded}} \tag{1.1}$$

which is mainly determined by the symbol rate R_{coded} at the FEC coder output. For a given source data rate R, the bandwidth W increases with decreasing coding rate of the FEC coding. In Fig. 1.9 we see the basic structure of an information transmission

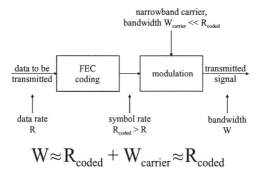

Figure 1.8 Bandwidth luxury by FEC coding

system, in which bandwidth luxury is brought about by spread spectrum techniques. In this system, the source data having rate R are immediately fed to the modulator. However, now a broadband carrier is chosen with a bandwidth W_{carrier} much larger than R. Then, the output signal of the modulator has the bandwidth

$$W \approx R + W_{\text{carrier}} \approx W_{\text{carrier}} \tag{1.2}$$

which approximately equals the bandwidth W_{carrier} of the broadband carrier. For the implementation of spectrum spreading, that is of the broadband carrier, various approaches are available, with the most important ones being direct sequencing, frequency hopping and multicarrier modulation. In practical systems often the two principles illustrated in Figs. 1.8 and 1.9 are combined so that the transmission bandwidth W is determined both by FEC coding and spectral spreading.

Historically, spread spectrum information transmission systems were mainly deployed with a view to their potentials to make the spectral power density low, and to reduce band limited interference, see Fig. 1.7. Obviously, such features are of paramount importance in military applications, where the avoidance of eavesdropping, the operation without being discovered, and combatting signals coming from unfriendly sources termed jammers are major issues. However, these potentials become more and more attractive also in the civil domain as will be elucidated by a few examples. In order to increase the efficiency with which a certain frequency band can be utilized, it is possible to overlay, over the signals of existing narrowband services, spread spectrum signals, which carry additional information. Thanks to their low spectral power density these spread spectrum signals do not significantly disturb the existing narrowband services. The spread spectrum potential of suppressing band limited interferers can

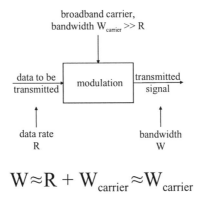

$$W \approx R + W_{carrier} \approx W_{carrier}$$

Figure 1.9 Bandwidth luxury by broadband carrier

be exploited in power line communications systems, in which usually such interfering signals occur. In satellite or land mobile radio systems, which are multiuser systems, the spread spectrum approach can be employed to solve the multiple access problem. The corresponding techniques are SSMA (Spread Spectrum Multiple Access) working with bandwidth luxury with respect to the total data rate of all users, and CDMA (Code Division Multiple Access) working with bandwidth luxury with respect only to the data rate of the individual user, and in principle these techniques avail themselves of the anti-jam potential of spread spectrum techniques. The classical multiple access schemes are FDMA (Frequency Division Multiple Access) and TDMA (Time Division Multiple Access), and for many decades one was quite happy with these schemes. However, FDMA and TDMA require a strict discipline when assigning frequency bands or time slots to the individual users, and, consequently, a significant signaling overhead. SSMA and CDMA require less overhead for organizing the multiple access, because the user signals are simultaneously active in the same frequency band, and this is one of the reasons why these multiple access schemes became interesting in the civil domain, especially for mobile radio applications, from the mid 1980s on [3]. Several proposals applying spread spectrum techniques were submitted in the course of the standardization of the pan european second generation digital mobile radio system GSM. The MATS-D concept uses CDMA in the downlink and FDMA in the uplink [4], taking advantage of frequency diversity in the downlink whereas antenna diversity at the base station could be used in the uplink. CD 900, another candidate concept for GSM, also uses spreading techniques for exploiting frequency diversity [5]. Another already mentioned reason for the application of CDMA in mobile radio communications is the increased interferer diversity. Unfortunately, the non-orderer multiple access of SSMA and CDMA is strongly connected with mutual disturbances of the individual users. However, due to the anti-jam potential of the spread spectrum approach this interference can be at least partly eliminated in the receivers.

The resistance of spread spectrum communications systems against interferers is purchased by bandwidth luxury and is only effective, if the interferers are band limited.

Also, interference suppression based on bandwidth luxury is more efficient, if this luxury is invested rather into FEC coding than into spreading the spectrum according to the scheme shown in Fig. 1.9. These efficiency differences when employing bandwidth luxury are illustrated by the curves shown in Fig. 1.10. In this figure, the SNRs (Signal to Noise Ratios) required for a bit error probability P_b equal to 10^{-3} are depicted versus the invested bandwidth luxury, which is defined as the ratio $B/B_{\min}$ of the bandwidth B of the system using bandwidth luxury and the bandwidth $B_{\min}$ of a narrowband reference system. The rate R of the data to be transmitted is kept constant. In our example, the reference system uses 4-ary orthogonal coding ($M = 4$) at the transmitter and coherent processing at the receiver. In the reference system, each coded symbol has the length $2/R$ and consists of four adjacent rectangular impulses with weight 1 or -1. Starting from the reference system, bandwidth luxury can be introduced by either increasing the parameter M of the orthogonal coding scheme, or by spectral spreading of the output signal of the reference system by direct sequencing. For a certain bandwidth luxury $B/B_{\min}$, both approaches lead to transmitted signals consisting of rectangular impulses which are shorter by a factor $B/B_{\min}$ than those of the reference system. In the case of introducing bandwidth luxury by FEC coding, only the discrete values $M/(2 \text{ ld } M)$, $M = 4, 8, 16, ...$, of $B/B_{\min}$ are possible, whereas in the case of the spread spectrum approach a practically continuous adjustment of $B/B_{\min}$ is possible. In the example of Fig. 1.10 two different interferer types are considered, namely broadband Gaussian noise with spectral power density N_0, and a band limited interferer with power N. We see the required SNR depending on the bandwidth luxury $B/B_{\min}$, where normalization is performed with respect to the SNRs required by the reference system. The dotted curves are valid for the spread spectrum approach and the solid curves for FEC. γ_b' represents the required normalized SNR per bit for broadband Gaussian interferers, and γ' represents the required ratio of the received desired and interfering powers for bandlimited interferers. It becomes evident from Fig. 1.10 that the spread spectrum approach is only effective against the bandlimited interferer, and that the FEC approach utilizes bandwidth luxury more efficiently to suppress interference. The latter approach is therefore said to have an additional coding gain.

As already mentioned above, another advantage of spread spectrum techniques applied in information transmission systems of both the military and civil type, frequency diversity, see Fig. 1.7, is important. Many radio channels, e.g. in land mobile radio communications or power line communications, are frequency selective, and spreading the spectrum is an obvious and reliable means to combat this type of fading.

1.4 CHANNEL IDENTIFICATION SYSTEMS

The task of channel identification systems consists in obtaining information on the properties of a radio channel connecting two locations. The most comprehensive information of this type would be the channel impulse response, but also less specific information like the attenuation or the power delay profile of the channel may be of interest. The problem of channel identification occurs for instance in radar, in communications systems and in mobile radio channel sounders. In the case of radar, the information gained on the channel forms the basis for further evaluations as e.g. the

10   CDMA TECHNIQUES FOR THIRD GENERATION MOBILE SYSTEMS

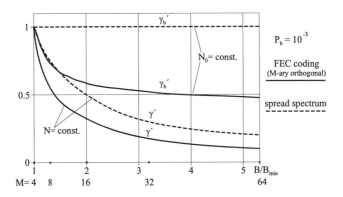

Figure 1.10 Decrease of required normalized SNR with increasing bandwidth luxury $B/B_{\min}$

detection and identification of radar targets. In the case of communications systems, information about the channel is absolutely necessary to adjust the channel equalizers and to regain the transmitted data at the receiver. In the case of mobile radio channel sounders, the obtained channel information helps to obtain a deeper understanding of the physical effects prevailing in mobile radio wave propagation, and to establish close-to-reality channel models.

As already shown in Fig. 1.3, the essential parts of a channel identification system are a transmitter generating a test signal, and a receiver performing the channel estimation. Depending on the local arrangement of these components, channel identification systems can be classified into monostatic, bistatic and multistatic systems, see Fig. 1.11, where we also find system examples for the different classes. In monostatic systems, the transmitter and receiver are virtually at the same location. In bistatic systems, the transmitter and receiver are at two different locations, which may be separated by a large distance. In multistatic systems, one transmitter and a set of receivers or one receiver and a set of transmitters are used, with each of these stations being at a different location. When specifying a channel identification system, the most important

system type	examples
monostatic	pulse radar laser range meter
bistatic	mobile radio channel sounder light barrier
multistatic	real time channel estimation in CDMA mobile radio systems

Figure 1.11 Classification of channel identification systems

parameters are the bandwidth W, within which the channel shall be observed, and the allowable observation time T. The larger the bandwidth W, the finer the temporal resolution of the channel measurement, which means for instance in the case of a pulse radar system that also closely spaced radar targets can be discerned. The observation time T is limited by the time variance of the channel. Each observation should be performed in a time interval T shorter than the time within which the channel changes its properties significantly.

The most straightforward approach to channel identification consists in transmitting a short impulse of width T, where T is related to the intended observation bandwidth W by the expression

$$T \approx 1/W. \qquad (1.3)$$

In this case, the time-bandwidth-product is close to 1. Temporal spreading in the sense of spreading techniques means that, instead of using a short test signal with bandwidth W, see (1.3), a temporally extended test signal with the same bandwidth W is used. In Fig. 1.12 such a temporal spreading is illustrated in the frequency domain. $\underline{A}(f)$ is the Fourier transform of a temporally non-spread impulse, see the upper run in Fig. 1.12, the time dependence of which is given by a sinc-function centered around $t = 0$ with the first zeros at $t = \pm 1$. Temporal spreading can be performed by fine-structuring the Fourier transform $\underline{A}(f)$, that is by multiplying $\underline{A}(f)$ by the spreading function $\underline{P}(f)$, see the middle run in Fig. 1.12. As the result, a time-spread function is obtained, the main part of which is extended along the time axis from $t = -31$ to $t = 31$, see the lower run in Fig. 1.12. Of course, the question arises why such a temporal spreading should be applied. In principle, the reasons were already mentioned in Section 1.2. The most obvious reason for temporal spreading consists in reducing the instantaneous transmitter power without reducing the bandwidth W and the energy of the test signal. It is more advantageous with respect to the implementation of transmitter power amplifiers to generate a certain signal energy by a temporally extended impulse instead of using a short impulse with high instantaneous power. Also, temporally extended impulses are less prone to interception, and they may be less harmful to other systems. As further advantages, the resistance against impulsive jammers and the frequency resolution of the channel measurement are increased.

1.5 TIME AND FREQUENCY ESTIMATION SYSTEMS

Often, either time or frequency is the unknown quantity to be estimated. Typical test signals with spectral spreading are the time synchronization bursts defined in the mobile radio standard GSM (Global System for Mobile Communications). Typical signals with temporal spreading are the frequency correction bursts defined in GSM. A straightforward implementation of the time and frequency estimator in the system structure shown in Fig. 1.4 would be a bank of matched filters and a maximum discriminator. Each of the filters would be matched for a different pair of the parameters t_0 and f_0, that is for a different time delayed and frequency shifted version of the test signal $\mathrm{Re}\{\underline{q}(t)\exp[\mathrm{j}2\pi ft]\}$. The parameters t_0 and f_0 of the filter exhibiting maximum output would be the estimates $\hat{t}_0$ and $\hat{f}_0$, respectively. A function which describes the resolution potential of the signal $\mathrm{Re}\{\underline{q}(t)\exp[\mathrm{j}2\pi ft]\}$ both with respect to time and

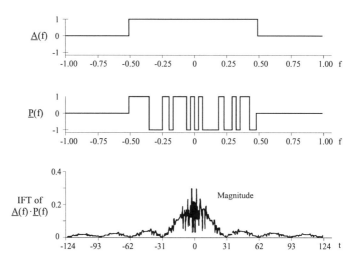

Figure 1.12 Temporal spreading

frequency is the ambiguity function [6]

$$\chi(\tau, f) = \int_{-\infty}^{+\infty} \underline{q}(t) \cdot \underline{q}^*(t+\tau)\exp(-j2\pi ft)dt \quad (1.4)$$

of the complex envelope $\underline{q}(t)$. Ideally, $\chi(\tau, f)$ has a spike for τ and f equal to zero and vanishes elsewhere. By increasing the temporal and spectral spreading of the test signal, the ambiguity function $\chi(\tau, f)$ can be brought closer to this ideal behavior.

1.6 APPLICATION EXAMPLES

1.6.1 Scope of Applications

The best known applications of spreading techniques are in the field of communications, with CDMA mobile radio systems being one of the most prominent representatives. In order to show the broad scope and potential of spreading techniques, many possible spreading applications from fields other than communications, will be presented in the following.

1.6.2 Thermal Flow Meter

Measuring the flow of a fluid through a pipe is a problem which is relevant in many technical areas as for instance fuel consumption monitoring or control of chemical processes. As long as the flow is laminar, the total flow can be calculated from the fluid velocity along the axis of the pipe. This velocity can be measured by inoculating heat impulses into the fluid and by detecting these impulses by a heat sensitive receiver which is positioned downstream relative to the transmitter. Fig.

1.13 shows such a measuring system which can be considered as a bistatic channel identification system. For obtaining a certain accuracy, one can either use short periodic heat impulses with high energy, which are transmitted with the repetition period T_p, or a PN (Pseudo Noise) sequence of short, low energy heat impulses with the same period T_p. The latter approach exerts less thermal stress on the fluid and is more easily implementable. A flow measuring system according to the design shown in Fig. 1.13 has been implemented for measuring e.g. the Diesel fuel consumption in motor vehicles [7]. The measuring range is 1 to 50 l/h with an accuracy of 5%.

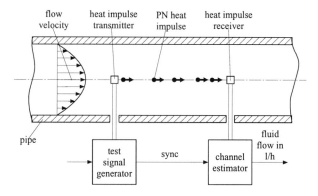

Figure 1.13 Thermal flow metering system

1.6.3 *Electrooptical Geodesic Range Meter*

A frequently used tool in geodesy are electrooptical range meters. In such systems, a transmitter sends an optical signal towards a prismatic corner reflector positioned at the terrain point the distance of which from the transmitter has to be determined. At the same location as the transmitter we have an optical receiver to evaluate the optical signal reflected by the reflector. Obviously, such a system is a monostatic channel identification system. Conventional geodesic electrooptical ranging systems work with periodic optical impulses, where first a low pulse repetition frequency is used to obtain a rough, but unambiguous estimate of the distance, and then successively one or several higher pulse repetition frequencies are used to increase accuracy. It was shown [8] that, instead of successively using several periodic impulse sequences having different pulse frequencies, a single optical PN impulse sequence can be advantageously applied.

1.6.4 *Antenna Diagram Spreading*

Airborne or satellite borne imaging systems are applied for taking images of the earth's surface. Besides optical imaging systems, radar based imaging systems are usual. An important issue when designing and operating imaging systems is the achievable resolution. The resolution potential of imaging radar systems is based on two physical grounds: The characteristic of the antenna and Doppler analysis. Concerning the antenna, a narrow beam pattern would be advantageous to have a good a priori separation of different points of the scenario to be imaged. Doppler analysis exploits the motion of the platform carrying the radar system and is based on the different Doppler histories of different points of the scenario. Concerning Doppler analysis, a large illumination time of the individual scenario points would be advantageous, which clearly contradicts the above mentioned requirement of a narrow antenna beam pattern. A solution to this dilemma can be found by using antennas with a pattern extended according to the spreading principle. Such antennas would have wide, but fine- structured patterns [9]. Thanks to this fine structure, the genuine resolution potential of the antenna is maintained, and, thanks to the increased width of the antenna pattern, a large illumination time can be achieved.

1.6.5 High Resolution Impulse Compression Radar

As mentioned earlier, pulse radar systems can be considered as monostatic channel identification systems. If temporal spreading is employed, the advantages mentioned in Sections 1.2 and 1.4 can be enjoyed. Pulse radar systems employing temporal spreading are usually termed impulse compression radars, because in the receiver the temporally spread radar impulses are compressed by passive correlation. Unfortunately, when doing so, correlation sidelobes arise. The correlation sidelobes resulting from large radar targets may be so high that the correlation mainlobes of small radar targets are masked. This undesired effect is especially large if impulse compression is performed by filters matched to the sent radar impulses. The correlation sidelobes can be reduced by using mismatched filters in the receiver. However, when using such filters the correlation mainlobes are broadened, which reduces the range resolution. A more recent approach to impulse compression consists in applying minimum variance unbiased estimation [10], which algorithmically eliminates the correlation sidelobes without widening the correlation mainlobes. In Fig. 1.14 this approach is illustrated. Fig. 1.14a shows a radar scenario comprising eight radar targets in different range gates. The radar targets are represented by the magnitudes $|\underline{x}_i|$ of their complex reflexion coefficients $\underline{x}_i$. Fig. 1.14b shows the magnitude $|\underline{\hat{x}}_i|$ of the output signal $\underline{\hat{x}}_i$ of a filter matched to the used temporally spread radar impulse, which, in this case, is a Frank Code having a time-bandwidth-product TW equal to 100. Evidently, the smaller targets disappear, because they are concealed by the correlation sidelobes of the larger targets. Also, a lot of non-existing targets are faked by the correlation sidelobes. In Fig. 1.14c we see the output signal of a minimum variance unbiased estimator. Now, also the smaller targets can be detected, and the only remaining disturbance is thermal noise. Presently, an impulse compression radar system for controlling the ground movements of airplanes on airports is being implemented following this approach [11].

1.6.6 Mobile Radio Channel Sounder

Radar systems and channel sounders for radio channels are channel identification systems. The main difference between these two system classes consists in the fact that the latter are bistatic, whereas radar systems are monostatic. Having the commonality of both system classes in mind, high resolution radio channel sounders can be imple-

SPREADING TECHNIQUES, A FAR-REACHING TECHNOLOGY   15

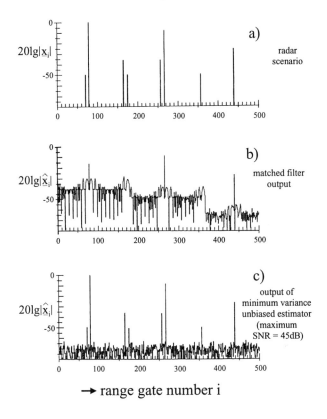

Figure 1.14 High resolution impulse compression radar

mented by applying the same principles which can be used to enhance the resolution of radar systems, see Section 1.6.5. An example is the mobile radio channel sounder SIMOCS 2000 [12], which uses temporally spread test signals and, at the receiver, minimum variance unbiased estimation. These test signals are optimized with respect to a low crest factor and to a low SNR degradation when performing the estimation.

1.6.7 Satellite Navigation Systems

The satellite navigation systems GPS (Global Positioning System) and GLONASS (Global Navigation Satellite System) can be considered as time and frequency estimation systems. In both systems [13] spread signals are used to enhance the accuracy of time and frequency estimates. The systems consist of a space segment with each 24 non-geostationary satellites, a control segment containing monitoring stations exclusively used for operational tasks, and a user segment. They are configured primarily to provide the user with the capability of determining his position. This is in principle accomplished by determining the distances measured to the satellites in view. The distance is computed by measuring the propagation time for a satellite-generated ranging code to reach the user receiver. Within the receiver, an identically coded ranging signal is generated and shifted in time until it achieves correlation with the satellite-generated ranging code. The larger the spectral spreading, the higher is the accuracy of determining the propagation time and finally the determining of the position. Besides time estimation, the system includes also information transmission. A low rate binary data signal containing navigation messages is spectrally spread by the aforementioned code sequence. The principal difference in the signaling of GPS and GLONASS is that GPS uses CDMA and GLONASS uses FDMA. Consequently, in GPS all carrier frequencies are equal, while each satellite uses a unique spreading code. In GLONASS each satellite uses the same spreading code but different carrier frequencies [13].

1.6.8 CDMA for UMTS and IMT-2000

In January 1998 ETSI (European Telecommunications Standards Institute) made a decision concerning the terrestrial radio air interface of UMTS (Universal Mobile Telecommunications System), the European third generation mobile radio system. It was decided that this standard will consist of two components, namely W- CDMA (Wideband CDMA) to be operated in the duplexing mode FDD (Frequency Division Duplex), and time TD- CDMA (Time Division CDMA) to be operated in the duplexing mode TDD (Time Division Duplexing). Both components are based on CDMA. It can be expected that the decision made by ETSI will also have an impact on ITU (International Telecommunications Union), which is expected to make a decision on the worldwide standard IMT-2000 (International Mobile Telecommunications 2000) in the near future. Consequently, all over the world third generation mobile radio systems will most probably apply CDMA and will therefore be based on the ideas of spreading techniques. Of special interest in this respect is TD-CDMA [14], which uses unbiased minimum variance estimation, see also Sections 1.6.5 and 1.6.6, for on-line channel estimation as well as for data estimation; in the case of data estimation, both ISI (Intersymbol Interference) and MAI (Multiple Access Interference) are eliminated in TD- CDMA. It is also interesting to combine CDMA with smart antennas [15, 16].

An important issue when designing air interface concepts for third generation mobile radio systems is the question – which spectrum efficiency and capacity can be achieved by different air interface proposals? The experience made, for instance, within COST231 (European Cooperation in the Field of Scientific and Technical Research) [14] and the EU project ACTS-(Advanced Communications Technologies and Services) FRAMES (Future Radio Wideband Multiple Access Systems) [17] clearly shows that binding and reliable statements in this respect are only possible on the basis of expensive simulations on the link and system levels. Nevertheless, the availability of rule-of- the-thumb formulae to basically judge the dependence of the system behaviour from various system parameters is desirable. Besides giving approximate indications of the system performance, such formulae can also be of use to obtain principal insights into the system functionality. In what follows, such formulae will be presented for the two modes W-CDMA and TD-CDMA.

When deriving such formulae, the cellular interference function $f(r)$ with r being the cluster size is of interest. $f(r)$ describes the average amount of interfering power arriving at a receiver in a reference cell of an infinite cell grid, if in each of the other

r	1	3	4	7
f(r)	0.59	0.07	0.04	0.009

ideal power control
cluster size r
attenuation coefficient $\alpha = 4$
standard deviation of the lognormal fading $\sigma = 8$ dB

Figure 1.15 Cellular interference function $f(r)$ (downlink)

infinitely many cells one user would be active, and if the desired powers at all receivers would be made equal to unity by perfect power control. Fig. 1.15 shows the values of $f(r)$ depending on r. These values have been obtained by A. Steil at the authors' labs by numerical simulations for an attenuation exponent α equal to 4 and a standard deviation σ of the lognormal fading equal to 8 dB. For instance, in the case r equal to 3 the average intercell interfering power arriving at a receiver in the reference cell would be 0.07, if in each of the other cells one user would be active. The function $f(r)$ can be used to calculate the available E_b/N_0 in the reference cell in the MAI limited case, i.e., if thermal noise can be neglected. The desired power impinging at a receiver in the reference cell equals 1. If K transmitters are active in each cell, in the case of W-CDMA, which is characterized by single user detection, the intracell MAI power takes the value $K - 1$, and the intercell MAI power is given by $f(r) \cdot K$. Consequently, in this case the total MAI power equals $K - 1 + f(r) \cdot K$, and the carrier-to-interference ratio becomes $1/(K - 1 + f(r) \cdot K)$. In the case of TD-CDMA, intracell MAI is eliminated by joint detection, and the total MAI power becomes $f(r) \cdot K$. In this case, the carrier-to-interference ratio takes the value $1/(f(r) \cdot K)$. With the total system bandwidth W, the cluster size r and the user data rate R the processing gain is given by the expression $W/(r \cdot R)$. Multiplication of the carrier-to-interference ratio by the processing gain yields the available E_b/N_0:

$$\frac{E_b}{N_0} = \frac{W/r}{R} \cdot \begin{cases} \frac{1}{K-1+f(r)\cdot K} & \text{for W-CDMA,} \\ \frac{1}{f(r)\cdot K} & \text{for TD-CDMA.} \end{cases} \tag{1.5}$$

The expression (1.5) for E_b/N_0 is visualized by the curves in Figs. 1.16 and 1.17 for W-CDMA and TD-CDMA, respectively. These curves show E_b/N_0 versus the number K of users per cell with r as the parameter. In both cases E_b/N_0 decreases with increasing K, which can be easily understood. In the case of W-CDMA, E_b/N_0 also decreases with increasing r, see Fig. 1.16, whereas in the case of TD-CDMA we have the inverse behaviour with respect to increasing r, see Fig. 1.17. This inverse behaviour in the case of TD-CDMA is due to the fact that the product $r \cdot f(r)$ appearing in the denominator of the above expression (1.5) decreases with increasing r, see also Fig. 1.15.

In order to obtain a certain quality of service, the available E_b/N_0, see Figs. 1.16 and 1.17, should come up to the E_b/N_0 required for said quality of service. In

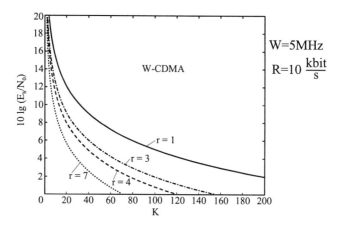

Figure 1.16 Available E_b/N_0 with single user detection

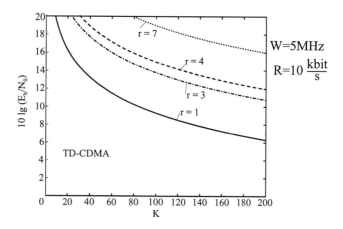

Figure 1.17 Available E_b/N_0 with multiuser detection

the case of W-CDMA, which is characterized by single user detection, the required E_b/N_0 equals a fixed value γ_0 independently of the number K of users per cell. Fig. 1.18 depicts the fixed required E_b/N_0 and the monotonous decreasing curves of the available E_b/N_0.

In the case of TD-CDMA, due to the properties of joint detection, the required E_b/N_0 exceeds γ_0 by the SNR degradation δ:

$$\left(\frac{E_b}{N_0}\right)_{\text{required}} = \gamma_0 \cdot \delta \tag{1.6}$$

According to simulations performed by M. Weckerle at the authors' labs, with the processing gain p_g per burst and the number K_{sim} of simultaneously active users per burst, the SNR degradation δ can be approximated by

$$\delta = \frac{p_g + 1}{p_g - K_{\text{sim}} + 1}, \qquad (1.7)$$

if ISI (intersymbol interference) is of inferior importance, as it is the case in small cells. With v the number of TDMA bursts per frame, for the processing gain

$$p_g = W/(r \cdot R \cdot v) \qquad (1.8)$$

and for the number of simultaneously active users per burst

$$K_{\text{sim}} = K/v \qquad (1.9)$$

hold. By substituting (1.7) to (1.9) into (1.6), one finally obtains the expression

$$\left(\frac{E_b}{N_0}\right)_{\text{required}} = \gamma_0 \cdot \frac{W/(r \cdot R) + v}{W/(r \cdot R) - K + v} \qquad (1.10)$$

in the case of TD-CDMA, which, of course, is only valid as long as the denominator is positive.

In Fig. 1.19 the required and the available E_b/N_0 in the case of TD-CDMA are depicted versus the number K of users per cell with r as the parameter and for the values of W, R, v and γ_0 given in the figure. The monotonous decreasing curves belong to the available E_b/N_0 whereas, according to (1.10), the required E_b/N_0 grows larger with increasing K and r.

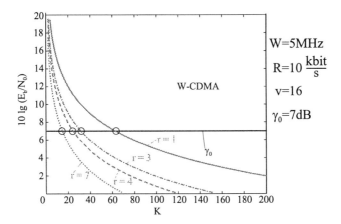

Figure 1.18 Required and available E_b/N_0 with single user detection

In order to achieve the required quality of service, the available E_b/N_0 and the required E_b/N_0 have to be matched to each other by choosing proper values for the parameters K and r, see Figs. 1.18 and 1.19. In general, different pairs (K, r) exist, for which such a match can be reached. For obtaining maximum spectrum efficiency

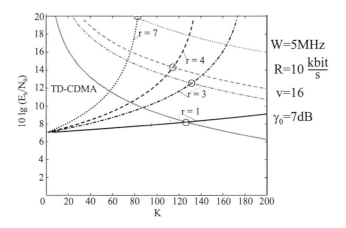

Figure 1.19 Required and available E_b/N_0 with multiuser detection

and capacity, this specific pair (K, r) should be chosen, which contains the maximum value K. For W-CDMA it can be concluded from Fig. 1.18 that r equal to 1 should be chosen, which leads to the maximum value 63 for K. For r equal to 3 the number K of supportable users would be much smaller than 63. For TD-CDMA it follows from Fig. 1.19 that the maximum achievable value of K is about 127 and can be reached with both r equal to 1 and r equal to 3.

1.7 PROMISING FIELDS OF FURTHER RESEARCH

As mentioned earlier, the origins of spreading techniques lie in the military field. Thanks to the worldwide political détente beginning in the late 1980s, a lot of previously classified knowledge and experience in spreading techniques step by step has become available also for civil use. However, it is still a challenge to identify new civil applications and to pave the way for new products employing spreading techniques. As illustrated in Section 1.6, these techniques can be utilized in many fields of modern information technology including totally different areas such as mobile radio systems and imaging radars. It would be rewarding to develop joint research programs for the applications of spreading techniques in different fields and to benefit from mutual stimulation. Sometimes, justifiably, it is emphasized that spreading techniques require a considerable amount of signal processing effort, which might be prohibitive especially for the deployment of low cost applications. Fortunately, the unbroken progress of microelectronics will lighten this burden, and conceptions like software radio are highly important in this respect. A special challenge has been set by the ETSI decision mentioned in Section 1.6.8. The efforts of many researchers and engineers will be needed to make the CDMA mobile radio systems envisaged by ETSI a reality. Spreading techniques are also interesting in power line communications. Power lines are a virtually omni-present and already installed medium. Therefore, it is highly rewarding to consider this medium for communications purposes such as the remote metering of the consumption of electrical energy or WLL (Wireless Local Loop). Already in the mid 1980s patents on the application of spread spectrum techniques in power line communications have been filed [18, 19] . This application of spread spectrum techniques is still a very active field of research [20, 21]. Spread spectrum techniques may also be useful to implement wireless multiuser microphone systems with CD (Compact Disc) quality [22].

1.8 CONCLUSIONS

The ideas behind spreading techniques can manifest themselves in temporal or spectral spreading and offer a series of advantages. Fields of applications are information transmission, channel identification as well as time and frequency estimation. Application examples underline the practical importance of the spreading principle in modern information techniques.

References

[1] E. Hüttmann, Verfahren zur Entfernungsmessung (ranging method). German Patent No. 768 068 (filed 22 March 1940, published 5 May 1955)

[2] Proc. IEEE Fourth International Symposium on Spread Spectrum Systems and Applications (ISSSTA'96), Mainz (1996), ISBN 0-7803-3567-8

[3] R.L. Pickholtz et al., Spread spectrum for mobile communications. IEEE Trans. on Vehicular Technology, vol. 40 (1991), pp. 313-322

[4] A. Eizenhöfer, Anwendung der Spread-Spectrum- Technik in dem hybriden Mobilfunksystem MATS-D (application of the spread spectrum technique in the hybrid mobile radio system MATS-D). Frequenz, vol. 40 (1986), pp. 255-259

[5] U. Langewellpott, Anwendung der Spread-Spectrum- Technik im Mobilfunk (application of spread spectrum techniques in mobile radio). Frequenz, vol. 40 (1986), pp. 249-254

[6] A.W. Rihaczek, Principles of high-resolution radar. Artech House, Boston, 1996

[7] W. Witte, Ein Beitrag zur korrelativen Volumenstrommessung mit pseudozufälligen Markierungen im Strömungsmittel (contribution to correlative volume flow metering with pseudo random marks in the fluid). Doctoral Dissertation, University of Kaiserslautern, Dept. Electr. Eng., 1983

[8] T. Zimmermann, Elektrooptische Entfernungsmessung mit Bandspreizverfahren (electrooptical range metering with spreading methods). Fortschrittberichte VDI, Reihe 8, Nr. 253, VDI Verlag, D~üsseldorf, 1991

[9] A. Löhner et al., Enhancement of the angular resolution of radar antennas by diagram spreading. Proc. IEEE Fourth International Symposium on Spread Spectrum Systems and Applications (ISSSTA'96), Mainz (1996), pp. 882-888

[10] T. Felhauer et al., A propositon to valuate expanded impulses in pulse radar systems using optimum unbiased estimation. European Trans. on Telcom. and Related Areas (ETT), vol. 4 (1993), pp. 175-181

[11] A.C. Schroth et al., Near-range radar network for ground traffic management: System concept and performance. International Journal on Air Traffic and Control Quarterly, vol. 5 (1996), pp. 69-90

[12] T. Felhauer, P. W. Baier, W. König, W. Mohr, Optimized wideband system for unbiased mobile radio channel sounding with periodic spread spectrum signals. IEICE Transactions on Communications (Special Issue on Spread Spectrum Techniques and Applications), vol. E76-B (1993), pp. 1016-1029

[13] E. D. Kaplan, Understanding GPS: Principles and applications. Artech House, Boston, 1996

[14] P.W. Baier, A. Klein, Flexible hybrid multiple access schemes for 3rd generation mobile radio systems. In E. Del Re (Editor) "Mobile and Personal Communications", Elsevier Science B.V., Amsterdam, 1995, pp. 31-43

[15] R. Kohno, Spatially and temporally joint optimum transmitter - receiver based on adaptive array antenna for multi-user detection in DS/CDMA. Proc. IEEE Fourth International Symposium on Spread Spectrum Systems and Applications (ISSSTA'96), Mainz (1996), pp. 365-369

[16] J.S. Thompson, P.M. Grant, B. Mulgrew, Smart antenna arrays for CDMA systems. IEEE Personal Communications, vol. 3 (1996), pp. 16-25

[17] T. Ojanperä, J. Sköld, J. Castro, L. Girad, A. Klein, Comparison of multiple access schemes for UMTS. Proc. 47th IEEE Vehicular Technology Conference (VTC'97), Phoenix, 1997, pp. 490-494

[18] P.W. Baier, K. Dostert, Verfahren und Einrichtung zum Empfang digitaler Informationen über Stromversorgungsnetze (method and device for the reception of digital information via electrical power networks). German Patent No. 40 01 265 (filed 18 January 1990, published 18 May 1995)

[19] P.W. Baier, K. Dostert, Verfahren und Sendeeinrichtung zur Übertragung digitaler Informationen über Stromversorgungsnetze (method and device for transmission of digital information via electrical power networks). German Patent No. 40 01 266 (filed 18 January 1990, published 27 July 1991)

[20] Y. Kanamori et al., Development of a spread spectrum transmission system for low-voltage distribution lines. Proc. 1998 International Symposium on Power-Line Communications and its Applications (PLC '98), Soka University, Tokyo (1998), pp. 175-178

[21] N. Suehiro et al., Cellular power line CDMA communications without co-channel interference. Proc. 1998 International Symposium on Power-Line Communications and its Applications (PLC '98), Soka University, Tokyo (1998), pp. 194-201

[22] F. Graf, Digitale drahtlose Mikrofonsysteme mit Vielfachzugriff (wireless digital microphone systems with multiple access). Doctoral Dissertation, University of Kaiserslautern, Dept. Electr. Eng., 1994

2 A LINEAR MODEL FOR CDMA SIGNALS RECEIVED WITH MULTIPLE ANTENNAS OVER MULTIPATH FADING CHANNELS

Lars K. Rasmussen, Paul D. Alexander and Teng J. Lim

Centre for Wireless Communications*
20 Science Park Road, #02-34/37 Teletech Park
Singapore Science Park II, Singapore 117674
cwclkr/cwcpa/cwclimtj@leo-nis.nus.edu.sg

Abstract:

This chapter shows how a simple linear model can be used to describe the signal received over a multiple-antenna, time-dispersive, fading code-division multiple-access (CDMA) channel. The realization that each of these channel effects can be absorbed into a general linear model validates the abstraction of the CDMA problem into a linear algebra problem which opens up mathematically elegant and powerful techniques to solve common problems such as detection and channel estimation. Based on the structural principles of practical systems, where all baseband processing is normally done in discrete time, we develop a corresponding matrix algebraic model. The model leads to the known convenient interpretation of the CDMA system as a time-varying convolutional encoder which allows for insight into the structure of the inherent interference. The model unifies many related models previously suggested in the literature. The novelty of this model is that it follows the structural principles of practical systems and leads to a convenient algebraic form which allows for powerful receiver design.

---

*The CWC is a National Research Centre funded by the National Science & Technology Board (NSTB), Singapore.

2.1 INTRODUCTION

As wideband CDMA has been accepted as one of the air interface technologies for UMTS/ IMT2000 in Europe [1] and Japan [2, 3], the development of practical multiuser detection techniques is becoming increasingly important. The use of Rake combining, beam forming and antenna diversity is also vital to efficiently combat channel impairments and multiple access interference (MAI) at high data rates and high system loads.

To develop efficient detection strategies for such elaborate systems, it is of paramount importance to develop an appropriate baseband model for the uplink communication. It is especially convenient and insightful for design to have the received signal represented by a sum of modulating waveforms, each directly associated with a transmitted symbol. In this chapter we show how such a linear model can be used to describe the signal received over a multiple-antenna, time-dispersive, fading CDMA channel. The realisation that each of these channel effects can be absorbed into the general linear model validates the abstraction of the CDMA problem into a linear algebra problem which opens up mathematically elegant and powerful techniques to solve common problems such as detection and channel estimation. We focus on modelling the oversampled received baseband signal coming out of the receiver A/D converter. Many practical aspects of the uplink are considered so the model facilitates the development of receiver structures based on the fundamental structural principles of practical systems.

In this chapter we first develop a model for a synchronous CDMA system to establish a conceptually simple benchmark from which to extend to more practical scenarios. This also serves to support the fact that the basic algebraic structure of the model remains the same regardless of the series of complications we later introduce. The development of the synchronous model includes Rayleigh fading, path loss, shadowing effects and discussions of channel estimation and power control.

The synchronous model is then extended to facilitate random access among users, typical for uplink transmission. Acquisition and tracking of the user dependent delays are discussed and modelling of delay estimation inaccuracies due to the finite delay resolution inherent to a discrete-time system is considered. This model is then extended to include multipath propagation. Additionally, we include a discussion of the resolvability of multipath components at the receiver.

The final extension incorporates multiple antenna elements into the model. We consider the cases of using the antenna elements either for diversity combining or for beam forming as well as hybrids thereof. We thus arrive at a linear model where each of the active users transmits over a time-dispersive multipath fading channel with path loss and shadowing and is received by multiple antenna elements.

Most models found in the literature are based on continuous-time transmitted and received signals. The received signal is then matched filtered to the spreading waveform to obtain a discrete-time baseband model at symbol rate. A few models are based on a discrete-time received signal at chip rate and are therefore similar in form to the one derived here, e.g., [4]. In our case however, we focus on modelling the sampled received signal coming out of the receiver A/D converter rather than considering the signal after chip matched filtering. This follows practical designs more closely and leaves the opportunity of investigating aspects of the receiver design which is normally performed prior to matched filtering.

The beauty of the developed model is that any additional system feature or channel impairment can be included without changing the fundamental algebraic form, The algebraic model retains the same structural form, where each transmitted symbol is associated with a specific modulating discrete-time waveform. This structure allows for intuitive insight into interference patterns and channel effects for systems of high complexity. It is also clear that detector structures developed for simple system models adhering to the generic form are equally suitable for practical systems.

The chapter is organised as follows. In Section II, structural block diagrams of a practical CDMA uplink are presented and discussed. These block diagrams are then used as the basis for the development of a general algebraic model for uplink communication in Section III. The concept of multiple antennas are incorporated into the model in Section IV. In Section V, the CDMA system is interpreted as a time-varying convolutional encoder and the inherent interference patterns are discussed. Concluding remarks are summarised in Section VI.

The notation to be used throughout this chapter is summarised below. Vectors and matrices are indicated as bold face lower case and bold face upper case, respectively. All vectors are defined as column vectors and subscripts k and m, $\mathbf{s}_{k,m}(i)$, indicate user number and path number while i denotes the symbol time interval. $\mathbf{0}_L$ represents a zero vector of length L, $*$ denotes discrete-time convolution and $\otimes$ denotes the Kronecker tensor product [5]. The operators $(\cdot)^{\mathrm{T}}$, $(\cdot)^*$ $(\cdot)^{\mathrm{H}}$, and $\|\cdot\|^2$ denote transposition, complex conjugate, complex conjugate transposition, and Euclidean norm, respectively.

2.2 CDMA UPLINK SYSTEM MODEL

Practical systems such as IS-95 [6], IS-665 [7] and the wideband CDMA proposals for IMT2000 [3] all have the same fundamental structure for the uplink. Block diagrams of the transmitter and receiver structures are shown in Fig. 2.1 and Fig. 2.2, respectively[1]. Both at the transmitter and receiver side, all baseband processing is done in discrete

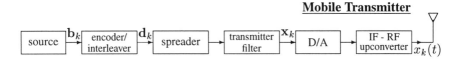

Figure 2.1 Block diagram of the mobile transmitter for user k.

time. The transmitter consists of an encoder, an interleaver, a spreader, a pulse shaping transmitter filter (root-Nyquist filter), a D/A converter and an IF-RF up-converter. The data stream $\mathbf{b}_k$ for an arbitrary user k is encoded and interleaved into the sequence $\mathbf{d}_k$ which is spread and digitally modulated onto a chip pulse shape[2]. The signal is then D/A converted and modulated onto an analogue carrier for transmission. In a system with K simultaneously active users, each user is represented by such a transmitter.

The base station receiver front end, common to all users, is depicted in Fig. 2.3. A total of B receiver antenna elements are assumed. First the signal from each

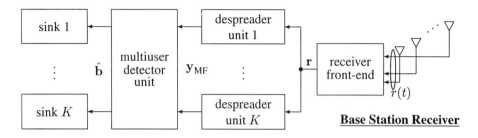

Figure 2.2 Block diagram of the base station receiver.

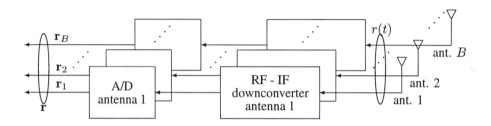

Figure 2.3 Block diagram of the base station front-end.

antenna element is down-converted to baseband and then digitised at a rate of Q/T_c samples per second where T_c is the chip interval and Q is the oversampling factor. The corresponding discrete-time received signal for each element, denoted $\mathbf{r}_b$, $b = 1, 2, ..., B$, is forwarded to a bank of K despreader units, one for each user. In the unit for user k, the multipath transmission delays are estimated through acquisition and tracked typically by delay-lock loop techniques [8]. These estimated and tracked delays are then used to control the timing for despreading as illustrated in Fig. 2.4. In a practical implementation only one receiver filter, working at a rate of Q/T_c is required. The despreading is done at chip rate where the tracking units provide the timing (delay estimates) for the appropriate sampling instances at the output of the receiver filter for each multipath finger. Since these delays can take on any continuous-time value, it is difficult to model the operation of the single receiver filter in a convenient manner. To avoid this complication we instead do acquisition, tracking and despreading based on the root-Nyquist filtered spreading sequence. This is in fact equivalent to including separate receiver filters in the acquisition unit as well as in each of the tracking and despreading units. It is later made clear that the two approaches lead to the same set of sufficient statistics coming out of the despreader.

In the detection unit, performance enhancing signal processing such as beam forming, Rake combining including channel estimation, antenna diversity combining[3], multiuser detection and error control decoding can be included. If perfect channel information is available, beam forming, Rake and antenna diversity combining merely represent linear transformations to obtain a minimal set of sufficient statistics.

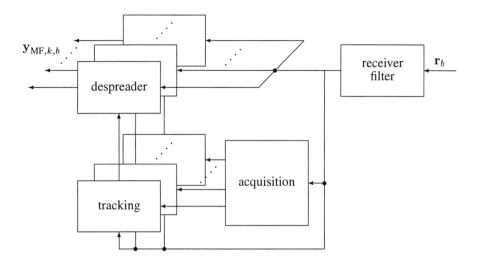

Figure 2.4 Block diagram of the base station despreading unit for each user.

From fundamental digital communications theory, this is in fact equivalent to filtering matched to the received waveform in order to maximise the signal-to-noise ratio (SNR) [9]. In case channel information is not available, some form of channel estimation must be done. This can be done separately to allow for coherent detection or adaptively allowing for either separate or joint channel estimation, Rake combining and beam forming. An example of separate adaption with antenna null steering is found in [10]. Traditionally detection and error control decoding are separated due to the inherent complexity of joint approaches, e.g., [8]. However, better performance is possible if they are designed jointly over all users [11]. Stepwise improvements can also be achieved by stepwise joint design [12]–[15].

2.3 DISCRETE-TIME BASEBAND UPLINK MODEL

In the following two sections we develop a discrete-time baseband uplink model derived from the structural block diagrams described above. The model is based on a single-cell environment and in all our discussions, we assume that K users are active simultaneously.

2.3.1 Synchronous CDMA

Referring to Fig. 2.1, user k in this multiuser CDMA system first encodes the information symbol stream $b_k(l)$, $l = 0, 1, 2, ...(\kappa L - n)/n$ at data rate $1/T_b$ symbols per second (sps) with a rate κ/n convolutional code followed by an interleaver. Denote the sequence output at rate $1/T_d = n/(\kappa T_b)$ sps as $d_k(i)$, where $i = 0, 1, 2, ...L - 1$ is the symbol interval index. At each symbol interval i, $d_k(i)$ is modulated by a rate $1/T_c = N/T_d = nN/(\kappa T_b)$ chips per second (cps) spreading sequence, $\mathbf{s}_k(i) = (s_k(iN + 1), s_k(iN + 2), ..., s_k((i+1)N))^\top$ with $s_k(j) \in \mathcal{S}$ where

$\mathcal{S}$ denotes the set of possible chip values. For binary spreading $\mathcal{S}_b = \{\pm 1/\sqrt{N}\}$, for quaternary spreading $\mathcal{S}_q = \{(\pm 1 \pm j)/\sqrt{2N}\}$. Examples of binary sequences are illustrated in Fig. 2.5. Note that the choice of chip values ensures that $\|\mathbf{s}_k(i)\|^2 = 1$

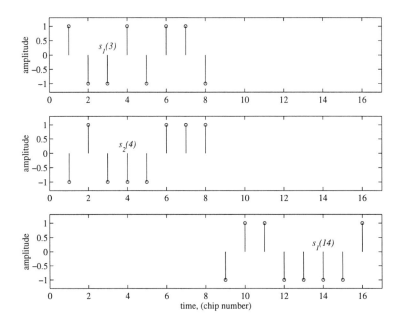

Figure 2.5 Examples of binary spreading sequences. From the top we have $\mathbf{s}_1(0)$, $\mathbf{s}_2(0)$ and $\mathbf{s}_1(1)$. Chips $s_1(3)$, $s_2(4)$ and $s_1(14)$ are specifically labelled.

so that modulation can be modelled as simple multiplication without modifying the energy contained in one symbol interval.

The bandwidth expansion of the spread signal is determined as

$$\text{PG} = \frac{1/T_c}{1/T_b} = \frac{nNT_b}{\kappa T_b} = \frac{nN}{\kappa},$$

which is traditionally termed the processing gain (PG). Both the encoding and the spreading contribute to the processing gain. The FEC code is usually quite powerful and relatively complicated to decode, while the spreading corresponds to a simple repetition code which can easily be decoded through correlation[4]. The spreading allows for fast techniques for acquisition and tracking and provides easily detectable initial multiuser separation, while the FEC provides error correction and, in some cases, refined user separation. In practice [6] the spreading sequences for each symbol interval are chosen pseudo-randomly and are known at the receiver. This models the use of spreading sequences with periods much longer than the data symbol duration.

As outlined in the previous section, for practical implementation all the baseband processing is usually done in discrete time. With this approach, we modulate each chip of $d_k(i)\mathbf{s}_k(i)$ by filtering the signal with a discrete-time pulse shaping filter[5]. In Fig.

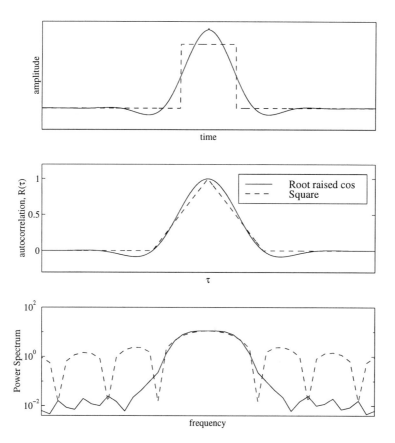

Figure 2.6 The filter impulse responses, autocorrelation functions and power spectra for a square pulse and a raised cosine pulse with a roll-off factor of 0.5..

2.6 examples of a square pulse and a raised cosine pulse with a roll-off factor of 0.5 are shown [9]. Included in the figure are the time functions (root-Nyquist filter impulse responses), their corresponding autocorrelation functions and their power spectra. It is clear that the square pulse only spans one chip interval while the raised cosine filter spans several chip intervals.

The filter operates at a rate of $1/T_s = Q/T_c$ where T_c is the chip interval. To express this operation algebraically, the spreading sequence must be zero padded before filtering to give $\mathbf{u}_1 \otimes \mathbf{s}_k(i)$ where $\mathbf{u}_1 = \left(1, \mathbf{0}_{Q-1}^\top\right)^\top$. The resulting transmitted discrete-time baseband signal due to symbol interval i for user k is then

$$\mathbf{x}_k(i) = d_k(i) \begin{pmatrix} \mathbf{0}_{iNQ} \\ (\mathbf{u}_1 \otimes \mathbf{s}_k(i)) * \mathbf{p} \\ \mathbf{0}_{(L-i-1)NQ} \end{pmatrix} = d_k(i) \begin{pmatrix} \mathbf{0}_{iNQ} \\ \hat{\mathbf{s}}_k(i) \\ \mathbf{0}_{(L-i-1)NQ} \end{pmatrix}, \qquad (2.1)$$

where $\mathbf{p}$ is the pulse shaping filter impulse response of length $P+1$, and $\hat{\mathbf{s}}_k(i)$ is the pulse shape filtered spreading sequence of length[6] $NQ + P$. This process is illustrated by way of example in Fig. 2.7. Here, $\mathbf{s}_k(i)$ and $\mathbf{x}_k(i)$ are depicted for $k = 1, 2$,

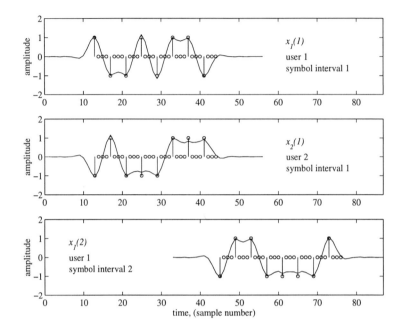

Figure 2.7 Examples of zero padded and pulse shape filtered spreading sequences, $\mathbf{x}_1(1)$, $\mathbf{x}_2(1)$ and $\mathbf{x}_1(2)$.

$i = 1$ and $k = 1$, $i = 2$, with $N = 8$, $Q = 4$ and $P = 6$. Of course all the users transmit continuously so the signal stream for user 1 in the example is represented by the sum of $\mathbf{x}_1(1)$ and $\mathbf{x}_1(2)$. When the two contributions are considered separately there is some overlap in the transition region at the symbol interval boundaries. These transient effects occur when the pulse shaping filter spans more than one chip interval. The extend of the transient is $P/2$ samples as observed in Fig. 2.7 at samples 42-44 ($P/2 = 3$). No intersymbol interference is encountered due to the symbol overlap as long as root-Nyquist pulses sampled at the correct instances are used. The length of $\mathbf{x}_k(i)$ is $LNQ + P$ which corresponds to the entire transmission interval of L symbols. Only the elements corresponding to symbol interval i are non-zero.

Now assume for simplicity that each user transmits over an AWGN channel with no phase or amplitude distortions. The received signal after A/D conversion is then

$$\mathbf{r} = \sum_{i=0}^{L-1} \sum_{k=1}^{K} \mathbf{x}_k(i) + \mathbf{n} = \sum_{i=0}^{L-1} \sum_{k=1}^{K} d_k(i) \begin{pmatrix} \mathbf{0}_{iNQ} \\ \hat{\mathbf{s}}_k(i) \\ \mathbf{0}_{(L-i-1)NQ} \end{pmatrix} + \mathbf{n}, \quad (2.2)$$

where $\mathbf{n}$ is a length $(LNQ + P)$ vector of independent, identically distributed (i.i.d.) complex Gaussian random variables of zero mean and variance $\sigma^2 = N_0$. The discrete-time model for generating the received signal is illustrated in Fig. 2.8. A

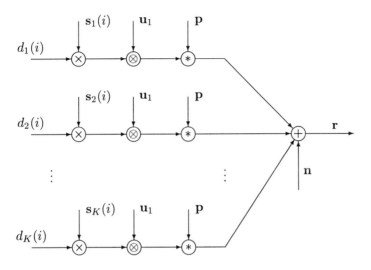

Figure 2.8 Discrete-time model of the uplink communication of a synchronous CDMA system without channel distortions.

more convenient form of (2.2) is

$$\mathbf{r} = \mathbf{S}\mathbf{d} + \mathbf{n}, \qquad (2.3)$$

where $\mathbf{S}$ is the matrix of transmitted spreading waveforms with column j expressed as

$$\hat{\mathbf{s}}_j = \begin{pmatrix} \mathbf{0}_{iNQ} \\ \hat{\mathbf{s}}_k(i) \\ \mathbf{0}_{(L-i-1)NQ} \end{pmatrix},$$

where j is uniquely related to k and i by $j = iK + k$. The structure of $\mathbf{S}$ is illustrated in Fig. 2.9 by placing dots representing the entries of $\hat{\mathbf{s}}_k(i)$. All other elements of $\mathbf{S}$ are zero. In this example $N = 4$, $Q = 2$ so $\mathbf{s}_k(i)$ has $NQ = 8$ samples and the transmitter filter has $P + 1 = 5$ taps, thus spanning 2.5 chip intervals. The synchronous structure is characterised by having all the spreading sequences aligned for any symbol interval i. Again we notice the overlap between symbol intervals as also observed in Fig. 2.7. This overlap represents the same transient effects. The data symbol vector is given by

$$\mathbf{d} = (d_1(0), d_2(0), \ldots, d_K(L-1))^\top = (d_1, d_2, \ldots, d_{LK})^\top,$$

where $d_j = d_k(i)$ with $j = iK + k$. In Eqns. (2.2) and (2.3), $\mathbf{r}$ represents a set of sufficient statistics for detection of $\mathbf{d}$. The length of $\mathbf{r}$ is however LNQ while only LK symbols were transmitted in total. A minimal set of sufficient statistics of length LK is obtained through correlation, matched to the received signal. This also ensures the maximisation of the SNR [9]. As discussed previously, in practice the received signal is firstly passed through a receiver filter $\mathbf{p}^*$ matched to the transmitter

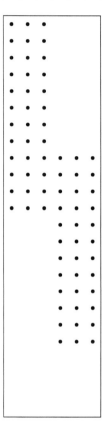

Figure 2.9 The fundamental structure of the matrix S for a synchronous system with $K = 3$, $L = 2$, $N = 4$, $Q = 2$ and $P = 4$. The entries representing $\hat{s}_k(i)$ are indicated with dots, while all other entries are zero.

filter and sampled at chip rate. Despreading or spreading sequence matched filtering with $s_k(i)$, $k = 1, ..., K$, is then done at chip rate with a bank of filters where the sampling instances for the individual filters are controlled by the tracking units. Instead of sampling the receiver filter at $1/T_c$, the output can be forwarded directly to the despreader where the matched filtering is now done with a bank of filters based on the zero-padded spreading sequences $\mathbf{u}_1 \otimes \mathbf{s}_k(i)$, $k = 1, ..., K$. The inserted zeros are in effect executing the sampling at $1/T_c$. Finally we can also despread the received signal directly with the relevant pulse shape filtered spreading sequence. In Fig. 2.10, it is made clear that these three approaches are equivalent. For modelling purposes the direct waveform matched filtering approach is the most convenient. This approach

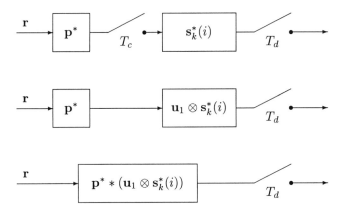

Figure 2.10 Equivalent matched filters for user k. The top diagram corresponds to the practical approach of one receiver filter while the bottom diagram depicts the approach adopted in the model.

corresponds algebraically to

$$\mathbf{y}_{\text{MF}} = \mathbf{S}^H \mathbf{r} = \mathbf{S}^H \mathbf{S} \mathbf{d} + \mathbf{S}^H \mathbf{n}, \tag{2.4}$$

where the subscript "MF" stands for matched filter.

In a mobile radio channel, each transmission path encounters temporal and spatial fading [16]. Furthermore, each user is transmitting at a specific power level. In our single-path K-user system this corresponds to each user being received with a random, time-dependent amplitude and phase, or equivalently, an arbitrary user k is affected by a random, time-dependent complex channel coefficient[7] $c_k(i)$. This is incorporated into the model to give us

$$\begin{aligned} \mathbf{r} &= \sum_{i=0}^{L-1} \sum_{k=1}^{K} c_k(i) \mathbf{x}_k(i) + \mathbf{n} \\ &= \sum_{i=0}^{L-1} \sum_{k=1}^{K} c_k(i) d_k(i) \begin{pmatrix} \mathbf{0}_{iNQ} \\ \hat{\mathbf{s}}_k(i) \\ \mathbf{0}_{(L-i-1)NQ} \end{pmatrix} + \mathbf{n}, \end{aligned}$$

which is illustrated in Fig. 2.11. The convenient matrix notation is now

$$\mathbf{r} = \mathbf{S}\mathbf{C}\mathbf{d} + \mathbf{n}, \tag{2.5}$$

where $\mathbf{C}$ is a $LK \times LK$ diagonal matrix containing the physical channel parameters with elements $c_j = c_k(i)$, $j = iK + k$.

The complex channel coefficient $c_k(i)$ contains all the fading and attenuation effects of the radio channel. Traditionally these effects are divided into small scale Rayleigh fading, large scale log-normal shadowing and path loss denoted for user k at symbol

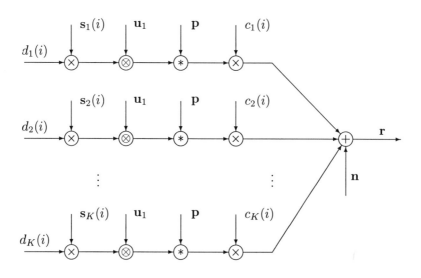

Figure 2.11 Discrete-time model of the uplink communication of a synchronous CDMA system over a mobile radio channel.

interval i by $h_k(i)$, ξ_k and $\varrho_k^{-\alpha_k}$, respectively. The variables ϱ_k, α_k and ξ_k are time-varying in practice. They are however, very slowly varying and can usually be assumed constant over L symbol intervals[8].

The Rayleigh coefficient $h_k(i)$ is usually assumed to have a Rayleigh distributed envelope, a uniformly distributed phase and a power spectral density determined by the surrounding environment. This can be modelled by a sequence of complex Gaussian random variables passed through a complex lowpass filter with a spectral shape determined by the Doppler spectrum and a cut-off frequency determined by the maximum Doppler shift [17]. The shadowing coefficient affecting user k, ξ_k is a log-normally distributed random variable while for the path loss, ϱ_k is the distance from the mobile to the base station and α_k is the path loss coefficient. Denoting the user controlled transmitted signal amplitude by $\beta_k(i)$, the diagonal elements of $\mathbf{C}$ can be expressed as

$$c_j = \beta_k(i)\varrho_k^{-\alpha_k}\xi_k h_k(i), \quad j = iK + k.$$

Notice that both (2.3) and (2.5) have the same fundamental algebraic form,

$$\mathbf{r} = \mathbf{Ad} + \mathbf{n}, \tag{2.6}$$

where $\mathbf{A} = \mathbf{S}$ and $\mathbf{A} = \mathbf{SC}$, respectively. Eqn. (2.6) clearly adheres to the aim of our model to associate a specific modulating waveform with each transmitted symbol.

A minimal set of sufficient statistics are obtained as in (2.4), i.e. $\mathbf{y}_{\text{MF}} = \mathbf{S}^H\mathbf{r}$. To allow for coherent detection and power control, the physical channel parameters must be known. The matrix $\mathbf{C}$ must therefore be estimated either based on $\mathbf{r}$ or $\mathbf{y}_{\text{MF}}$. In all the practical systems employing coherent uplink detection, channel estimation is aided by either pilot channels [3] or pilot symbols [10]. A variety of strategies can then be applied to obtain reliable channel estimates. For the purpose of further developing the model, we assume perfect knowledge of the channel parameters since estimates can easily be substituted in. Coherent detection is therefore assured from the statistics

$$\mathbf{y} = \mathbf{C}^H \mathbf{y}_{MF} = \mathbf{C}^H \mathbf{S}^H \mathbf{S} \mathbf{C} \mathbf{d} + \mathbf{C}^H \mathbf{S}^H \mathbf{n}, \qquad (2.7)$$

which is equivalently described by

$$\mathbf{y} = \mathbf{A}^H \mathbf{r} = \mathbf{A}^H \mathbf{A} \mathbf{d} + \mathbf{A}^H \mathbf{n} = \mathbf{R} \mathbf{d} + \mathbf{z}, \qquad (2.8)$$

where $\mathbf{y} = (y_1(0), y_2(0), ..., y_K(L-1))^T = (y_1, y_2, ..., y_{LK})^T$ with $y_j = y_k(i)$, $j = iK + k$, $\mathbf{R} = \mathbf{A}^H \mathbf{A}$ and $\mathbf{z} = \mathbf{A}^H \mathbf{n}$. This representation of $\mathbf{y}$ proves useful later on.

Transmission power control (TPC) is an integral part of any mobile communication system. The target for TPC is to maintain the received signal power for each user fixed at a pre-determined level. This also corresponds to a fixed, pre-determined SNR level for each user. Perfect TPC (symbol-by-symbol) can be accomplished by adjusting the transmitter amplitudes to allow for $\text{diag}(\mathbf{A}^H \mathbf{A}) = \text{diag}(\mathbf{P}_{\text{set}})$ where $\mathbf{P}_{\text{set}}$ is a diagonal matrix with the target received power levels. This leads to a transmitter amplitude for user k at symbol interval i determined as

$$\beta_k(i) = \frac{\sqrt{P_{\text{set},k}}}{\|\varrho_k^{-\alpha_k} \xi_k h_k(i) \hat{s}_k(i)\|}.$$

As we will discuss again later, for multipath propagation, a perfect TPC strategy requires instantaneous adjustments for each symbol interval due to the randomly changing spreading codes. Even if individual multipath components are not faded, the sum of the components can still result in virtually no received signal for an unfortunate choice of the specific spreading sequence. For any other choice of spreading sequence, the received signal encounters no attenuating effects. Instantaneous adjustments are impractical so instead TPC typically compensates for the effects of the physical channel only and not the variations due to the random spreading sequences. The strategy is now to adjust the transmitter amplitudes according to $\text{diag}(\mathbf{C}^H \mathbf{C}) = \text{diag}(\mathbf{P}_{\text{set}})$, which leads to

$$\beta_k(i) = \frac{\sqrt{P_{\text{set},k}}}{\|\varrho_k^{-\alpha_k} \xi_k h_k(i)\|}.$$

Both strategies for perfect TPC are based on symbol-by-symbol adjustments whether the adjustments are done instantaneously or with a fixed delay. This is not practical either so instead TPC adjustments are done periodically (slot-by-slot). As long as the maximum Doppler frequency is much smaller than the TPC frequency, then slot-by-slot adjustments work well. When the fading becomes faster than the TPC rate, a slot-by-slot strategy is no longer effective and instead only the average power level should be adjusted. In this case the strategy is described by $\text{diag}(\mathrm{E}\{\mathbf{C}^H \mathbf{C}\}) = \text{diag}(\mathbf{P}_{\text{set}})$, where the expectation is over the Rayleigh fading. So for user k, interval i,

$$\beta_k(i) = \frac{\sqrt{P_{\text{set},k}}}{\|\varrho_k^{-\alpha_k} \xi_k\|}.$$

2.3.2 Asynchronous CDMA

Random access, usually employed on the uplink, is one of the factors that makes a synchronous system impractical. When we allow the users to have random access, then each user encounters a transmission delay relative to other users. The delay is measured against an arbitrary reference selected such that all transmission delays are constrained to $0 < t_k < T_d$. A specific user can be chosen as the reference but this is not necessary.

Assume first for simplicity that all delays are constrained to be integer multiples of the sampling interval. The relative delay normalised to the sampling interval is then $\tau_k = t_k/T_s \in \{0, 1, ..., NQ - 1\}$. Introducing a transmission delay for user k leads to a modification of $\mathbf{x}_k(i)$. The position of $\hat{\mathbf{s}}_k(i)$ in $\mathbf{x}_k(i)$ of (2.1) corresponds to a delay of zero. For a delay of τ_k we get instead

$$\hat{\mathbf{x}}_k(i) = d_k(i) \begin{pmatrix} \mathbf{0}_{iNQ+\tau_k} \\ \hat{\mathbf{s}}_k(i) \\ \mathbf{0}_{(L-i)NQ-\tau_k-1} \end{pmatrix},$$

i.e., the position of $\hat{\mathbf{s}}_k(i)$ has been moved downwards by τ_k and the length of $\hat{\mathbf{x}}_k(i)$ is now $(L+1)NQ + P - 1$. This operation is in fact just a filtering of $\mathbf{x}_k(i)$ with a linear filter $\mathbf{g}_k = (g_{k,1}, g_{k,2}, ..., g_{k,NQ})^T$ of length NQ where

$$g_{k,j} = \begin{cases} 1 & \text{if } j = \tau_k \\ 0 & \text{otherwise} \end{cases}.$$

Defining for the time being $\mathbf{c}_k(i) = \mathbf{g}_k c_k(i)$ then we can obtain $\hat{\mathbf{x}}_k(i)$ as indicated in Fig. 2.12 with $\mathcal{M} = NQ$. The received signal is then expressed as

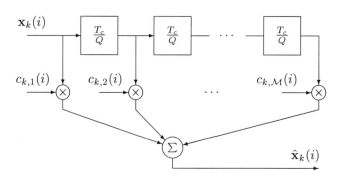

Figure 2.12 Diagram of the tapped-delay line filtering of $\mathbf{x}_k(i)$ with $\mathbf{c}_k(i)$.

$$\mathbf{r} = \sum_{i=0}^{L-1} \sum_{k=1}^{K} \mathbf{c}_k(i) * \mathbf{x}_k(i) + \mathbf{n} \qquad (2.9)$$

$$= \sum_{i=0}^{L-1} \sum_{k=1}^{K} \mathbf{c}_k(i) d_k(i) \begin{pmatrix} \mathbf{0}_{iNQ+\tau_k} \\ \hat{\mathbf{s}}_k(i) \\ \mathbf{0}_{(L-i)NQ-\tau_k-1} \end{pmatrix} + \mathbf{n}.$$

This is illustrated in Fig. 2.13 and the matrix formulation remains as Eqn. (2.5) where

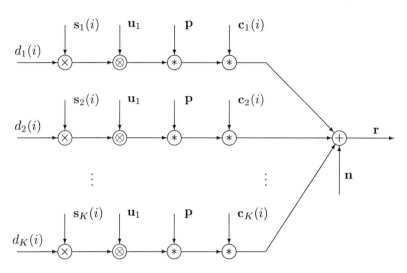

Figure 2.13 Discrete-time model of the uplink communication of an asynchronous CDMA system over a mobile radio channel.

S is now a matrix of shifted transmitted spreading waveforms with column j expressed as

$$\hat{\mathbf{s}}_j = \begin{pmatrix} \mathbf{0}_{iNQ+\tau_k} \\ \hat{\mathbf{s}}_k(i) \\ \mathbf{0}_{(L-i)NQ-\tau_k-1} \end{pmatrix}.$$

The asynchronous structure of **S** is illustrated in Fig. 2.14. The same basic example as used in Fig. 2.9 is considered. Here, we observe the characteristics of an asynchronous system as a staggered set of spreading sequences. The spreading sequences however, do not have to be ordered according to increasing transmission delays. We also note that the columns for user k, symbol intervals 1 and 2 overlap with the same amount as for the synchronous system due to the same transient effects.

The transmission delays are obviously continuous-time parameters. Since the system is based on a discrete-time representation, inaccuracies are introduced due to the finite resolution of delays. With an oversampling factor of Q, the maximum resolution error is limited to $\pm T_s/2 = \pm T_c/2Q$. This means that the A/D converter at the receiver can have a timing error in the sampling instances of up to $\pm T_c/2Q$ for each user. This timing error inherently leads to a loss in the received signal power. This is illustrated in Fig. 2.15 where the dashed curve represents the "actual" received waveform and the solid curve represents the estimated received waveform used for matched filtering at the receiver. This "aquired" waveform is based on delay estimates quantised to the nearest sampling instance, which differs from the actual delay by $T_s/2$ in the figure. The pulse shape is a raised cosine with a roll-off factor of 0.5,

38 CDMA TECHNIQUES FOR THIRD GENERATION MOBILE SYSTEMS

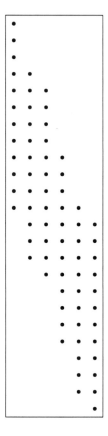

Figure 2.14 Fundamental structure of the matrix S for asynchronous CDMA with $K = 3$, $L = 2$, $N = 4$, $Q = 2$, $P = 4$ and $\tau_1 = 0$, $\tau_2 = 4T_s$ and $\tau_3 = 5T_s$.

$N = 8$, $Q = 2$, where the circles represent sample points at the receiver and the vertical lines denote the spreading sequence. The matched filtered statistic is then the sum of products of the two curves at sample points. It is clear that even for $Q = 2$ the loss is minimal. As long as Q is sufficiently large, the loss is negligible as confirmed in Table 2.1 for a few different pulse shapes.

The loss can however be included into the model by allowing each user to use an appropriately sampled pulse shaping filter where the sampling instances account for the non-resolvable delay $\gamma_k = \tau_k - \lfloor \tau_k \rfloor$. Let $p(t)$ be the analogue pulse shape. Now let us define,

$$p_k(n, \gamma_k) = p\left((n + \gamma_k)\frac{T_c}{Q}\right), \quad n = -\frac{P}{2}, -\frac{P-1}{2}, \ldots, \frac{P}{2},$$

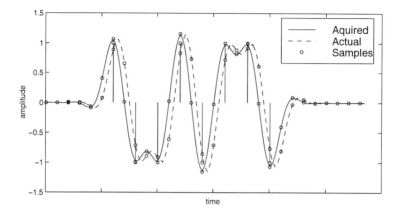

Figure 2.15 An example of the transmitter and receiver mismatched waveforms for a maximum timing error of $T_s/2$ caused by off-sets between the receiver clock and the transmitter clock.

Table 2.1 Table of the quantisation loss for different pulse shapes in terms of the maximum percentage energy loss due to sampling. Rc denotes a raised cosine filter while LPF - Hamming denotes a lowpass filter designed with a Hamming window.

| Pulse | Q | | | |
Shape	2	4	8	16
Rc ($h = 0.25$)	19.53	5.209	1.324	0.3322
Rc ($h = 0.5$)	21.28	5.727	1.459	0.3664
Rc ($h = 0.75$)	24.12	6.585	1.683	0.4232
Square	43.75	23.44	12.11	6.152
LPF - Hamming	20.37	5.46	1.388	0.3485

where obviously $\gamma_k \in [-1/2; 1/2]$. We therefore now have a user dependent root-Nyquist filter impulse response, $\mathbf{p}_k(\gamma_k)$ which may be used in place of $\mathbf{p}$ in (2.1) to give a new $\mathbf{x}_k(i)$.

It is clear from above that we need knowledge about the time delays. Initial acquisition is attempted based on $\mathbf{r}$, typically through a sliding correlator. The sliding correlator conducts an exhaustive search among all possible spreading sequence phases in order to acquire the right timing to within $\pm T_c/2$. The delay estimate is further refined and locked on to through tracking, an operation that also attempts to compensate for phase drifts between the oscillator in the transmitter and in the receiver. Usually some form of a delay-lock loop is applied which controls the sample timing of the despreading operation.

2.3.3 Multipath Propagation

As we are considering a discrete-time baseband model, we will represent the multipath fading channel as a discrete-time tapped-delay line [9] with a tap-spacing of T_c/Q as illustrated in Fig. 2.12 with $\mathcal{M} = \hat{M}$. The impulse response is denoted $\mathbf{c}_k(i) = \left(c_{k,1}(i), c_{k,2}(i), ..., c_{k,\hat{M}}(i)\right)^T$ where $\hat{M}$ indicates the maximum delay spread normalised to the sampling interval. By having a tap-spacing of T_c/Q, path delays can be resolved to within $\pm T_c/2Q$. As discussed above, for a sufficiently large Q, the modelling error due to quantisation is negligible. However, arbitrary continuous-time delays can be included as outlined in the previous section by having separate pulse shaping filters for each path for each user $\mathbf{p}_{k,m}(\gamma_{k,m})$. For notational simplicity, we will assume that Q is sufficiently large to ignore the quantisation error.

We will here also assume that there are only M resolvable multipath components per user so among the $\hat{M}$ filter taps, only M represent non-zero, time varying complex channel coefficients[9]. The sequence of non-zero channel coefficients are collected in a vector defined as $\hat{\mathbf{c}}_k(i) = \left(\hat{c}_{k,1}(i), \hat{c}_{k,2}(i), ..., \hat{c}_{k,M}(i)\right)^T$. Each non-zero channel coefficient has an associated relative path delay denoted $\tau_{k,m} = t_{k,m}/T_s \in \{0, 1, ..., \hat{M} - 1\}$, where $t_{k,m}$ is the absolute path delay. In practice, the multipath profile is user dependent and slowly time-varying, i.e., the number of paths and their locations in time change. Usually it can be assumed though that M and $\tau_{k,m}$ are constant over L symbol intervals[10], while the channel coefficients $\hat{c}_{k,m}(i)$ are independent over m and k and constant over one symbol interval[11]. As for the single-path channel, the coefficients contain all the radio channel effects,

$$\hat{c}_{k,m}(i) = \beta_k(i)\varrho_k^{-\alpha_k}\xi_k \hat{h}_{k,m}(i).$$

The coefficient $\hat{h}_{k,m}(i)$ is again assumed to have a Rayleigh distributed envelope, a uniformly distributed phase and a power spectral density determined by the Doppler properties of the path. The power level for each path can be regulated by adjusting the mean value of the Rayleigh distribution through the Doppler filtering. Specific tapped-delay line multipath models are available from various standardisation bodies, e.g., Vehicular B [18], COST207 [19, 20].

Transmitting $\mathbf{x}_k(i)$ over the multipath fading channel, we get a resulting signal similar to the asynchronous case where now M coefficients (one for each path) are non-zero instead of only one.

$$\hat{\mathbf{x}}_k(i) = \mathbf{c}_k(i) * \mathbf{x}_k(i) = d_k(i)\begin{pmatrix}\mathbf{0}_{iNQ}\\\mathbf{c}_k(i)*\hat{\mathbf{s}}_k(i)\\\mathbf{0}_{(L-i-1)NQ}\end{pmatrix}$$

$$= d_k(i)\begin{pmatrix}\mathbf{0}_{iNQ}\\\sum_{m=1}^{M}\hat{c}_{k,m}(i)\hat{\mathbf{s}}_{k,m}(i)\\\mathbf{0}_{(L-i-1)NQ}\end{pmatrix}. \quad (2.10)$$

This signal is of length $(L+\hat{M})NQ+P-1$ which again spans the entire transmission interval of L symbols. Here, $\hat{s}_{k,m}(i)$ is a length $(\hat{M}+1)NQ+P-1$ vector determined from $s_k(i)$ and $\tau_{k,m}$ as,

$$\hat{s}_{k,m}(i) = \begin{pmatrix} 0_{\tau_{k,m}} \\ \hat{s}_k(i) \\ 0_{\hat{M}NQ-\tau_{k,m}-1} \end{pmatrix}.$$

Substituting (2.10) appropriately into (2.9), the received discrete-time baseband signal is

$$\mathbf{r} = \sum_{i=0}^{L-1} \sum_{k=1}^{K} \sum_{m=1}^{M} \hat{c}_{k,m}(i) d_k(i) \begin{pmatrix} 0_{iNQ} \\ \hat{s}_{k,m}(i) \\ 0_{(L-i-1)NQ} \end{pmatrix} + \mathbf{n}. \qquad (2.11)$$

The discrete-time model for generating the received signal is the same as in Fig. 2.13 and the matrix formulation is still the same as in (2.5) where now $\mathbf{S}$ is the matrix of received spreading waveforms with column j expressed as

$$\hat{\mathbf{s}}_j = \begin{pmatrix} 0_{iNQ} \\ \hat{s}_{k,m}(i) \\ 0_{(L-i-1)NQ} \end{pmatrix},$$

where j is uniquely related to k, i and m by $j = iKM + (k-1)M + m$. We now have M multipath columns for each user for each symbol interval. The entire matrix $\mathbf{S}$ is of dimensions $(LNQ + \hat{M} + P - 1) \times KLM$, and consequently, the length of $\mathbf{r}$ is $LNQ + \hat{M} + P - 1$. The structure of $\mathbf{S}$ is illustrated in Fig. 2.16 for the same case as in Fig. 2.14. Now, however, each user has $M = 2$ paths with a multipath delay spread of $\hat{M} = 5T_s$. The multipath spreading sequences for user k, symbol interval i are inserted side by side as shown in the figure, again leading to a band-diagonal structure.

The channel matrix $\mathbf{C}$ also needs to be modified to reflect the multipath profile. The matrix is $KLM \times KL$ block diagonal,

$$\mathbf{C} = \begin{pmatrix} \hat{\mathbf{c}}_1(0) & 0 & \cdots & 0 \\ 0 & \hat{\mathbf{c}}_2(0) & & \vdots \\ & & \ddots & \\ 0 & 0 & \cdots & \hat{\mathbf{c}}_K(L-1) \end{pmatrix},$$

where

$$\hat{\mathbf{c}}_k(i) = \beta_k(i) \varrho_k^{-\alpha_k} \xi_k \hat{\mathbf{h}}_k(i).$$

Due to the increased dimensionality of $\mathbf{S}$ and $\mathbf{C}$, the minimal set of sufficient statistics in (2.7), restated here for convenience,

$$\mathbf{y} = \mathbf{C}^H \mathbf{S}^H \mathbf{S} \mathbf{C} \mathbf{d} + \mathbf{C}^H \mathbf{S}^H \mathbf{n}, \qquad (2.12)$$

42 CDMA TECHNIQUES FOR THIRD GENERATION MOBILE SYSTEMS

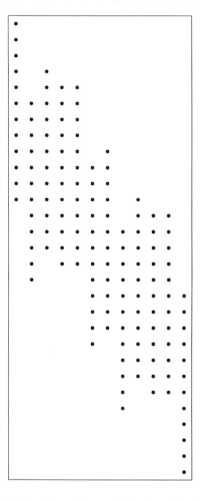

Figure 2.16 The fundamental structure of the matrix S for multipath CDMA with $K = 3$, $L = 2$, $N = 4$, $Q = 2$, $P = 4$, $M = 2$ and $\hat{M} = 5T_s$. The delays for the three users are $(0, 5T_s)$, $(3T_s, 4T_s)$ and $(4T_s, 9T_s)$, respectively.

now represents coherent multipath combining, also known as Rake reception [9]. The principle of Rake reception is to collect all the power in the multipath components through weighted coherent combining. As the multipath case is still described by Eqn. (2.5) and thus also by Eqn. (2.6), then it is clear that Rake combining is essentially traditional matched filtering where the filter is matched to the received waveform rather than the transmitted waveform,

$$\mathbf{y} = \mathbf{A}^H \mathbf{r} = \mathbf{A}^H \mathbf{A} \mathbf{d} + \mathbf{A}^H \mathbf{n}.$$

This is made clear by our model which specifically associates one modulating waveform to each transmitted symbol.

In Eqn. (2.12) we combine all the received multipath components. Some components however are received at a very low power level, making channel estimation considerably inaccurate. Such components are usually excluded from Rake combining. In fact, usually only the $\tilde{M}$ strongest paths are combined in practice.

In a CDMA system, multipath components can only be resolved at the receiver if they are separated by more than 1-2 chip intervals [17] depending on the autocorrelation function of the Nyquist pulse. The delays between paths however can take on any continuous value so unresolvable multipaths are possible. In Fig. 2.17 we consider an example for user k. The delay for the first path is used as a reference so the path

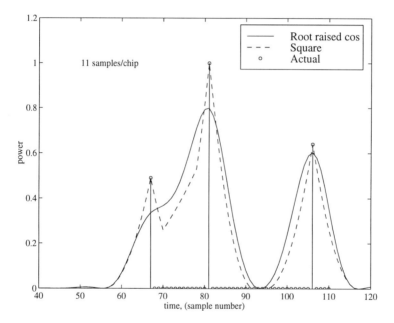

Figure 2.17 Example of the resolution of paths in a multipath channel using a sliding correlator energy detector based on a square pulse shape and a raised cosine pulse shape, respectively. The actual channel is included for reference.

delays take on the values $\tau_{k,1} = 0$, $\tau_{k,2} = 14T_s = 1.3T_c$ and $\tau_{k,3} = 39T_s = 3.5T_c$. In the figure, the outputs of a sliding correlator energy detector based on a square pulse and a raised cosine pulse with a roll-off of 0.5, respectively, are included. For the square pulse, all the paths can be resolved. For the raised cosine pulse however, paths 1 and 2 cannot be distinguished. In such cases we can for example assume that unresolvable paths combine to form an equivalent single path with a delay and complex channel coefficient determined by (unknown) functions of the path delays and channel coefficients. Very few results however, are currently available addressing this problem. Preliminary work suggesting some strategies for Rake finger assignment can be found in [21].

The $\tilde{M}$ strongest paths must obviously be acquired. The continually acquisition of the multipath components is known as multipath searching and must be done periodically to ensure that the $\tilde{M}$ strongest paths are always used for combining. Tracking and channel estimation is done for each acquired component independently. Denoting the spreading code and channel matrices corresponding to the $\tilde{M}$ resolvable, acquired paths by $\tilde{\mathbf{S}}$ and $\tilde{\mathbf{C}}$, respectively, the equivalent set of matched filtered and Rake combined statistics is

$$\tilde{\mathbf{y}} = \tilde{\mathbf{C}}^H \tilde{\mathbf{S}}^H \mathbf{SCd} + \tilde{\mathbf{C}}^H \tilde{\mathbf{S}}^H \mathbf{n} = \tilde{\mathbf{A}}^H \mathbf{Ad} + \tilde{\mathbf{A}}^H \mathbf{n}.$$

Perfect transmission power control in the multipath case requires instantaneous adjustments for each symbol interval due to the randomly changing spreading codes. This is due to the fact that an unfortunate combination of the multipath profile and the autocorrelation function for a specific spreading sequence can lead to severe attenuation of the received signal as illustrated in Fig. 2.18. Here, a two path case is considered

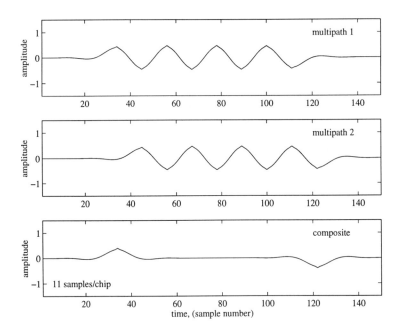

Figure 2.18 Example of the severe attenuation due to a destructive combination of multipath delays, coefficients and spreading sequence. The composite signal is practically annihilated.

where the composite signal is received with practically no power. At first sight this is surprising since the individual multipath components are not in deep fades. In this case however, the path delays, the complex channel coefficients and the spreading sequence combine destructively to virtually annihilate the received signal. As mentioned earlier, instantaneous adjustments are not practical so instead TPC only compensates for the effects of the physical channel. The techniques described earlier also applies in the multipath case with obvious complications. In this case the TPC maintains the power level of the coherently combined multipath components at a set target.

2.4 MULTIPLE-ANTENNA

In a practical system, the use of multiple antennas has proven a powerful concept to improve performance [22, 23]. In this section, we extend the single-antenna model from the previous section to conveniently incorporate the use of multiple antennas. We consider the use of multiple antennas for either diversity combining or beam forming as well as hybrids thereof.

2.4.1 Multiple-Antenna, Independent-Multipath Channel Model

With an antenna array consisting of B elements, there will generally be B different multipath channels per user, per symbol interval (see Fig. 2.19). Specifically we have

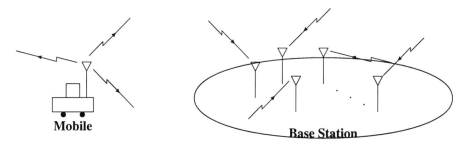

Figure 2.19 Illustration of the transmitted signal for one user arriving at B antennas at the receiver over B different multipath channels.

the channel impulse response $\hat{\mathbf{c}}_{k,b}(i)$ as the vector of multipath coefficients for user k received at antenna b in symbol interval i.

There will also be B received signal vectors $\mathbf{r}_1$ to $\mathbf{r}_B$, which when stacked one on top of the other gives the (very large) vector

$$\mathbf{r} = \begin{pmatrix} \mathbf{r}_1 \\ \vdots \\ \mathbf{r}_B \end{pmatrix} = \begin{pmatrix} \mathbf{S}_1 \mathbf{C}_1 \\ \vdots \\ \mathbf{S}_B \mathbf{C}_B \end{pmatrix} \mathbf{d} + \mathbf{n} = \begin{pmatrix} \mathbf{S}_1 & & \mathbf{0} \\ & \mathbf{S}_2 & \\ & & \ddots \\ \mathbf{0} & & & \mathbf{S}_B \end{pmatrix} \begin{pmatrix} \mathbf{C}_1 \\ \mathbf{C}_2 \\ \vdots \\ \mathbf{C}_B \end{pmatrix} \mathbf{d} + \mathbf{n}$$

$$= \mathbf{SCd} + \mathbf{n}. \qquad (2.13)$$

The matrices $\mathbf{S}_b$ and $\mathbf{C}_b$ are respectively the $\mathbf{S}$ and $\mathbf{C}$ matrices defined earlier for antenna b. As the continuing argument throughout this chapter, we note that again the system can conveniently be described by Eqn. (2.6) and thus, one specific modulating waveform can be associated with each transmitted symbol.

With this model for a multiple-antenna system, Rake combining is still represented by pre-multiplication with the matrix $\mathbf{C}^H\mathbf{S}^H$ (assuming perfect multipath acquisition), i.e., the KL decision statistics will again be described by Eqn. (2.7). In this system, Rake reception is identical to Rake combining of the signal received at each antenna element, and then an addition of the resulting B signals. Since the resulting signal at each antenna has already been appropriately weighted through Rake combining, the final addition is in fact performing maximal ratio antenna diversity combining.

The antennas can be organised in different ways. The above is the general model for arbitrary antenna positions. The following two scenarios are special cases of the above.

2.4.2 Multiple Antennas in a Uniform Linear Array

If the distance d between neighbouring elements of an antenna array is of the order of a wavelength λ, the multipath channels between a transmitter and the individual elements will no longer be independent as assumed in the previous section. The antenna array is now said to be capable of beam forming with a directional response that can be shaped by complex weighting of the antenna outputs. The beam forming scenario is significantly different from the previous space-diversity one because of the simple relationship between the channel responses seen at each antenna, as explained presently.

For simplicity, we consider a uniform linear array (ULA) in which antenna elements are placed in a straight line at a distance of d from each other (see Fig. 2.20). A plane

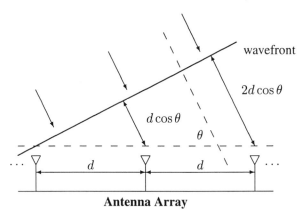

Figure 2.20 An illustration of the ULA with the direction of arrival θ indicated.

wave of wavelength λ arriving at an angle θ from broadside experiences a phase difference of

$$\phi = \frac{2\pi d \cos\theta}{\lambda},$$

between adjacent antennas. Assume further that the multipath component m of user k arrives at an angle[12] $\theta_{k,m}$. It is also assumed that all users are located in the far field, so that their signals arrive at the antenna array on a plane wave. The second antenna will then have a channel vector from the same user given by

$$\mathbf{c}_{k,2}(i) = \begin{pmatrix} e^{j\phi_{k,1}} & 0 & \cdots & 0 \\ 0 & e^{j\phi_{k,2}} & & \\ \vdots & & \ddots & \\ 0 & \cdots & 0 & e^{j\phi_{k,M}} \end{pmatrix} \mathbf{c}_{k,1}(i) = \mathbf{\Phi}_k \mathbf{c}_{k,1}(i),$$

where $\phi_{k,m} = 2\pi d \sin\theta_{k,m}/\lambda$ and $\mathbf{c}_{k,1}(i)$ is the channel for user k, symbol interval i at antenna 1. Clearly, $\mathbf{\Phi}_k$ relates not just the first two antenna elements but any adjacent elements b and $b+1$, i.e.,

$$\mathbf{c}_{k,b+1}(i) = \mathbf{\Phi}_k \mathbf{c}_{k,b}(i) = \mathbf{\Phi}_k^b \mathbf{c}_{k,1}(i),$$

for $b = 1, \ldots, B-1$.

In this case, the $\mathbf{C}$ matrix in (2.13) becomes

$$\mathbf{C} = \begin{pmatrix} \mathbf{C}_0 \\ \mathbf{\Phi}\mathbf{C}_0 \\ \vdots \\ \mathbf{\Phi}^{B-1}\mathbf{C}_0 \end{pmatrix} = \begin{pmatrix} \mathbf{I} \\ \mathbf{\Phi} \\ \vdots \\ \mathbf{\Phi}^{B-1} \end{pmatrix} \mathbf{C}_0 = \mathbf{\Psi}\mathbf{C}_0.$$

Here $\mathbf{\Phi} = \mathrm{diag}(\mathbf{\Upsilon}, \mathbf{\Upsilon}, \ldots, \mathbf{\Upsilon})$, with

$$\mathbf{\Upsilon} = \begin{pmatrix} \Phi_1 & & & \bigcirc \\ & \Phi_2 & & \\ & & \ddots & \\ \bigcirc & & & \Phi_K \end{pmatrix},$$

where $\mathbf{\Phi}$ is of dimension $KLM \times KLM$. The channel matrix has the following form,

$$\mathbf{C}_0 = \begin{pmatrix} \mathbf{c}_1(0) & & & \bigcirc \\ & \mathbf{c}_2(0) & & \\ & & \ddots & \\ \bigcirc & & & \mathbf{c}_K(L-1) \end{pmatrix}.$$

Since each antenna now sees the same path delays from each user, the sub-matrices $\mathbf{S}_b$ in (2.13) are now all equal to $\mathbf{S}_1$, so that $\mathbf{S} = \mathrm{diag}(\mathbf{S}_1, \ldots, \mathbf{S}_1)$. The received signal vector is therefore

$$\mathbf{r} = \mathbf{S}\mathbf{\Psi}\mathbf{C}_0\mathbf{d} + \mathbf{n} = \mathbf{A}\mathbf{d} + \mathbf{n},$$

and the Rake-combined output, following the objective of the model, is again

$$\mathbf{y} = \mathbf{A}^H\mathbf{A}\mathbf{d} + \mathbf{A}^H\mathbf{n}.$$

The resulting modulating waveform for user k with two antenna elements in the ULA is shown in Fig. 2.21. It is clear that the two contributions only differ by a constant phase offset.

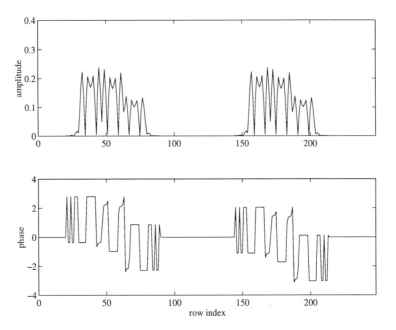

Figure 2.21 Example of the resulting modulating waveform for user k with two antenna elements in a ULA.

2.4.3 Combined Antenna Diversity and Beam Forming

The B antenna elements can be divided into B_d diversity locations with B_u elements in a ULA at each location, i.e., $B = B_d B_u$. Combining the results from the previous two subsections, we get a received signal of the form,

$$\mathbf{r} = \mathbf{S}\mathbf{\Psi}\mathbf{C}\mathbf{d} + \mathbf{n} = \mathbf{A}\mathbf{d} + \mathbf{n},$$

where

$$\mathbf{S} = \begin{pmatrix} \text{diag}(\mathbf{S}_1, ..., \mathbf{S}_1) & & & \mathbf{O} \\ & \text{diag}(\mathbf{S}_2, ..., \mathbf{S}_2) & & \\ & & \ddots & \\ \mathbf{O} & & & \text{diag}(\mathbf{S}_{B_u}, ..., \mathbf{S}_{B_u}) \end{pmatrix},$$

with $\text{diag}(\mathbf{S}_b, ..., \mathbf{S}_b)$ consisting of B_d replicas of $\mathbf{S}_b$.

$$\mathbf{\Psi} = \begin{pmatrix} \mathbf{\Psi}_1 & & & \mathbf{O} \\ & \mathbf{\Psi}_2 & & \\ & & \ddots & \\ \mathbf{O} & & & \mathbf{\Psi}_{B_d} \end{pmatrix},$$

where $\boldsymbol{\Psi}_b$ is the matrix corresponding to the ULA at location b, and

$$\mathbf{C} = \begin{pmatrix} \mathbf{C}_1 \\ \mathbf{C}_2 \\ \vdots \\ \mathbf{C}_{B_d} \end{pmatrix}.$$

Again, Rake combining can be done by

$$\mathbf{y} = \mathbf{C}^H \boldsymbol{\Psi}^H \mathbf{S}^H \mathbf{r} = \mathbf{A}^H \mathbf{r}.$$

2.4.4 Processing Block Diagram

The order in which we have chosen to do the signal processing in this model suggests a functional separation in the multiuser detection unit. This separation is depicted in Fig. 2.22. It is however, not necessary to separate the functionalities this way or do

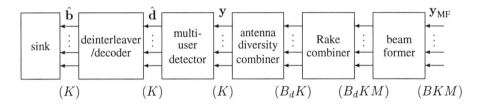

Figure 2.22 Block diagram of the base station multiuser detection unit. The numbers in parentheses indicate the number of signals passed between blocks in one symbol interval.

the signal processing in this order. Other strategies can lead to an identical minimal set of sufficient statistics $\mathbf{y}$. As mentioned earlier, the functional separation of detection and error control decoding does not lead to optimal performance. Better performance is achieved through joint operation as shown in [11, 15].

2.5 DISCUSSION

In this section we discuss some features of the developed model. The model allows for an intuitive insight into the interference patterns encountered in a CDMA system. The interpretation of the CDMA channel as a time-varying convolutional encoder becomes obvious based on this representation. We emphasise the specific characteristics of CDMA interference and then discuss the encoding equivalence of the channel. Finally, we briefly discuss the motivation behind and strength of iterative techniques for detection and decoding.

2.5.1 Interference Structure

We have so far provided a detailed model of a multiple access channel. Several signals are fed into the channel $\mathbf{d}$ and several signals emerge from the channel $\mathbf{y}$. When we form **y** from **r** by matched filtering for **A** we get precisely one output for every input as we originally intended regardless of the included system concepts.

$$\mathbf{y} = \mathbf{R}\mathbf{d} + \mathbf{d}, \qquad (2.14)$$

where **y** and **d** are column vectors of length LK and **R** is a band-diagonal square matrix of dimensions $LK \times LK$ for a single antenna system. The fundamental band-diagonal structure of **R** is illustrated in Fig. 2.23 for the multipath case of Fig. 2.16 with 6 symbol intervals instead of 2.

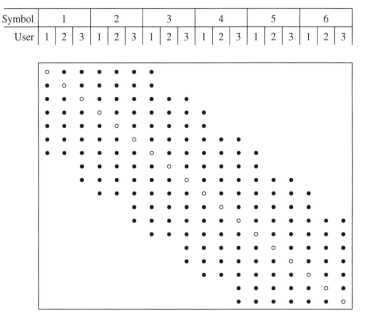

Figure 2.23 The fundamental structure of **R** for the multipath case in Fig. 2.16 where $K = 3$ and $L = 6$. The circles on the diagonal represent the autocorrelations. The remaining dots represent non-zero cross correlations. The band-diagonal structure is obvious.

Interference arises since, in general, every output has a contribution from every input. Under ideal conditions of orthogonal columns of **A** there would be no interference, but such a scenario is virtually impossible in practice since we would require an improbable coincidence of multipath channel, spreading code and direction of arrival. In the general case where any output y_j potentially depends on every input $\{d_j : j = 1, \cdots, LK\}$, we need not even make the direct association between y_j and d_j. Indeed, in pathological cases, some other observation, y_i say, may contain more information about d_j than y_j. In practice however we find that d_j is among the strongest contributors to y_j and that the other contributors are localised, in terms of the subscript $j = iK + k$ around j as indicated by the band-diagonal structure of **R** illustrated in Fig. 2.23. Most of the interference in y_j comes from symbols transmitted within a symbol interval or two of the symbol interval corresponding to j.

In order to provide a feeling for the components constituting each observation y_j we consider the components of (2.14) pertaining to y_j. Assuming perfect power control, we can write, for the noiseless case,

$$\mathbf{y} = \mathbf{Rd} = (\mathbf{P}_{\text{set}} + \mathbf{M})\mathbf{d}, \qquad (2.15)$$

where $\text{diag}(\mathbf{R}) = \text{diag}(\mathbf{P}_{\text{set}})$. Then for y_j

$$y_j = P_{\text{set},k} d_j + \mathbf{m}_j^H \mathbf{d},$$

where $\mathbf{m}_j^H$ is row[13] j of the matrix $\mathbf{M}$. It is of interest to compare the average size of the two components, one being the signal component and the other being the interference component. Assuming perfect knowledge of the channel matrix $\mathbf{A}$, and therefore perfect matched filtering, the power in the interference term (K/NB) is shown in Fig. 2.24 as a function of K.

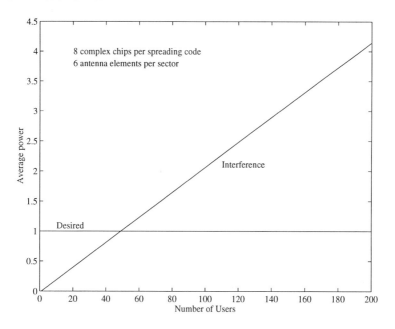

Figure 2.24 Interference power versus desired signal power as a function of the number of active users.

The desired signal component is the horizontal line at 1. We see that the interference power should not be ignored and moreover should be removed as effectively as possible. If not, effective signal to interference ratios well below 0 dB are possible. It is apparent from Fig. 2.24 that if the interference is treated as random noise that cannot be estimated then the channel capacity per user will go to zero as the number of users increases [24]. Although this is also true when the structure of the interference is exploited through joint detection, the degradation is more rapid in the former case. It is true that the CDMA channel is in some sense interference limited[14], but this limit is imposed by the adder channel nature of CDMA, not by excessive noise in correlator outputs.

2.5.2 Encoding Interpretation

The encoding process for convolutional coding can be described as

$$\mathbf{d} = \mathbf{G}^\mathsf{T}\mathbf{b}, \tag{2.16}$$

where $\mathbf{b}$ is the stream of information symbols to be encoded, $\mathbf{d}$ is the stream of encoded output symbols and $\mathbf{G}$ is the so-called generator matrix for the code [26]. The structure of $\mathbf{G}$ is illustrated below for a rate 1/2, constraint length 3 code with generator polynomials $\mathbf{g}_1 = (g_1(0), g_1(1), g_1(2))^\mathsf{T}$ and $\mathbf{g}_2 = (g_2(0), g_2(1), g_2(2))^\mathsf{T}$,

$$\mathbf{G} = \begin{pmatrix} g_1(0) & g_2(0) & g_1(1) & g_2(1) & g_1(2) & g_2(2) & & & \\ & & g_1(0) & g_2(0) & g_1(1) & g_2(1) & g_1(2) & g_2(2) & \\ & & & & g_1(0) & g_2(0) & g_1(1) & g_2(1) & g_1(2) & g_2(2) \\ & & & & & & \ddots & & & & \ddots \end{pmatrix}$$

For a trivial rate one code (e.g., a scrambler [27] or an ISI channel [28]), $\mathbf{G}^\mathsf{T}$ is a lower left triangular matrix.

For our CDMA model, Eqn. (2.15) is of the same form as (2.16). Furthermore, Fig. 2.23 indicates that $\mathbf{R}$ has a band-diagonal structure similar to $\mathbf{G}^\mathsf{T}$. The structure of $\mathbf{R}$ is not necessarily lower left triangular though. As long as $\mathbf{R}$ is band-diagonal however, it can conveniently be interpreted as a generator matrix where the non-causalities introduced by the band-diagonal structure are absorbed into the metric calculation. We can therefore consider $\mathbf{R}$ to represent a rate one time-varying, convolutional encoding of the input stream $\mathbf{d}$ having at least q^{K-1} states for q-ary modulation formats. The impact of this realisation is that the maximum likelihood (ML), or Maximum-A-Posteriori (MAP) sequence detectors for $\mathbf{d}$ given $\mathbf{y}$ and $\mathbf{R}$ requires the use of a trellis decoder. Unfortunately such a trellis has a number of states that is exponential in the number of users.

2.5.3 Iterative Detection Techniques

We have seen that each output symbol y_j consists of a desired signal component and a MAI component. From the convolutional code point of view we know that the output symbol of the encoder is determined equally by the state (previous inputs) and the new input. In the CDMA case we have argued above that the new input symbol (the desired symbol) predominantly determines the output symbol. The state consists of the interfering symbols which have, on average, a secondary influence on the output symbol. This fact can be used to great advantage in iterative detection strategies such as interference cancellation.

Although we know how to implement ML or MAP sequence detection, it is prohibitively complex, so perfect[15] interference cancellation is an attractive alternative since it efficiently removes all of the interference in each observation y_j, leaving behind the desired contribution from d_j in the presence of truly random noise. The problem with such an approach is that the receiver must know all of the interfering symbols. In practice, possibly erroneous estimates can be used instead, however, the quality of such estimates is paramount for the resulting performance.

Based on minimum mean squared error considerations and convenient Gaussian approximations, it is possible to derive an optimal tentative decision function for the intermediate estimates [30]. For BPSK modulation formats, the function is a hyperbolic tangent of the matched filtered statistics y, while for QPSK it is the same function applied independently to the in-phase and quadrature components. As a good approximation to the hyperbolic tangent, a piecewise linear clipped soft decision can be used. In [31], close to ML performance is reported for a multi-stage successive interference cancellation based on the linear clipper.

In general, successive cancellation provides better performance than parallel cancellation at the expense of a greater detection delay. The intermediate estimates for parallel interference cancellation (PIC) can however be improved through weighted cancellation. Significant improvements have been observed both for a weighted PIC based on the hyperbolic tangent as well as the linear clipper and the linear soft decision [30]. For the non-linear cases, the weighting factors are found heuristically as opposed to the linear case where it is possible to analytically derive the optimal weighting factors with respect to the mean squared error [32]. The weighted linear PIC then effectively implements the MMSE detector for long–code CDMA at a remarkably low complexity [33].

The quality of the estimates can be further improved by incorporating error control coding within the cancellation stages. Embedding Viterbi decoding within the cancellation structure, significant improvements are achieved at the expense of detection delay [34]. An all together new level of performance has been reached by interpreting the bank of K single-user convolutional encoders at the transmission side together with the time-varying convolutional encoder representing the CDMA channel as a serially concatenated turbo-coding system. This system can be iteratively decoded based on the turbo-decoding principles as illustrated in Fig. 2.25 [14].

However, as we have already established, in its pure form, the CDMA MAP decoder, is prohibitively complex for practical systems. Since the outputs of the CDMA encoder are predominantly determined by the corresponding input symbol, it is possible however to sub-optimise the corresponding MAP decoder into a parallel interference canceller based on probabilistic information [15]. Only marginal performance degradation as compared to optimal MAP is observed at a substantially lower complexity. This is due to the inherent power of the iterative exchange of reliability information rather than the optimality of each step. Performance approaching the single-user bound is observed even for overloaded systems [15]. Indeed, such a receiver is seen to provide levels of service within a dB of the information theoretic capacity of the CDMA channel.

2.6 CONCLUDING REMARKS

In this chapter we have developed a linear model for CDMA transmission over a multipath, time-dispersive fading channel usually encountered in cellular mobile communications. The model is based on the structural principles of practical systems where all baseband processing is done in discrete-time. For each additional system concept introduced, the corresponding algebraic model retains the same fundamental

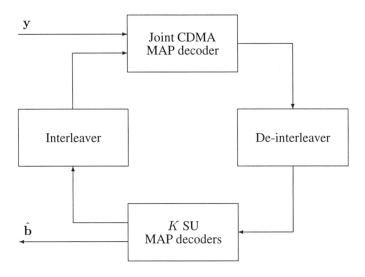

Figure 2.25 The fundamental structure of a powerful iterative receiver for joint detection and decoding.

form where each transmission is associated with a specific modulating discrete-time waveform.

An expression for the oversampled, received baseband vector allows for realistic investigations and design of channel estimation, transmit power control algorithms, acquisition, multipath searching and tracking as well as detection and decoding. The model invites the interpretation of the CDMA channel as a time-varying convolutional encoding which then in turn leads to the idea of modelling the K single-user encoders and the CDMA channel as a concatenated convolutional coding scheme. This structure, developed originally for an asynchronous CDMA system, encourages the application of iterative decoding techniques in which the inner decoder creates reliability information for the outer decoder which in turn creates reliability information for the inner decoder. Performance within a dB of the information theoretic capacity of the CDMA channel is achieved. Since the model retains the same algebraic structure for all additionally included system concepts or channel impairments, the same decoding principles can be used and performance in reach of the single-user bound obtained for very practical scenarios.

Notes

1. For system designs where the uplink and the downlink are structurally equivalent, the received signal model developed here is, with obvious minor alterations, equally applicable for both links.

2. For the single user transmitters, we label the data streams with a subscript, e.g., $\mathbf{b}_k$ and $\mathbf{d}_k$, while the data streams at the receiver consisting of all transmissions, possibly interlaced, are denoted by $\mathbf{b}$ and $\mathbf{d}$, respectively.

3. Here we distinguish between antenna elements in a phased array used for beam forming and antenna elements used for space diversity.

4. In the single user case (or when all users are orthogonal) correlation or matched filtering is in fact performing maximum likelihood detection.

5. This discrete-time pulse shaping filter is merely an appropriately sampled version of the corresponding continuous-time filter.

6. The last $Q-1$ elements of $\hat{\mathbf{s}}_k(i)$ are zero since $\mathbf{u}_1 \otimes \mathbf{s}_k(i)$ is terminated by $Q-1$ zeros. Strictly speaking $\hat{\mathbf{s}}_k(i)$ is then of length $(N-1)Q+1+P$. We will however, for notational ease, define $\mathbf{u}_1 \otimes \mathbf{s}_k(i)$ to be of length NQ and $\hat{\mathbf{s}}_k(i)$ to be of length $NQ+P$ regardless of the value of Q.

7. It is commonly assumed that the channel time variations are slow enough to consider the channel time-invariant over one symbol interval.

8. Path loss and shadowing are large scale channel effects and are very location dependent. They can be considered constant for at least up to 5-10 s.

9. This is merely a convenient assumption. The general case of $M_k(i)$ paths for user k can also be incorporated through appropriate adjustment of the size of the involved vectors and matrices.

10. The number of multipaths and the corresponding path delays change relatively slowly with time compared to a symbol time interval. As long as LT_b is small as compared to the time variations of $M_k(i)$ and $\tau_{k,m}(i)$, then our assumptions hold true. Assuming that the profile is constant over 100-300 ms, then for a symbol rate of 10 ksps, $L \simeq$ 1000-3000 symbols.

11. This holds true for any realistic terrestrial Doppler rate.

12. The direction of arrival (DOA) is connected to the location of the mobile and the multipath profile. The location of the mobile changes very slowly while the multipath profile changes more rapidly. It is reasonable to assume that the DOA remains constant for 100-300 ms periods.

13. Column j of $\mathbf{M}$ is denoted $\mathbf{m}_j$. Since both $\mathbf{R}$ and $\mathbf{M}$ are hermitian, then row j of $\mathbf{M}$ is $\mathbf{m}_j^H$.

14. The capacity of CDMA, expressed as bits per channel use per user goes to zero as the number of users goes to infinity [25, pg. 406].

15. The information theoretic capacity of the CDMA channel places fundamental constraints on the number of users that can be perfectly cancelled [29].

References

[1] M. Sundelin, W. Granzow and H. Olofsson, "A test system for evaluation of the WCDMA technology," in the proceedings of the *IEEE Vehicular Technology Conference*, pp. 983–987, May 1998, Ottawa, Ontario.

[2] F. Adachi and M. Sawahashi, "Coherent multi-rate wideband DS-CDMA for next generation mobile radio access: Link design and performance," in the proceedings of the *Third Asia-Pacific Conference on Communications '97*, pp. 1479–1483, Dec. 1997, Sydney, Australia.

[3] K. Okawa, Y. Okumura, M. Sawahashi and F. Adachi, "1.92 Mbps data transmission experiments over a coherent W-CDMA mobile radio link," in the proceedings of the *IEEE Vehicular Technology Conference*, pp. 1300–1304, May 1998, Ottawa, Ontario.

[4] P. Jung and J. Blanz, "Joint detection with coherent receiver antenna diversity in CDMA mobile radio systems," *IEEE Trans. Veh. Technol.*, vol. 44, pp. 76–88, Feb. 1995.

[5] R. A. Horn and C. R. Johnson, *Topics in Matrix Analysis*, Cambridge University Press, 1991.

[6] *TIA/EIA IS-95A: Mobile Station-Base Station Compatibility Standard for Dual-Mode Wideband Spread Spectrum Cellular System*, Mar. 95.

[7] *JTC(Air): W-CDMA (Wideband Code Division Multiple Access) Air Interface Combatibility Standard for 1.85 to 1.99 GHz PCS Applications*, JTC(Air)/95.02.02-037R1, T1P1/95-036, Feb. 95.

[8] A. Viterbi, *CDMA - Principles of Spread Spectrum Communication*, Addison-Wesley, 1995.

[9] J. G. Proakis, *Digital Communications*, 3rd edition, McGraw-Hill, 1995.

[10] S. Tanaka, M. Sawahashi and F. Adachi, "Pilot symbol-assisted decision-directed coherent adaptive array diversity for DS-CDMA mobile radio reverse link," *IEICE Trans. Fund. Elec, Comp. Comp. Sc.*, vol. 12, pp. 2445–2454, Dec. 1997.

[11] T. R. Giallorenzi and S. G. Wilson, "Multiuser ML sequence estimator for convolutionally coded asynchronous DS-CDMA systems," *IEEE Trans. Commun.*, vol. 44, pp. 997–1008, Aug. 1996.

[12] T. R. Giallorenzi and S. G. Wilson, "Suboptimum multiuser receivers for convolutionally coded asynchronous DS-CDMA systems," *IEEE Trans. Commun.*, vol. 44, pp. 1183–1196, Aug. 1996.

[13] P. D. Alexander, L. K. Rasmussen, and C. B. Schlegel, "A linear receiver for coded multiuser CDMA," *IEEE Trans. on Commun*, vol. 45, pp. 605–610, May 1997.

[14] M. C. Reed, C. B. Schlegel, P. D. Alexander and J. A. Asenstorfer, "Iterative multiuser detection for CDMA with FEC," in the proceedings of the *International Symposium on Turbo Codes & Related Topics*, pp. 162–165, Sept. 1997, Brest, France.

[15] P. D. Alexander, A. J. Grant and M. C. Reed, "Iterative detection in code-division multiple-access with error control coding," *European Trans. Telecommun.*, Vol. 9, July-Aug. 1998.

[16] W. C.Jakes, *Microwave Mobile Communications*, IEEE Press Classic Reissue, 1974.

[17] T. S. Rappaport, *Wireless Communications - Principles & Practice*, Prentice-Hall, 1996.

[18] *ITU-R TG8-1: Procedures for evaluation of transmission technologies for FPLMTS*, 8-1/TEMP/233-E, Sept. 95.

[19] European Cooperation in the Field of Scientific and Technical Research EURO-COST 207, "Digital land mobile radio communications," final rep., Luxemburg. Office for Official Publications of the European Communities, 1989.

[20] European Cooperation in the Field of Scientific and Technical Research EURO-COST 231, "Urban transmission loss models for mobile radio in the 900 and 1800 MHz bands," Revision 2, The Hague, The Netherlands, Sept. 1991.

[21] G. E. Bottomley, E. Sourour, R. Ramèsh and S. Chennakeshu, "Optimizing the performance of limited complexity RAKE receivers," in the proceedings of the *IEEE Vehicular Technology Conference*, pp. 968–972, May 1998, Ottawa, Ontario.

[22] L. C. Godara, "Application of antenna arrays to mobile communications, part I: Performance improvement, feasibility, and system considerations," *Proceedings of the IEEE*, vol. 85, pp. 1031–1060, July 1997.

[23] L. C. Godara, "Application of antenna arrays to mobile communications, part II: Beamforming and direction-of-arrival considerations," *Proceedings of the IEEE*, vol. 85, pp. 1195–1245, Aug. 1997.

[24] S. Verdú, "The capacity region of the symbol-asynchronous Gaussian multiple access channel," *IEEE Trans. Inform. Theory*, vol. 35, pp. 733–751, July 1989.

[25] T. M. Cover and J. A. Thomas, *Elements of Information Theory*, Wiley & Son, 1991.

[26] S. B. Wicker, *Error Control Coding for Digital Communication and Storage*, Prentice Hall, 1995.

[27] P. K. Gray and L. K. Rasmussen, "Bit error rate reduction of TCM systems using linear scramblers," in the proceedings of the *IEEE International Symposium on Information Theory '95*, p. 64, Sept. 1995, Whistler, Canada.

[28] G. D. Forney, "Maximum-likelihood sequence estimation of digital sequences in hte presence of intersymbol interference," *IEEE Trans. Inform. Theory*, vol. 18, pp. 363–378, May 1972.

[29] A. J. Grant and P. D. Alexander, "Random sequence multisets for synchronous code-division multiple-access channels," to appear in the *IEEE Trans. Inform. Theory*.

[30] D. Divsalar, M. Simon and D. Raphaeli, "Improved parallel interference cancellation for CDMA," *IEEE Trans. Commun.*, vol. 46, pp. 258–268, Feb. 1998.

[31] L. K. Rasmussen, T. J. Lim and H. Sugimoto, "Fundamental BER-floor for long-code CDMA," to appear in the proceedings of the *IEEE International Symposium on Spread Spectrum Techniques and Applications '98*, Sept. 1998, Sun City, South Africa.

[32] D. Guo, L. K. Rasmussen, S. Sun, T. J. Lim and C. Cheah, "MMSE-based linear parallel interference cancellation in CDMA," to appear in the proceedings of the *IEEE International Symposium on Spread Spectrum Techniques and Applications '98*, Sept. 1998, Sun City, South Africa.

[33] D. Guo, *Linear Parallel Interference Cancellation in CDMA*, M.Eng. Thesis, National University of Singapore, 1998.

[34] M. R. Koohrangpour and A. Svensson, "Joint interference cancellation and Viterbi decoding in DS-CDMA," in the proceedings to the *IEEE International Symposium on Personal, Indoor and Mobile Radio Communications '97*, pp. 1161–1165, Sept. 1997, Helsinki, Finland.

3 ANTENNA ARRAYS FOR CELLULAR CDMA SYSTEMS

P.M. Grant, J.S. Thompson and B. Mulgrew

Signals and Systems Group,
The Department of Electronics and Electrical Engineering,
The University of Edinburgh,
Kings Buildings,
Edinburgh EH9 3JL, UK.

Peter.Grant@ee.ed.ac.uk

3.1 INTRODUCTION

This chapter provides a tutorial review of cell site antenna array processing techniques for wireless code division multiple access (CDMA) networks. The basic ideas behind CDMA and antenna array processing are explained, followed by an introduction to channel models for antenna arrays. A number of algorithms for operating a base station antenna array receiver are then described. Major issues in the selection of a suitable algorithm are highlighted through simulation results. The potential increase in uplink capacity is analysed briefly and finally the extension of antenna array processing to downlink transmission is described.

3.2 BACKGROUND

Current cellular or personal communication service (PCS) networks are often based on time division multiple access technology, such as the European GSM system and the IS-54 standard in North America (NA) [1]. However, the NA IS-95 standard is based on code division multiple access (CDMA) techniques and is currently being deployed widely in NA and Asia. The third generation (3G) of cellular systems is also being developed across the world, to provide higher data rate services than are currently available on cellular networks. Most of the candidate systems will be based on CDMA techniques, such as the European UTRA system [2] and the NA IS-95 3G [3].

In general, the most complex and expensive part of the radio path for these systems is the base station. As a result, manufacturers have been designing networks which have high efficiency in terms of the bandwidth occupied and the number of users per base station [1]. This trend has been at the expense of high-power transmitters and receivers which employ very computationally expensive signal processing techniques. Recent studies have shown that considerable system capacity gains are available from exploiting the different spatial locations of cellular users: see [4] for a general overview of antenna arrays for cellular systems. There are a number of methods to achieve capacity improvements, from simple sectorisation schemes [5] to complex adaptive antenna array techniques [6]. These methods will be discussed in more detail below. Firstly, however, the basic principles of CDMA communications will be described. Then, the motivations for using antenna array techniques in cellular systems will be addressed.

3.3 DIRECT-SEQUENCE CDMA

CDMA techniques are based on spread spectrum communications, which were originally developed for military applications. A simple definition of a spread spectrum signal is that its transmission bandwidth is much wider than the bandwidth of the original signal [7]. This can be achieved in a number of ways, e.g. frequency hopping, time hopping and the use of multiple radio frequency (RF) carriers. This paper will focus on direct-sequence spread spectrum techniques which are used in the NA IS-95 and European UTRA based systems.

In a CDMA system, it is often possible for the same RF frequencies to be shared between multiple users in each cell, so that complete frequency re-use is obtained throughout all cells [8]. To distinguish one user's transmission from another, each mobile modulates the voice data symbols by a pseudo-noise (PN) code, as shown in part (a) of Figure 3.1. Each symbol is composed of W binary "chips" which have a period T_c that is much shorter than that for the original data symbols, T_s, so that the signal bandwidth is considerably increased. Now consider the ith active user in a cell, communicating on the uplink. They use a PN code $c_i(t)$, such as that shown in Figure 3.1 (a), to modulate their baseband data $d_i(t)$ and the resulting signal $s_i(t)$ is transmitted at RF. At the base station, the received signal is correlated with the PN code to provide a delayed estimate of $s_i(t)$. A typical PN code auto-correlation function observed at the correlator output is shown in Figure 3.1 b): it takes on significant values only within one chip of the code arrival time. An example cross-correlation function between two codes for different users is shown in Figure 3.1(c): it should be small to minimise multiple access interference (MAI).

The reverse link of a CDMA system such as that specified by IS-95 has a number of essential characteristics for effective multiple access communication. A detailed introduction to spread spectrum and CDMA techniques can be found in [8, 9], but here only points relevant to this discourse will be addressed.

- **Spread Spectrum Bandwidth:** The chip rate of the spread spectrum signal is an important parameter, and is inversely proportional to the chip period T_c. The chip period of so-called *narrowband* IS-95 CDMA system is roughly 800 ns

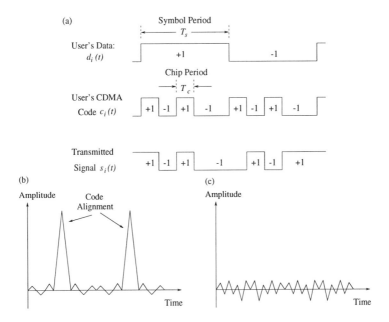

Figure 3.1 (a) Direct sequence modulation of a data sequence by a PN code, (b) A typical autocorrelation function for a PN code, (c) A typical cross-correlation function for two different users' PN codes.

(chip rate 1.25 MHz). Proposals for third generation systems based on *wideband CDMA*, usually have a higher chip rate, in the range of 4 − 16 MHz, to permit higher data rate services.

- **Modulation Scheme:** The uplink of an IS-95 system employs 64-ary orthogonal data modulation, transmitted using offset QPSK [8]. This permits non-coherent detection at the base station. However, the downlink of IS-95, in common with both links for third generation wideband CDMA systems, employs coherent BPSK or QPSK modulation, with a pilot signal or training symbols used for channel estimation purposes.

- **Multipath diversity:** In urban areas, multipath propagation is common, whereby the receiver observes a large number of copies of the transmitted signal, each with a different time delay. The noise like autocorrelation function of a PN code, shown in Figure 3.1 (b), means that the correlator receiver can resolve multipath components which are spaced by 1 chip period or more. This provides a form of multipath diversity, which can be exploited by using a RAKE filter to combine the multipath components at the outputs of the code correlators [10].

- **Multiple access interference:** The uplink of a CDMA system is usually asynchronous, in the sense that the arrival times for each user's code are different. This means that the receiver for each user will observe interference from all other users in the system, as the received codes will not be orthogonal. On the downlink, transmissions to each user from one cell sector can be made orthogonal provided there is no multipath. However, each user will still observe asynchronous interference from other cells. As a result, the capacity of a CDMA system is interference-limited in general.

- **Power control:** A corollary to the above is that power control is essential on both links, to minimise multiple access interference. A particular problem on the uplink is to prevent the case where mobile transmitters far away from the cell's base station are swamped by interference generated by users closer to the receiver. Provided that all signals arrive with the same power, a receiver's tolerance to CDMA interference is proportional to the *processing gain* [9] $W = T_s/T_c$.

The obvious way to increase the capacity of a CDMA system is to reduce the levels of multiple-access interference. In the following, an approach based on antenna array receivers will be considered. The next section will discuss the motivations for using antenna arrays and some of the issues involved in receiver design.

3.4 MOTIVATIONS FOR USING ANTENNA ARRAYS

An antenna array consists of M antenna receivers, whose operation and timing is usually controlled by one central processor. The geometry of the antenna locations can vary widely, but the most common configurations are to place the antennas round a circle (circular array) or along a straight line (linear array). Antenna arrays have been proposed in the past for the operation of Radar and military communications systems [11]: it is possible to perform direction finding tasks and to null out enemy jammers. However, in the context of civilian cellular systems, the main aim of the antenna array receiver is to maintain an acceptable error performance on the radio link and hence maximise the signal-to-interference and noise ratio (SINR) for each user in the system. An antenna array containing M-elements can provide a mean power gain of M over white noise, but suppression of interference from other cellular users is dependent on the form of the received data.

If the processing gain W of the CDMA system is small, the interference contribution from one other user on the same RF channel will be significant. An antenna array can null out up to $(M-1)$ users on the uplink [12], so significant performance improvements may be possible. However, if W is large, a good model for the received signal in a power controlled CDMA system is a strong desired signal corrupted by a large number of small cross-correlation interference terms, which arrive with a uniform distribution from throughout the cell[1]. In this case nulling out $(M-1)$ interferers

---

[1]This assumption permits general results to be obtained for system capacity. However, the validity of this assumption will depend on the layout of the cell for practical situations.

is unlikely to significantly improve the received SINR [6] because of the very large number of interference components [2]. A better methodology here is to estimate the form of the received signal and determine the matched filter solution [13]. This form of receiver can exploit any spatial diversity present, while suppressing the mean level of CDMA interference by a factor proportional to M.

Assuming that the antenna array provides significantly improved SINR levels at the base station receiver, the number of channel errors, measured by the bit error ratio (BER), will reduce. This provides the cellular operator with some degrees of freedom which may be used for the following purposes [6, 14]:

- To increase the number of active users for a given BER quality threshold
- To improve the BER performance for a given number of users within a cell
- To reduce the SINR required at each antenna to achieve a target BER; this would reduce the transmit power required by the mobile handset for the reverse link
- To increase the range of the base station receiver and thus the cell size

One additional application of antenna arrays is in position location applications, such as for emergency 911 calls in NA. Antenna arrays located at three separate base stations could be used for direction finding, in order to permit a mobile position to be determined. However, there are a number of other candidate solutions to this problem, as described in [15].

While antenna arrays provide many advantages, these must be offset against the cost and complexity of their implementation. There are a number of points which must be taken into account here [16]:

- The hardware/software requirements increase in order to control M sets of receivers for each user
- Practical antenna arrays may be adversely affected by channel modelling errors, calibration errors, phase drift and noise which is correlated between antennas

With these points in mind, this paper will move on to consider channel modelling aspects of antenna arrays. This will motivate a discussion on the candidate algorithms for a CDMA antenna array base station receiver.

3.5 CHANNEL MODELLING CONSIDERATIONS

The effects of the RF channel on both the uplink and downlink are very complex. This section will consider the impact of these on CDMA network design and look particularly at how channel models may be derived for antenna array receivers. Radio propagation affects the transmitted waveforms in different ways over short and long timescales. This section will begin by considering the latter.

---

[2] An exception is if a mobile undergoes a power control error and transmits at a very high power: nulling out the resulting interference is clearly desirable in this case.

As a mobile unit moves through its environment, the average received power at the base station is affected by two major phenomena, namely shadowing and pathloss. Shadowing is caused by the transmitted signal being attenuated at certain times due to obstructions between the mobile and the base station. A typical model for this effect is that the received power follows a log-normal distribution, with a standard deviation of 6–12 dB for urban environments. Pathloss is the signal attenuation due to the mobile-base station distance r. The attenuation is typically proportional to r^{-n}: the exponent of r will vary with the environment, but $n = 4$ is often used for urban environments. Third generation CDMA systems will employ fast power control on both links to equalise these effects whilst minimising the required transmit power.

For smaller timescales, the received signal on either link is corrupted by the effects of *multipath fading*. In most environments there are multiple paths by which the mobile's RF signal can reach the base station. Different paths will have different time delays, amplitudes and phases, so that the received signal is the summation of all of these components. In a CDMA system, the receiver can separately resolve multipath components with a relative time delay of at least T_c. However, in dense multipath environments, each resolved path will consist of a number of path components with similar delays. Depending on the phases of these components, the instantaneous power can fluctuate enormously, giving rise to the phenomenon of fading.

A typical low-pass model for the RF channel between the mobile and the M-element array consists of a L tap delay line [17, Chapter 14], where L is the number of resolved multipath components. The $M \times 1$ channel vector for the ith user may be written in the following form:

$$\mathbf{h}_i(t) = \sum_{l=0}^{L-1} \mathbf{h}_{i,l}(t - \tau_i - lT_c)\delta(t - \tau_i - lT_c).$$

The notation τ_i denotes the propagation delay of the RF channel for the ith user and $\delta(t)$ is the Dirac delta function. The mth entry of the vector $\mathbf{h}_{i,l}(t)$ is a complex scalar, representing the amplitude and phase of the channel between the mobile and the mth antenna element of the receiver for the lth channel tap.

The multipath channel tap vectors $\mathbf{h}_{i,l}(t)$ are typically modelled as the sum of a large number of multipath components that cannot be separately resolved. A reasonable model for the Rayleigh fading channel that results is Raleigh's model [18]:

$$\mathbf{h}_{i,l}(t) = \sqrt{(r^{-4}\zeta)} \sum_{b=1}^{B} \alpha_b \exp\{j(\phi_b + 2\pi\nu_b t)\}\mathbf{a}(\theta_b), \qquad (3.1)$$

where r^{-4} represents the pathloss due to the mobile-base station distance r and ζ denotes a log-normal variable modelling shadowing. The received waveform is made up of B multipath rays with amplitudes α_b, phases ϕ_b, Doppler frequencies ν_b and angles-of-arrival θ_b. The steering vector $\mathbf{a}(\theta)$ is the impulse response of the array to a plane wave arriving with angle θ. For a uniform linear array (ULA) with antenna spacing D, it has the form:

$$\mathbf{a}(\theta) = [1, \exp\{j\psi_1 \cos(\theta)\}, \ldots \exp\{j\psi_{M-1} \cos(\theta)\}]^T,$$

where ^{T} is the matrix transpose operation. The scalar $\psi_m = (2\pi mD/\lambda)$ and λ is the RF carrier wavelength. The ULA broadside (i.e. perpendicular) is 90°: for 120° sectorisation, the user's angles of arrival will be restricted to $[30°, 150°]$.

Five important issues relating to the model of equation (3.1) will now be addressed:

- **Fading distribution:** Two different models are often used for the amplitude of each entry of $\mathbf{h}_{i,l}(t)$. If B is large and the amplitude components $\{\alpha_b\}$ are roughly equal, the sum follows a Rayleigh distribution. This distribution gives rise to catastrophic *deep fades* of the signal, where communication is impossible. To overcome this problem [17], space and multipath diversity techniques are used to provide multiple independent copies of the signal. In the case where there is a dominant line-of-sight path present, the amplitude follows a Rician distribution which is a less severe form of signal fading.

- **Multipath power profile:** The mean power present at the different time delays will be a function of the type of environment encountered. Some example profiles, taken from the COST-207 report [19], are shown in Figure 3.2. The various profiles illustrate the point that the extent of multipath delay spread generally increases from suburban areas to urban areas to hilly areas.

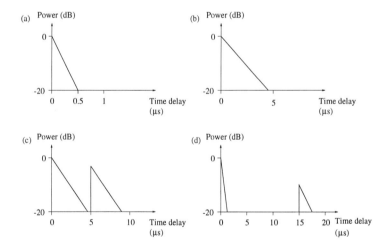

Figure 3.2 Typical multipath profiles, drawn from the COST-207 report [19], for (a) rural area, (b) typical urban area, (c) bad urban area and (d) hilly terrain.

- **Doppler spectra:** When the mobile is in motion with velocity v, individual propagation paths will experience a Doppler shift in the range $[-f_m, f_m]$, where $f_m = v/\lambda$. The important parameter here is the time-frequency product $T_s f_m$. The larger its value, the more rapidly the channel must be tracked, compared to the symbol rate. The worst case will be for mobiles being operated in rapidly moving trains or vehicles. For example, a mobile moving at 110 kph will give rise to a Doppler shift of 200 Hz for a 2 GHz carrier system. A typical voice data rate for a system like IS-95 is 10 kbps, only 50 times the maximum Doppler shift.

- **Angular width:** The spatial spreading of multipath energy is an important parameter, as it specifies the array element spacing need to obtain uncorrelated fading at each antenna. As the angular width increases, so the required antenna spacing reduces. Some typical values for different environments from [4] are given in table 3.1. As the multipath becomes more dense and the antenna height is reduced (e.g. from macro-cell to micro-cell), the angular width increases. In this paper, the model of [20] will be used, where multipath energy is uniformly distributed over the angular width Δ. Other statistical and deterministic spatial channel models are described in [18].

Table 3.1 Table of typical angular widths for different environments (after [4]).

Cell Type	Flat Rural Macro-cell	Urban Macro-cell	Hilly Macro-cell	Dense Urban Micro-cell	Indoor Pico-cell
Angular Width	0.5°	20°	30°	120°	360°

- **Antenna Spacing:** In order to maximise diversity gain, the antenna spacing should be set according to the angular widths encountered in the environment, as described above. However, in practice the antenna spacing may be restricted by the cell site. Also if angular information is to be preserved, the antenna spacing must be restricted to $\lambda/\sqrt{3}$ for a 120° sector.

3.5.1 Receiver Signal Modelling

A diagram of a typical space-time receiver structure for one user in a 120° cell sector, using a ULA configuration, is shown in Figure 3.3. The received baseband signal at the array from the ith user is simply the convolution of $s_i(t)$ with the channel $\mathbf{h}_i(t)$. The antenna array observes the sum of the P users' waveforms, corrupted by thermal noise. The $M \times 1$ antenna array vector output at time t is thus:

$$\mathbf{y}(t) = \sum_{p=1}^{P} \sum_{l=0}^{L-1} s_p(t - \tau_p - lT_c)\mathbf{h}_{p,l}(t - \tau_p - lT_c) + \mathbf{v}(t),$$

where $\mathbf{v}(t)$ represents spatially and temporally white Gaussian noise of zero mean. It will be assumed that the channel vectors $\{\mathbf{h}_{p,l}\}$ are time invariant, so that the time index may be dropped.

The receiver is interested in the waveform of user i. Therefore, the vector $\mathbf{y}(t)$ is subject to rectangular pulse shape matched filtering and sampling at the chip rate T_c

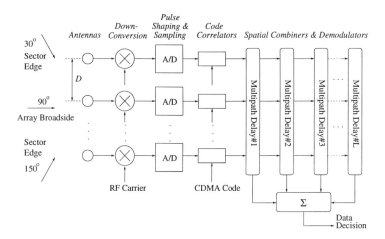

Figure 3.3 The space-time receiver structure for one user, operating with a uniform linear array in a single $120°$ sector.

to obtain a discrete time representation with index k:

$$\mathbf{y}(k) = \sum_{p=1}^{P} \sum_{l=0}^{L-1} s_p(k - l + (\tau_i - \tau_p))\mathbf{h}_{p,l} + \mathbf{v}(k).$$

In order to process $\mathbf{y}(k)$ to recover the original data sequence $d_i(n)$, it is necessary to consider the different multipath components present in $\mathbf{y}(k)$. To this end, define the $ML \times 1$ received data vector $\mathbf{y}_L(k)$ as:

$$\mathbf{y}_L(k) = [\mathbf{y}^T(k-L+1), \mathbf{y}^T(k-L+2) \ldots \mathbf{y}^T(k)]^T.$$

By using this "oversampling" formulation, $\mathbf{y}_L(k)$ may be written as:

$$\mathbf{y}_L(k) = \mathbf{h}_i s_i(k - L + 1) + \mathbf{i}_1 + \mathbf{i}_2 + \mathbf{v}_L(k). \tag{3.2}$$

The size $ML \times 1$ vector $\mathbf{i}_1$ represents self-noise interference between the L multipath taps, due to non-ideal code-correlation properties. Similarly, the $ML \times 1$ vector $\mathbf{i}_2$ represents MAI from the $(P - 1)$ other users. Finally, the $ML \times 1$ vectors $\mathbf{h}_i$ and $\mathbf{v}_L(k)$ are defined as follows:

$$\mathbf{h}_i = [\mathbf{h}_{i,0}^T, \mathbf{h}_{i,1}^T, \ldots \mathbf{h}_{i,L-1}^T]^T,$$
$$\mathbf{v}_L(k) = [\mathbf{v}^T(k-L+1) \; \mathbf{v}^T(k-L+2) \ldots \mathbf{v}^T(k)]^T.$$

In order to estimate the nth transmitted symbol for the ith user, the vector $\mathbf{y}_L(k)$ is correlated with the code c_i as follows, to form the $ML \times 1$ vector $\mathbf{x}_{i,L}(n)$:

$$\mathbf{x}_{i,L}(n) = [\mathbf{x}_i^T(0, n), \mathbf{x}_i^T(1, n), \ldots \mathbf{x}_i^T(L-1, n)]^T,$$

$$= \sum_{w=1}^{W} \mathbf{y}_L((n-1)W + (L-1) + w)c_i((n-1)W + w).$$

The notation $\mathbf{x}_i(l,n)$ denotes the post-correlation signal for the lth multipath and nth symbol.

In a Rayleigh fading environment, the base station should try to exploit as many sources of diversity as possible [21], in order to minimise the probability of deep fades. Typically, the receiver will choose an $ML \times 1$ vector $\mathbf{g}_i$, which comprises L filters of size M for the L multipath delays. Then $d_i(n)$ is estimated as the inner product $\mathbf{g}_i^H \mathbf{x}_{i,L}(n)$ (where ^{H} is the Hermitian transpose). Equation (3.2) indicates that the filter $\mathbf{g}_i$ should provide a good estimate of $\mathbf{h}_i$, whilst minimising the effects of self-noise ($\mathbf{i}_1(k)$), MAI ($\mathbf{i}_2(k)$) and Gaussian noise ($\mathbf{v}_L(k)$). Different approaches to estimating $\mathbf{g}_i$ will now be described.

3.6 RECEIVER ALGORITHMS

In this section, a number of techniques for operating antenna array receivers will be introduced. The general approach and advantages/disadvantages for each technique will be discussed.

3.6.1 Fixed Beam Receivers

The first approach considered here is to apply increased sectorisation using fixed beam antennas. There is a ULA of M elements in each 120° sector with maximum antenna element spacing $D = \lambda/\sqrt{3}$, so that there is a unique steering vector for each possible bearing θ in the range $[30°, 150°]$. In this case, the receiver contains a set of J fixed steering vectors $\{\mathbf{a}(\theta_j)\}$ which can be set to cover the cell. It is possible to generate a set of $J = M$ beams which are orthogonal — however, there may be cusps of 3–4 dB between the beams. To minimise this effect, the receiver might employ a set $J = 2M$ overlapping beams to minimise cusping effects.

The receiver operates by measuring the power from each steering vector, for each multipath $\mathbf{x}_i(l,n)$ separately, perhaps averaged over multiple symbols. It then selects the largest or largest two outputs, for combining with the equivalent beam outputs from the other $(L-1)$ multipath delays. However, it should be noted that when Δ is very large, signal energy will appear in multiple beams so that all the beams must be combined to avoid loss of signal power. The receiver might employ a RAKE filter [17, Chapter 14] to combine the outputs from the selected fixed beams. Results for fixed beam receivers [22] suggest that the available diversity gain may be inferior to that obtained through using space diversity with widely spaced antennas. Therefore a fixed beam receiver will perform best when additional diversity is present. One example is the use of a "diversity" antenna placed at a distance of perhaps $10\lambda - 20\lambda$ from the array [23].

3.6.2 Bearing Estimation Techniques

The idea behind this type of receiver is similar to fixed beam receivers. Again, the antenna spacing D must be limited to $\lambda/\sqrt{3}$ to ensure that the steering vector for each value of θ is unique. Each multipath component $\mathbf{x}_i(l,n)$ may be used as an input to a bearing estimation algorithm. The algorithm outputs a set of H bearings $\{\theta_h\}$ which provide information on the angles of arrival of the multipath components present. The spatial filters $\{\mathbf{a}(\theta_h)\}$ corresponding to the bearings can be applied to the multipath vector and the outputs from all L multipath delays combined in a RAKE filter.

There are a large number of possible techniques that can be used for bearing estimation, see [24] for example. Simple techniques such as the conventional beamforming tend to have poor angular resolution. More sophisticated techniques, such as the MUSIC algorithm [25], have been proposed to resolve closely spaced sources. However, these algorithms can be sensitive to coloured noise and also fail in the presence of coherent multipath. The coherent multipath problem can be overcome by spatial smoothing [26] or by using more complex multi-dimensional search techniques, such as weighted subspace fitting [24].

However, even if all the coherent multipath components can be correctly identified, the performance of bearing estimation techniques is at best about the same as channel estimation techniques [27]. The complexity and sensitivity of these techniques makes them less attractive than other receiver structures. The main reason for considering them is that they may be useful for mobile position determination, such as for locating users placing emergency 911 calls in NA [15].

3.6.3 Channel Estimation

In this case, the receiver attempts to estimate the channel vector $\mathbf{h}_i$. The filter $\mathbf{g}_i$ is then set equal to the channel estimate $\hat{\mathbf{h}}_i$. The receiver therefore operates as a matched filter in both time and space.

The simplest approach to estimating the channel [28] is to average the received data over N consecutive symbols (or *snapshots* in array processing terminology):

$$\hat{\mathbf{g}}_i = \frac{1}{N} \sum_{n=1}^{N} \mathbf{x}_{i,L}(n) d_i(n). \tag{3.3}$$

Unfortunately, this requires that the data sequence $d_i(n)$ is known. This can be achieved if the mobile transmits a pilot signal with the data, or by using periodic training sequences combined with decision feedback schemes.

In order to avoid estimating $d_i(n)$, methods based on principal components (PC) analysis may be used. The receiver can form the following matrices:

$$\mathbf{R}_{\mathbf{y},L} = \sum_{n=0}^{N-1} \sum_{w=1}^{W} \mathbf{y}_L(nW+w) \mathbf{y}_L(nW+w)^H,$$

$$\mathbf{R}_{\mathbf{x},L} = \sum_{n=1}^{N} \mathbf{x}_{i,L}(n) \mathbf{x}_{i,L}(n)^H. \tag{3.4}$$

The principal eigenvector of $(\mathbf{R}_{\mathbf{x},L} - \mathbf{R}_{\mathbf{y},L})$ is used as an estimate of the multipath channel $\mathbf{h}_i$ [12]. The disadvantage of this approach is the increased computational complexity compared to equation (3.3). In addition, the eigenvector cannot provide an estimate of the absolute phase of the channel, so that non-coherent data detection may be required. Similar channel estimation approaches have been proposed [28], where smaller covariance matrices are formed separately for each multipath delay.

Channel estimation techniques provide good performance when the MAI is spatially or temporally white. They are likely to be a good choice for large processing gain systems like IS-95. However, when W is small, the MAI from other users in the cell is likely to be spatially or temporally coloured. In this case, interference suppression techniques may be the better choice.

3.6.4 Interference Suppression Techniques

It is possible to design a filter $\mathbf{g}_i$ which is optimal in the minimum mean squared error (MMSE) sense [12]. The filter thus obtained has the following form:

$$\mathbf{g}_i = \mathbf{R}_{\mathbf{y},L}^{-1}\mathbf{h}_i, \qquad (3.5)$$

where $\mathbf{R}_{\mathbf{y},L}^{-1}$ denotes the matrix inverse operation. In practice, the vector $\mathbf{h}_i$ is obtained from a channel estimation algorithm and $\mathbf{R}_{\mathbf{y},L}$ is obtained from equation (3.4). This filter can null out $(M-1)$ users in low noise environments [12], provided that the equaliser length is large enough. Therefore, it may be of use for the small processing gain case, where nulling out a small number of users ($< M$) can significantly reduce the level of MAI. However, the performance of this technique may not be much different from channel estimation for the large processing gain case, where nulling out $(M-1)$ users is likely to have little effect on the level of MAI. The channel vector $\mathbf{h}_i$ may be estimated using equation (3.3) or the PC method with equation (3.4). The latter approach will be used below and denoted as the PC-MMSE method.

A similar, but suboptimal technique has been described in [29] which calculates the MMSE equaliser for each multipath tap separately. The L equaliser outputs are then combined using a non-coherent RAKE filter.

3.6.5 Multi-user Detection

It is possible to combine antenna array techniques with interference cancellation or multi-user detection techniques. The optimum receiver for CDMA antenna arrays has been derived in the case of no multipath in [30] and extended to the multipath case in [31]. It consists a receiver that combines multipath energy in space, followed by a Viterbi decoder, containing enough states to compensate for all the multipath time lags of all the CDMA users present on the same channel. In practice, this detector has exponentially increasing complexity with the number of users, so that simpler suboptimal detectors are likely to be preferred. For a thorough review of multi-user detection techniques, see [32]. These type of techniques require to cancel a large number of users in order to improve the performance of systems with large W, and in any case are likely to be limited by MAI from other cells. However, if W is small, cancelling even the MAI from one user may significantly improve the receiver performance.

3.7 UPLINK SIMULATION WORK

In this section, the results of a number of simulations to analyse the performance of the algorithms will be presented. All the simulations consider a single 120° sectored cell, with a M-element ULA in the base station. Unless otherwise stated, the antenna spacing $D = 0.5\lambda$. All the simulations assume that uplink power control is in operation and that all users are received with the same mean power.

The remainder of this section is split into four parts. The first part will consider general comparisons of the different receiver algorithms. Then the convergence properties of channel estimation and interference suppression methods will be analysed. Thirdly, some initial simulations addressing multi-user detection techniques are presented. The last part will discuss the uplink capacity increase that is available with antenna array techniques.

3.7.1 General Comparisons

Firstly, a simple scenario is presented to highlight the operation of different algorithms. There are two users in a sector, with the desired user's channel consisting of two multipaths arriving at the same time delay with amplitude/bearings: $(1.0, 100°)$ and $(0.5, 60°)$. The interferer's signal arrives from $40°$ and the array size $M = 4$. The beam patterns for fixed beam[3], channel estimation and PC-MMSE equaliser techniques are shown in Figure 3.4. The fixed beam receiver simply picks the beam with the largest power, in this case with angle $90°$. The channel estimation approach combines both multipath components, but ignores the interferer. Finally, the PC-MMSE equaliser is also able to combine both multipaths, but in addition it places at null at $40°$ to remove the interferer's signal.

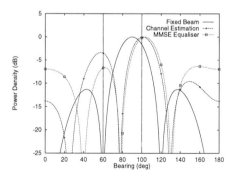

Figure 3.4 The beam patterns for the fixed beam, channel estimation and PC-MMSE equaliser techniques, for a two users received at an $M = 4$ element array. The bearings of the desired user's two multipath components are shown as vertical lines.

---

[3]The fixed beam receiver is assumed to use $M = 4$ orthogonal beams.

Next, some simple simulations have been performed to compare the performance of a fixed beam receiver, the conventional beamformer bearing estimation algorithm and channel estimation. Two versions of the fixed beam have been used, using the two sets of M beams which have the best and worst performance for $\Delta = 0°$. The array size $M = 8$ and the receiver is tracking a single slow Rayleigh fading tap, with an SINR of 5 dB per antenna element. The maximum SINR gain is 8 (9 dB) and the algorithms estimate the channel from $N = 50$ symbols. Figure 3.5 presents the SINR gain achieved by the three algorithms as a function of the tap's angular width Δ. The results show that both the bearing estimation and fixed beam techniques degrade as the angular width increases. This is because of the the fixed beamwidth of the associated spatial filters. However, the channel estimation algorithm is able to track the fading channel and always combines all the signal energy.

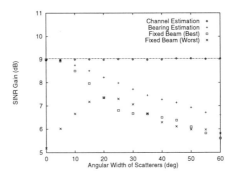

Figure 3.5 The achieved SINR gain for the fixed beam (best and worst beam sets), bearing estimation and channel estimation algorithms, plotted against the angular width Δ. The array size $M = 8$ and the algorithms estimate the channel from $N = 50$ symbols.

A second set of simulations compare the performance of the PC channel estimation and PC-MMSE methods (equation (3.5)). The array size is again $M = 8$ and each user has a three equal tap Rayleigh fading multipath channel. The antenna element spacing D is set to be several wavelengths in this case, in order to achieve uncorrelated fading at all the antennas. The number of symbols N is assumed to be large enough for both algorithms to have converged. Two different scenarios have been considered: (a) $W = 8$ and background noise equivalent to -20 dB (relative to the desired signal) and (b) $W = 64$ and and background noise level -10 dB. The SINR of the two algorithms for both cases is plotted against the number of users in the cell sector in Figure 3.6.

The results show that in case (a), the PC-MMSE equaliser can suppress the MAI to obtain substantial gains in performance. This is because the processing gain is small and the noise level is relatively low. However, for case (b), there is little difference between the two methods. In this case, the interference from each user is small due to the large processing gain and the background noise is the dominant source of interference.

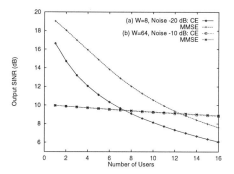

Figure 3.6 The output SINR vs number of users plotted for the channel estimation and PC-MMSE equaliser methods. Two different scenarios are considered: (a) $W = 8$ and background noise -20 dB and (b) $W = 64$ and and background noise -10 dB.

3.7.2 Algorithm Convergence

The next set of results look more closely at the convergence performance of the PC channel estimation technique. An antenna array of M elements again tracks a single channel tap, but this time the angular width $\Delta = 0°$ and the channel amplitude is fixed (no fading). In part (a) of Figure 3.7, the convergence performance of this algorithm is shown for maximum achievable SINRs of 0, 5 and 10 dB with an array size of $M = 4$. As the SINR is reduced, the algorithm requires a longer data length N to converge to the maximum achievable SINR. In Figure 3.7(b), the SINR is fixed at 5 dB and the array size M varies from 2–16 elements. Increasing M also increases the noise at each antenna, so the receiver again requires a larger value of N to achieve a certain SINR performance. When designing an antenna array receiver, the algorithm convergence performance will limit the capacity gain of the system and the antenna array size that can be used.

In a fast fading environment, the channel estimation method must be rapidly updated to track channel variations. The convergence performance of the PC method has been measured for the case of a single Rayleigh fading tap that varies in time according to the Classical Doppler fading model. The data rate of the system is 10 kbps and the maximum fading frequency of the channel tap is 50 Hz. Results have been obtained for an $M = 8$ array with three angular widths $\Delta = 0°$, $20°$ and $60°$ and are shown in Figure 3.8. The results show that as the angular width increases, the algorithm's channel estimate becomes more rapidly out of date. The optimum value of the data length N is only 16 symbols for $\Delta = 20°$ or $60°$.

Figure 3.9 compares the convergence performance of the channel estimation PC method and the PC-MMSE method. The array size is $M = 4$ and results are shown in Figure 3.9(a) for a CDMA system with $W = 64$, 10 users and a background noise level of -10 dB. The maximum achievable SINR is shown as a horizontal line. In this case, there is little to choose between the two algorithms as the interference contribution from each user is very small. Figure 3.9(b) shows similar results for a system with $W = 8$,

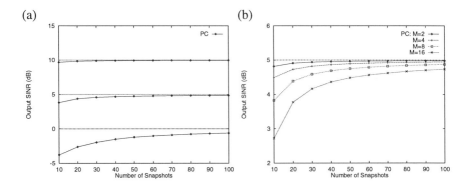

Figure 3.7 The convergence performance of the PC method, plotted as output SINR vs the number of averaged symbols (snapshots) N. Results are shown for (a) array size $M = 4$ and maximum SINRs of 0, 5 and 10 dB (b) maximum SINR of 5dB and array size $M = 2, 4, 8$ or 16.

5 users and a background noise level of -20 dB. The maximum achievable SINRs for channel estimation and interference suppression methods are shown as horizontal lines. This time, the noise level is much lower and the interference from each user is more significant due to the small value of W. The result is that the PC-MMSE method achieves a performance gain of about 1 dB over the PC algorithm. This is because the multi-user interference is not spatially white for this case.

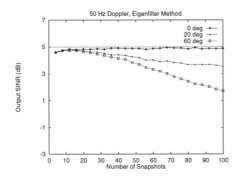

Figure 3.8 The output SINR convergence performance of the PC method with an $M = 8$ element array and angular widths of $0°$, $20°$ and $60°$.

3.7.3 Multi-user Detection

Some initial simulation studies has been performed for antenna arrays operating using the PC approach or using the PC method combined with a multi-user detection technique called the decision feedback decorrelator (DFD) [32]. The antenna array size was $M = 1$ or 4 and each user's channel consisted of a single tap with fixed

ANTENNA ARRAYS FOR CELLULAR CDMA SYSTEMS 75

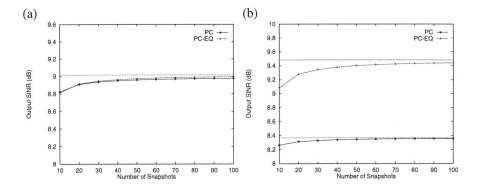

Figure 3.9 The output SINR plotted against number of averaged symbols (snapshots) N for a scenario with array size $M = 4$. The noise variance σ^2 is set to achieve an SINR ignoring CDMA interference of (a) 10 dB with processing gain $W = 64$ and 10 users; (b) 20 dB with processing gain $W = 8$ and 5 users. The maximum SINR and matched filter bounds are shown as horizontal lines.

amplitude (no fading). These receivers have been tested for a multiple user scenario with processing gain $W = 31$. The resulting bit error ratio (BER) for a desired user is plotted in Figure 3.10 against the number of active users in the cell.

The results for $M = 1$ show a substantial performance advantage for the DFD method, compared to using channel estimation alone. When the array size M is increased to 4, the performance improvement is greater for the CE method than the DFD method. This is because the multi-user detector has already cancelled a lot of the MAI affecting the desired user. However, for $M = 4$, the performance of the DFD method is almost independent of the number of users, indicating that DFD combined with antenna array processing has cancelled almost all the MAI.

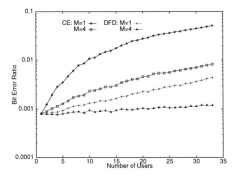

Figure 3.10 The BER performance of a base station receiver plotted against the number of users in a cell. The antenna array size was $M = 1$ or 4; the receiver uses the single user CE method or the multi-user DFD approach. The processing gain of the system $W = 31$.

3.8 CAPACITY IMPROVEMENT WITH CDMA ANTENNA ARRAYS

The performance of antenna array base stations in a CDMA network has been analysed in a number of publications, including [13, 33, 34]. In this section, some basic analysis is presented to highlight how antenna arrays can improve performance. Some simulation results, first presented in [35], will then be used to illustrate the points made.

A simple equation for the uplink capacity U of a single CDMA cell is given by [36]:

$$U = 1 + \frac{WG}{E_b/N_0} - \frac{\sigma^2}{S}$$

where the value of E_b/N_0 represents that required for adequate link performance. The scalar σ^2 is the background noise power and S is the received signal power for each user. Finally, G is the ratio of the antenna gain for the desired user to that for the other interfering users in the cell. The value of G depends on the beam pattern for each user, but will be roughly proportional to the array size M.

As a result, antenna arrays can improve the capacity in two ways:

1. Increasing the antenna gain G and hence the array size M. This reduces the average level of interference from each user in the cell, permitting a capacity increase. However, this gain factor can be reduced by user clustering in one part of the cell.

2. Reducing the required E_b/N_0. Antenna arrays can provide increased space diversity at the base station, which can permit the receiver to operate at a lower signal power. This increases the tolerance of the receiver to multiple access interference. However, in dense multipath environments, there may be a lot of multipath diversity already, so that increased space diversity does not improve performance significantly.

Here, some results first presented in [35] are summarised, to highlight these points. A single 120° sector was considered, with users uniformly distributed in angle over the sector. The CDMA processing gain was $W = 256$ and each user had a 4 tap Rayleigh fading multipath profile, with mean power levels: 1, 0.5, 0.25 and 0.125. Three different mobile ranges were considered, giving rise to three values of the angular width Δ: 53°, 14° and 3.5°. The receiver used the channel estimation approach, with perfect estimates available.

The maximum number of users for which a BER of 10^{-2} can be sustained is plotted against the array size M and the angular width Δ in table 3.2. The results show a large increase in capacity as the array size M increases. The capacity also increases with the angular width Δ, which is mainly due to improved diversity gain and reduced variability in the interference suppression capability of the array. In practice, these gains will be offset by combining losses in the receiver algorithm and the clustering of users within one part of the cell.

3.9 DOWNLINK TECHNIQUES

Table 3.2 Maximum number of users for different array sizes M and angular widths Δ for a BER threshold of 10^{-2}.

M	Angular Width Δ		
	53°	14°	3.5°
1	30	30	30
2	73	59	58
4	169	114	105
8	366	230	189

The downlink is more challenging than the uplink, when considering antenna array techniques for cellular systems. One difficulty occurs due to the fact that most major cellular standards currently employ frequency division duplex. The separation between the uplink and downlink frequencies may typically be on the order of tens of MHz. Therefore, the downlink is outside the coherence bandwidth of the uplink channel, which means that the instantaneous fading on the two links will be uncorrelated. As a result, the downlink channel cannot always be directly inferred from the uplink channel.

In addition, the IS-95 standard uses a high power pilot signal on the downlink, to permit coherent data demodulation [36]. In this case, the pilot must be transmitted over the same channel as each user's data. Transmitting a separate pilot signal for each user will increase the downlink interference, so it is better to share the pilot between multiple users. One feasible technique in this case is simply to use increased sectorisation. Alternatively, transmit diversity techniques [37], such as soft-handoff, may be used to assist mobiles to obtain additional diversity gain.

If a coherent pilot is not used on the downlink, beamforming techniques may be applied to each user separately. It is possible to use bearing estimates from the uplink to choose weights for downlink beamforming [38]. Alternatively, the spatial filter $\mathbf{g}_i$ obtained on the uplink can be used directly to re-transmit on the downlink. However, this will only be effective when the antenna spacing $D \leq \lambda/\sqrt{3}$ (so that the array steering vectors are unique for each θ) and the multipath angular width is small. If these conditions are not met, downlink beamforming is unlikely to provide much performance benefit [39]. Another approach is "transmitting with feedback" [40]. In a training phase, the base station array transmits a "probing signal" to the user, switching between a number of different beamformer weight vectors. Information fed back from the user can be used to choose the best set of weights for subsequent transmission of data to that user. This scheme clearly involves some delay, so it is perhaps most suited to scenarios where the channel changes very slowly, e.g. indoor communication systems.

An alternative approach to downlink beamforming is to use time division duplex transmission techniques [41], so that the uplink and downlink are conducted on the same RF frequencies. By alternating uplink and downlink transmissions, it is possible to use channel estimates obtained on the uplink to retransmit to the mobile on the downlink. In this case, the RF channel must not change significantly between the uplink and downlink time slots, so that the time slot length must be carefully chosen with maximum Doppler shifts in mind.

3.10 SUMMARY

This paper has provided an introduction to the subject of antenna arrays for CDMA base station receivers. A number of points have been discussed, and are summarised below:

- The topic of antenna arrays has been introduced, noting that they can reduce cellular interference levels and improve capacity. Results in this paper suggest that employing M antennas can multiply the reverse link capacity by a factor of roughly M. However, this requires additional base station hardware and software.

- Channel modelling aspects have been described: in urban areas, several channel taps are often resolvable. An antenna array receiver must be able to handle the angular width of each tap and should also be able to track channel variations due to Doppler effects.

- Several different receiver algorithms have been described, including fixed beam, bearing estimation, channel estimation, interference suppression and multi-user detection techniques. Selecting one method for a practical system will require comparison of the complexity and performance of the different algorithms.

- The capacity increases due to the use of an antenna array receiver has been described. An array of M elements can provide an average capacity increase proportional to M. However, the capacity can be degraded by user clustering or poor algorithm convergence.

- A number of downlink algorithms have been introduced. However, further research is required to assess how well they perform.

Acknowledgments

The authors gratefully acknowledge the sponsorship of these studies by the UK EPSRC and Nortel. They thank Catherine Hassell Sweatman for permission to use Figure 3.10. Finally, they also thank Chris Ward, Damian Bevan and John Hudson of Nortel for many enlightening discussions.

References

[1] D.C. Cox, "Wireless personal communications: What is it?", *IEEE Personal Communications Magazine*, vol. 2, pp. 20–35, April 1995.

[2] E. Nikula, A. Toskala, E. Dahlman, L. Girard, and A. Klein, "FRAMES multiple access for UMTS and IMT-2000", *IEEE Personal Communications Magazine*, vol. 5, pp. 16–24, April 1998.

[3] E. G Tiedemann, Y.C. Jou, and J.P Odenwalder, "The evolution of IS-95 to a third generation system and to the IMT-2000 era", *in Proc. ACTS Mobile Communications Summit, Aalborg (Denmark)*, pp. 924–9, October 1997.

[4] A.J. Paulraj and C.B. Papadias, "Space-time processing for wireless communications", *IEEE Signal Processing Magazine*, vol. 14, pp. 49–83, November 1997.

[5] G.K. Chan, "Effects of sectorisation on the spectrum efficiency of cellular radio systems", *IEEE Transactions on Vehicular Technology*, vol. 41, pp. 217–25, August 1992.

[6] J.H. Winters, J. Salz, and R.D. Gitlin, "The impact of antenna diversity on the capacity of wireless communication systems", *IEEE Transactions on Communications*, vol. 42, pp. 1740–50, February/March/April 1994.

[7] J.L. Massey, "Information theory aspects of spread spectrum communications", *in Proc. 3rd IEEE International Symposium on Spread Spectrum Techniques and Applications (ISSSTA), Oulu (Finland)*, pp. 16–21, July 1994.

[8] R. Padovani, "Reverse link performance of IS–95 based cellular systems", *IEEE Personal Communications Magazine*, vol. 1, pp. 28–34, Third Quarter 1994.

[9] M.K. Simon, J.K. Omura, R.A. Scholtz, and B.K. Levitt, *"Spread Spectrum Communications Handbook (Revised Edition)"*, McGraw–Hill, 1994.

[10] R. Price and P.E. Green, "A communications technique for multipath channels", *Proceedings of the IRE*, vol. 2, pp. 555–70, March 1958.

[11] B.D. Van Veen and K.M. Buckley, "Beamforming: A versatile approach to spatial filtering", *IEEE Acoustics, Speech and Signal Processing Magazine*, vol. 5, pp. 4–24, April 1988.

[12] H. Liu and M.D. Zoltowksi, "Blind equalization in antenna array CDMA systems", *IEEE Transactions on Signal Processing*, vol. 45, pp. 161–172, January 1997.

[13] A.F. Naguib, A. Paulraj, and T. Kailath, "Capacity improvement with base–station antenna arrays in cellular CDMA", *IEEE Transactions on Vehicular Technology*, vol. 43, pp. 691–8, August 1994.

[14] A.F. Naguib and A. Paulraj, "Performance enhancement and trade–offs of smart antennas in CDMA cellular networks", *in IEEE Vehicular Technology Conference (VTC), Chicago (USA)*, pp. 40–4, July 1995.

[15] T.S. Rappaport, J.H. Reed, and B.D. Woerner, "Position location using wireless communications on highways of the future", *IEEE Communications Magazine*, vol. 34, pp. 33–41, October 1996.

[16] J.E. Hudson, *"Adaptive Array Principles"*, Peter Peregrinus (Stevenage), 1981.

[17] J.G. Proakis, *"Digital Communications (3rd Edition)"*, McGraw-Hill, New York, USA, 1995.

[18] R.B. Ertel, P. Cardieri, K.W. Sowerby, T.S. Rappaport, and J.H. Reed, "Overview of spatial channel models for antenna array communication systems", *IEEE Personal Communications Magazine*, vol. 5, pp. 10–22, February 1998.

[19] Commission of the European Communities, *"Digital Land Mobile Radio Communications: COST-207 Final Report"*, chapter 2, 1988.

[20] J. Salz and J.H. Winters, "Effect of fading correlation on adaptive arrays in digital mobile radio", *IEEE Transactions on Vehicular Technology*, vol. 43, pp. 1049–57, November 1994.

[21] B. Sklar, "Rayleigh fading channels in mobile digital communications systems, Part I: Characterisation and Part II: Mitigation.", *IEEE Communications Magazine*, vol. 35, pp. 136–55, September 1997.

[22] D.D.N. Bevan, C.R. Ward, and J.E. Hudson, "Performance comparison of switched beam and adaptive beam architectures for intelligent antenna systems", *in Proc. 1st Surrey Symposium on Intelligent Antenna Technology, Guildford (UK)*, August 1997.

[23] T. Moorti and A. Paulraj, "Performance of switched beam systems in cellular base stations", *in Proc. 29th IEEE Asimolar Conference on Signals, Systems and Computers, Pacific Grove (USA)*, pp. 388–92, October–November 1995.

[24] H. Krim and M. Viberg, "Two decades of array signal processing research", *IEEE Signal Processing Magazine*, vol. 13, pp. 67–94, July 1996.

[25] R.O. Schmidt, "Multiple emitter location and signal parameter estimation", *IEEE Transactions on Antennas and Propagation*, vol. 34, pp. 276–280, March 1986.

[26] T. Shan, M. Wax, and T. Kailath, "On spatial smoothing for direction-of-arrival estimation of coherent signals", *IEEE Transactions on Acoustics, Speech and Signal Processing*, vol. 33, pp. 806–811, August 1985.

[27] D. Pal and B.H. Khalaj, "A RAKE type receiver structure for narrow band wireless systems can be designed using multiple antennas at the receiver", *in Proc. IEEE International Conference on Communications (ICC), New Orleans (USA)*, pp. 1701–5, May 1994.

[28] B.H. Khalaj, A. Paulraj, and T.Kailath, "Spatio–temporal channel estimation techniques for multiple access spread spectrum systems with antenna arrays", *in Proc. IEEE International Conference on Communications (ICC), Seattle (USA)*, pp. 1520–4, June 1995.

[29] A.F. Naguib and A. Paulraj, "Recursive adaptive beamforming for wireless CDMA", *in Proc. IEEE International Conference on Communications (ICC), Seattle (USA)*, pp. 1515–9, June 1995.

[30] S.L. Miller and S.C. Schwartz, "Integrated spatial–temporal detectors for asynchronous Gaussian multiple–access channels", *IEEE Transactions on Communications*, vol. 43, pp. 396–411, February/March/April 1995.

[31] A.F. Naguib, "Space–time receivers for CDMA multipath signals", *in Proc. IEEE International Conference on Communications (ICC), Montreal (Canada)*, pp. 304–8, June 1997.

[32] S. Moshavi, "Multi–user detection for DS–CDMA communications", *IEEE Communications Magazine*, vol. 34, pp. 124–36, October 1996.

[33] J.C. Liberti and T.S. Rappaport, "Analytical results for capacity improvements in CDMA", *IEEE Transactions on Vehicular Technology*, vol. 43, pp. 680–90, August 1994.

[34] A.F. Naguib and A. Paulraj, "Performance of wireless CDMA with M-ary orthogonal modulation and cell site antenna-arrays", *IEEE Journal On Selected Areas In Communications*, vol. 14, pp. 1770–83, December 1996.

[35] J.S. Thompson, P.M. Grant, and B. Mulgrew, "Analysis of CDMA antenna array receivers with fading channels", *in Proc. 4th IEEE International Symposium on Spread Spectrum Techniques and Applications (ISSSTA), Mainz (Germany)*, pp. 297–301, September 1996.

[36] K.S. Gilhousen, I.M. Jacobs, R. Padovani, A.J. Viterbi, L.A. Weaver, and C.E. Wheatley, "On the capacity of a cellular CDMA system", *IEEE Transactions on Vehicular Technology*, vol. 40, pp. 303–12, May 1991.

[37] G.W. Wornell and M.D. Trott, "Efficient signal processing techniques for exploiting transmit antenna diversity on fading channels", *IEEE Transactions on Signal Processing*, vol. 45, pp. 191–205, January 1997.

[38] P. Zetterberg and B. Ottersten, "The spectral efficiency of a basestation antenna array system for spatially selective transmission", *in Proc. IEEE Vehicular Technology Conference (VTC), Stockholm (Sweden)*, pp. 1517–21, June 1994.

[39] G. Auer, J.S. Thompson, and P.M. Grant, "Performance of antenna array transmission techniques for CDMA", *Electronics Letters*, vol. 33, pp. 369–70, February 1997.

[40] D. Gerlach and A. Paulraj, "Adaptive transmitting antenna arrays with feedback", *IEEE Signal Processing Letters*, vol. 1, pp. 150–2, October 1994.

[41] V. Kezys, J. Litva, T. Todd, and B. Currie, "Characterising the propagation channel for evaluating smart antenna performance", *in Proc. 4th Workshop on Smart Antennas in Wireless Mobile Communications, Stanford University (USA)*, July 1997.

4 SPATIAL FILTERING AND CDMA

Michiel P. Lötter *, Pieter van Rooyen † and Ryuji Kohno‡

* Alcatel Altech Telecoms
P.O. Box 286, Boksburg, South Africa
† Department of Electrical and Electronic Engineering
University of Pretoria, Pretoria, South Africa*
‡ Department of Electrical and Computer Engineering
Yokohama National University, Japan

m.p.lotter@ieee.org, pvrooyen@postino.up.ac.za, kohno@kohnolab.dnj.ynu.ac.jp

Abstract: A number of smart antenna techniques will form an important part of the new Wideband Code Division Multiple Access (WCDMA) standard that will realize the Universal Mobile Telephone System (UMTS). One such technique is High Sensitivity Reception (HSR) which is implemented in the uplink of cellular systems. This chapter addresses a few issues of importance when HSR techniques are used in a cellular CDMA system. Firstly, a brief overview of smart antenna techniques is presented followed by a theoretical analysis of a HSR/CDMA system. The analysis is focused on both micro and macro multi cell, multipath Rayleigh fading scenario's with imperfect power control. As a system performance measure, Bit Error Rate (BER) is used to investigate the influence of user location, number of antennas and power control error. An important parameter in the design of a HSR system is the antenna array element spacing. In the analysis a Uniform Linear Array (ULA) is considered and a measure is defined to determine the optimal antenna element spacing in a CDMA cellular environment. Normally, the mobile users in a cell are assumed to be uniformly distributed in cellular performance calculations. To reflect a more realistic situation, a novel probability density function is proposed to model the non-uniform distribution of the mobile users in a cell. It is shown that multipath and imperfect power control, even with antenna arrays, reduces the system performance substantially.

---

*This work was partially funded by Alcatel CIT and Alcatel Altech Telecoms through the Alcatel Research Unit for Wireless Access at the University of Pretoria

4.1 INTRODUCTION

As wireless technology advances, new, improved services at lower costs are created, resulting in an increase in the number of mobile users, which further stimulates progress in wireless technologies. As a result, it is predicted that by the turn of the century, some half a billion people worldwide will subscribe to wireless networks, introducing the most challenging demand on spectrum efficiency[1]. The definition of a Universal Mobile Telephone System (UMTS) is currently receiving a great deal of attention from standardization bodies such as the ITU, and current evolution paths predict the availability of UMTS services early in the next century. This process has been the driving force behind the design of new coding, modulation, planning and access methodologies, which are desperately needed for its success.

Wireless multiple access networks can be either fixed or mobile. In fixed applications, also known as wireless local loop (WLL), voice and data access is provided to non-mobile subscribers. This type of service is used extensively in developing countries, but is not limited to developing countries only. Mobile networks can be divided into high mobility, which offer voice and data services to high speed vehicle borne users, and low mobility to serve pedestrians. With the advent of the Internet and other applications that require high data rate services, such as video transmission etc., it is necessary to provide voice at low data rates, together with data transfer at high data rates. With these different criteria, more dimensions are added to the cellular engineering problem. Starting with Frequency Division Multiple Access (FDMA) systems, cellular networks have seen the introduction of Time Division Multiple Access (TDMA), CDMA and smart antenna techniques such as HSR and space division multiple access (SDMA) systems. Systems employing SDMA enable multiple users within the same radio cell to be accommodated on the same frequency, at the same time and using the same code, and are normally used in combination with one of the other multiple access schemes mentioned. Realization of this technique is accomplished by using an adaptive antenna array. Adaptive antenna arrays are effectively antenna systems capable of modifying their temporal, spectral and spatial responses by means of amplitude and phase weighting. As a subset of SDMA techniques, HSR systems make use of null steering to reduce excessive interference and/or the correlation between received signals in the uplink to improve system performance.

To reliably estimate the system performance of a HSR system, various aspects of the system should be considered. Two of the most important aspects are the distribution of the users in the cell and the direction of the arriving signals. The direction of arrival of the signals is related to the environment in which the system operates, which is in turn related to the size of the cell, i.e. macro-, micro- or pico cell. This chapter addresses the effect on the uplink when an adaptive antenna array is implemented at the base station and where it is assumed that each mobile is surrounded by a local scattering area where the scatterers are assumed to have a Gaussian distribution around the mobiles. This is a valid assumption, justified by measurements of the radio propagation channel [2]. In our analysis, it is further assumed that the subscribers are non-uniformly distributed throughout the cell.

As background to the numerical performance evaluation of a HSR/CDMA system, the basic operating principles of various smart antenna concepts are recapped in section

4.2. Then in section 4.3, the uplink BER performance of a cellular system consisting of a reference cell, surrounded by a single tier of interfering cells is derived. The base station employs an antenna array, and it is assumed that this array is a Uniform Linear Array (ULA). Furthermore, each received signal at the base station is assumed to be subject to Rayleigh fading. Following this derivation, it is shown in section 4.4 how the optimum antenna spacing for a ULA can be derived for the general case of signals arriving at the base station with a non-uniform angle of arrival. As a special case, the optimum spacing for a ULA with uniform AOA is derived. Section 4.5 provides a discussion on the modeling of the angle of arrival of signals at the base station and this information is combined with the BER analysis to generate some performance results for a cellular HSR/CDMA system (section 4.6). Finally some conclusions are presented in section 4.7.

4.2 SMART ANTENNA TECHNIQUES

It is interesting to note that in both the W-CDMA and TD-CDMA proposals accepted by ETSI for the implementation of UMTS, the use of smart antenna and SDMA [3, 4, 5] concepts is fully supported in order to increase the capacity of third generation mobile systems. Specifically, SDMA techniques will be implemented to yield [6]:

Increased cell coverage areas to reduce high base station site cost in low traffic areas.

Reduction of interference to improve service quality and/or increase the frequency re-use factor. This point is especially important in CDMA based systems which are interference limited.

Extend system traffic capacity.

Cellular access systems rely on the fact that users of a single resource - the Base Station (BS) - will be separable in one or more domain, that is frequency (viz FDMA), time (viz TDMA) or code (viz CDMA). Thus, in a FDMA system (for example AMPS [7] and CT-2 [8]), simultaneous transmissions to a BS will have different carrier frequencies and will therefore not interfere with one another. Similarly, in a TDMA system (for example the access method employed in one frequency allocation in an IS-54 system [8]), transmissions to the BS are separated in time to prevent interference. These multiple access techniques can also be combined to form, for example, TDMA/FDMA (IS-54) or FDMA/CDMA (IS-95 [8]) systems or any combination thereof. All of the above mentioned multiple access techniques do however share one common trait, being the non-homogeneous geographical distribution of their subscribers. This means that all of the mentioned multiple access systems can exploit another dimension, viz the spatial dimension, of the cellular problem to increase system capacity or cellular spectral efficiency.

In [9, 10, 11], cellular spectral efficiency is defined as a basis to rate the performance of a cellular system. Many definitions for cellular spectral efficiency have been proposed, including $bit/s/Hz$ [9, 11] (with the data rate measured at some predefined Bit Error Rate (BER)), $Erlang/MHz/km^2$, equivalent telephone Erlangs per square kilometer [12] and even Mbit/s-per-floor for indoor environments [12]. Because SDMA and smart antenna systems rely on spatial parameters, a spatial parameter is included in the definition of spectral efficiency in order to evaluate cellular system performance:

Definition 1: Cellular Spectral Efficiency (η)
The cellular spectral efficiency of a system is defined as the sum of the maximum data rates that can be delivered to users affiliated to all base stations in a re-use cluster of cells, occupying a defined physical area.

Mathematically, cellular spectral efficiency, η is defined as

$$\eta = \frac{\sum_{j=1}^{r} \sum_{k=1}^{K} R_{kj}}{B} \frac{1}{A_{cluster}} \quad \text{bit/s/Hz/km}^2 \quad (4.1)$$

where r denotes the number of cells in a re-use cluster, R_{kj} denotes the data rate measured in $bits/s$ at some predefined BER available to user k in cell j of the re-use cluster, B denotes the total bandwidth measured in Hz allocated to all cells in the re-use cluster and $A_{cluster}$ denotes the physical area, measured in km^2, occupied by the re-use cluster. Clearly, the concept of the re-use cluster is fundamental in the determination of η. Definitions for a re-use cluster can be found in [11, 13]. In this chapter, a re-use cluster will be defined as follows:

Definition 2: Re-use Cluster (r)
A set of cells which have access to the total Time/Frequency and Code (T/F/C) resources available in the cellular system.

Figure 4.1 shows this scenario, for the case where $r = 3$. Thus, each set of cells forming a re-use pattern exists totally independent of the other cells in the area (as far as T/F/C resources go).

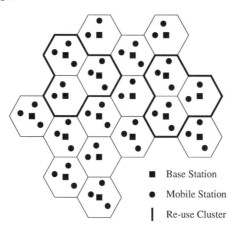

Figure 4.1 Cellular re-use concept.

With the above definitions in mind, the task of a smart antenna or SDMA system is clear - optimize the cellular spectral efficiency of the network, or in other words, increase the re-use of the available cellular resources. Examining (4.1) it is clear that reducing the size of the re-use cluster will increase the cellular spectral efficiency. However, a reduced cell size will significantly increase the interference present in the cell, limiting the capacity of the network in the case of a CDMA system. The increased amount of interference, specifically in the uplink, can be overcome using a smart antenna technique called High Sensitivity Reception (HSR) [6].

Definition 3: High Sensitivity Reception
High Sensitivity Reception refers to the use of adaptive antenna arrays in the uplink of a cellular network to focus the antenna beam on a specific user, thereby increasing the antenna gain in the direction of the user and suppressing transmissions received from interfering users.

This concept is depicted in Figure 4.2. In the case of TDMA and FDMA systems the HSR system may use pencil antenna beams [14] to focus on the active users whereas in CDMA systems, the HSR system can increase the signal-to-noise ratio (SNR) in the uplink by introducing nulls in the antenna pattern in the direction of strong interfering signals. The antenna gain of HSR with an M-element antenna is equal to $10 \log_{10} M$, where M is the number of elements in the array.

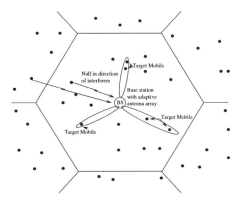

Figure 4.2   SDMA system implemented using adaptive antenna arrays.

In a manner similar to HSR, Spatial Filtering for Interference Reduction (SFIR) [6] can be used in the downlink of a cellular system to focus all the energy radiated by the base station onto a single user or cluster of users. Therefore, SFIR is defined as:

Definition 4: Spatial Filtering for Interference Reduction
SFIR reduces the interference experienced by mobile communication systems in the downlink by concentrating all radiated electromagnetic energy in the direction of a user or group of users, avoiding geographical areas where no users are active.

Because the uplink of a cellular network is in general the capacity limiting factor, it might seem that HSR systems will yield greater capacity advantages than SFIR systems. However, the increased downlink quality afforded by SFIR techniques may lead to less dropped calls during handovers (because of the better signal quality estimates available to the mobile), increasing the overall quality of service. Also, due to the dynamic nature of an adaptive antenna array, a SFIR system can facilitate the tracking of a user across cell boundaries, increasing the chances of a successful handover to the next cell.

Whereas HSR and SFIR techniques increase the cellular spectral efficiency by decreasing the total co-channel interference levels in a cell, SDMA techniques increase cellular spectral efficiency by decreasing $A_{cluster}$ with the consequent decrease in the physical size of the re-use cluster. In other words, the same physical cellular network resources can be re-used more often. Various definitions have been proposed to define SDMA techniques [3, 5, 4]. We summaries these as follows:

Definition 5: Space Division Multiple Access
A SDMA system is a multiple access technique which enables two or more subscribers, affiliated to the same base station, to use the same Time and Frequency and Code (T/F/C) resources on the grounds of their physical location or spatial separation.

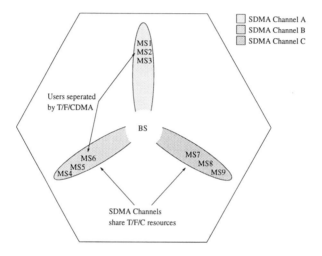

Figure 4.3 Space Division Multiple Access (SDMA): Allowing users in the same cell to share time/frequency and code resources.

This scenario is depicted in Figure 4.3, where Mobile Stations (MS) MS1, MS2 and MS3 share a same set of T/F/C resources with MS4, MS5, MS6 and MS7, MS8 and MS9. For example, MS4 and MS1 may both be allocated carrier frequency f_1, time slot T_1 and code c_1 although they are affiliated to the same BS, because of their spatial separation. In [6] it is shown analytically and by measurements that gains in the order of 5-9 dB can be obtained using antenna arrays with 8 elements.

Most often, the introduction of smart antenna techniques into cellular systems follows a migration path from HSR to SFIR and finally to SDMA. In the following section, the first step of this migration process, HSR, is analysed.

4.3 BER PERFORMANCE CALCULATION

Figure 4.4 is a representation of the cellular system under discussion, with K_l subscribers in the reference cell and a total of K subscribers in the reference and adjacent first tier of interfering cells. The output of the transmitter for user k can then be written as

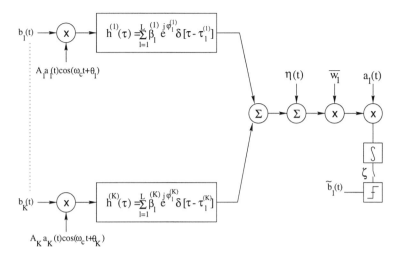

Figure 4.4 Basic block diagram of cellular CDMA system.

$$s_k(t) = A_k b_k(t) a_k(t) \cos(\omega_c t + \theta_k) \tag{4.2}$$

where A_k denotes the transmitted signal amplitude, $b_k(t)$ denotes binary data with bit period T seconds, $a_k(t)$ denotes a random binary spreading sequences with chip period T_c seconds and length $N = T/T_c$. Also, standard Binary Phase Shift Keying (BPSK) modulation is used with a carrier frequency of ω_c rad/s and unknown carrier phase θ_k, a random variable uniformly distributes over $[0, 2\pi)$. The transmitted signal propagates over a radio channel modeled as a Rayleigh fading, time invariant, discrete multipath channel with equivalent low-pass response

$$h^{(k)}(\tau) = \sum_{l=1}^{L} \beta_l^{(k)} e^{j\varphi_l^{(k)}} \delta[\tau - \tau_l^{(k)}]. \tag{4.3}$$

Each path is characterized by the variables $\beta_l^{(k)}$, a Rayleigh distributed random variable denoting the strength of path l from user k, $\varphi_l^{(k)}$ uniformly distributed over $[0, 2\pi)$ and denoting the phase shift of path l from user k and $\tau_l^{(k)}$, uniformly distributed over $[0, T)$ and denoting the propagation delay of path l from user k. Assuming that $L = \left\lfloor \frac{T}{T_c} \right\rfloor$ multipath components are present, the received signal can be written as

$$r(t) = \sum_{k=1}^{K} \sum_{l=1}^{L} A_k b_k(t) a_k(t) \beta_l^{(k)} \cos(\omega_c t - \omega_c \tau_l^{(k)} + \theta_k + \varphi_l^{(k)}) + \eta(t) \tag{4.4}$$

where $\eta(t)$ denotes AWGN with a two-sided power spectral density of $N_0 T/4$.

Since coherent demodulation is assumed, the receiver coherently recovers the carrier phase and delay locks to the desired signal. Assuming that this signal is that of subscriber $k = 1$, the output of the receiver after correlation and demodulation can be written as

$$\begin{aligned}
\zeta &= \text{Re}\left\{\int_0^T \frac{\bar{w}_1^H}{\|\bar{w}_1\|} r(t) a_1(t) \cos(\omega_c t) dt\right\} \\
&= \|\bar{w}_1\| A_k \frac{T}{2} b_0^{(1)} \beta_1^{(1)} \\
&+ \sum_{l=2}^{L} \frac{A_1}{2} \beta_l^{(1)} \cos(\theta_l^{(1)}) \|\bar{w}_1\| \bar{\mathcal{R}}_{11} \left\{ b_{-1}^{(1)} R_{11}(\tau_l^{(1)}) + b_0^{(1)} \hat{R}_{11}(\tau_l^{(1)}) \right\} \\
&+ \sum_{k=2}^{K} \sum_{l=1}^{L} \frac{A_k}{2} \beta_l^{(k)} \cos(\theta_l^{(k)}) \|\bar{w}_k\| \bar{\mathcal{R}}_{k1} \left\{ b_{-1}^{(k)} R_{k1}(\tau_l^{(k)}) + b_0^{(k)} \hat{R}_{k1}(\tau_l^{(k)}) \right\} \\
&+ \eta(t)
\end{aligned} \quad (4.5)$$

where θ is a random variable uniformly distributed on $[0, 2\pi)$ and

$$\bar{\mathcal{R}}_{k1} = \frac{\text{Re}[\bar{w}_1^H \bar{w}_k]}{\|\bar{w}_1\| \|\bar{w}_k\|} \quad (4.6)$$

with $(\cdot)^H$ denoting the Hermitian transpose and $\bar{w}_k$ the array manifold vector or steering vector optimizing the response of the antenna array for user k. Furthermore, $b_0^{(1)}$ denotes the current information bit detected for user $k = 1$ and $b_{-1}^{(1)}$ denotes the previous information bit. Also,

$$\begin{aligned}
R_{k1}(\tau_l^{(k)}) &= \int_0^{\tau_l^{(k)}} a_k(t - \tau_l^{(k)}) a_1(t) dt \\
\hat{R}_{k1}(\tau_l^{(k)}) &= \int_{\tau_l^{(k)}}^{T} a_k(t - \tau_l^{(k)}) a_1(t) dt.
\end{aligned} \quad (4.7)$$

It should be clear that the detection variable is made up of four distinct elements, each represented by a term in (4.5). The first term represents the contribution of the main received path of the reference user, the second term all multipath components of this user or self-interference, the third term all transmissions from interfering users within the reference cell and the surrounding interfering cells and the final term the effect of AWGN. Therefore, (4.5) can be rewritten as the sum of the desired signal and some interference term, i.e.

$$\zeta = \frac{A_1 T}{2} \left\{ \beta_1^{(1)} \|\bar{w}_1\| b_0^{(1)} + \alpha \right\} + \eta \quad (4.8)$$

where

$$\alpha = \sum_{l=2}^{L} \frac{\beta_l^{(1)}}{T} \cos(\theta_l^{(1)}) \|\bar{w}_1\| \bar{\mathcal{R}}_{11} \left\{ b_{-1}^{(1)} R_{11}(\tau_l^{(1)}) + b_0^{(1)} \hat{R}_{11}(\tau_l^{(1)}) \right\}$$

$$+ \sum_{k=2}^{K_I} \sum_{l=1}^{L} \frac{A_k \beta_l^{(k)}}{A_1 T} \cos(\theta_l^{(k)}) \|\bar{w}_k\| \bar{\mathcal{R}}_{k1} \left\{ b_{-1}^{(k)} R_{k1}(\tau_l^{(k)}) + b_0^{(k)} \hat{R}_{k1}(\tau_l^{(k)}) \right\}$$

$$+ \sum_{k=K_I+1}^{K} \sum_{l=1}^{L} \frac{A_k \beta_l^{(k)}}{A_1 T} \cos(\theta_l^{(k)}) \|\bar{w}_k\| \bar{\mathcal{R}}_{k1} \left\{ b_{-1}^{(k)} R_{k1}(\tau_l^{(k)}) + b_0^{(k)} \hat{R}_{k1}(\tau_l^{(k)}) \right\}$$

(4.9)

To calculate the error performance of the system, and making use of the Gaussian assumption, that is assuming that α is Gaussian distributed, it is then necessary to calculate the variance of α. In order to simplify the analysis it is assumed that β, θ and τ are all independent random variables. Furthermore, from inspection of (4.9) it is clear that the expected value of α, $E\{\alpha\}$ must be equal to zero as $E\{\cos(\theta)\} = 0$. Therefore, the variance of α is equal to $E\{\alpha^2\}$ or,

$$E\{\alpha^2\} = \sum_{l=2}^{L} \left(\frac{E\{\beta_l^{(1)^2}\}}{E\{T^2\}} E\{\|\bar{w}_1\|^2\} E\{\bar{\mathcal{R}}_{11}^2\} \right.$$

$$\cdot \left. E\left\{ \left((b_{-1}^{(1)} R_{11}(\tau_l^{(1)}) + b_0^1 \hat{R}_{11}(\tau_l^{(1)})) \cos(\theta_l^{(1)}) \right)^2 \right\} \right)$$

$$+ \sum_{k=2}^{K_I} \sum_{l=1}^{L} \left(\frac{E\{A_k^2\} E\{\beta_l^{(k)^2}\}}{E\{A_1^2\} E\{T^2\}} E\{\|\bar{w}_k\|^2\} E\{\bar{\mathcal{R}}_{k1}^2\} \right.$$

$$\cdot \left. E\left\{ \left((b_{-1}^{(1)} R_{k1}(\tau_l^{(k)}) + b_0^1 \hat{R}_{k1}(\tau_l^{(k)})) \cos^2(\theta_l^{(k)}) \right)^2 \right\} \right)$$

$$+ \sum_{k=K_I+1}^{K} \sum_{l=1}^{L} \left(\frac{E\{A_k^2\} E\{\beta_l^{(k)^2}\}}{E\{A_1^2\} E\{T^2\}} E\{\|\bar{w}_k\|^2\} E\{\bar{\mathcal{R}}_{k1}^2\} \right.$$

$$\cdot \left. E\left\{ \left((b_{-1}^{(1)} R_{k1}(\tau_l^{(k)}) + b_0^1 \hat{R}_{k1}(\tau_l^{(k)})) \cos^2(\theta_l^{(k)}) \right)^2 \right\} \right) \quad (4.10)$$

Each term in (4.10) is discussed below.

Let us first consider the term $E\{\beta_l^{(k)^2}\}$. The envelope of a fading mobile signal can be described by a Rayleigh fading process. Using this assumption, the received signal strength of the multipath component l of user k has the well-known pdf

$$p_{\beta_l^{(k)}}(\beta) = \left(\frac{2\beta^2}{\Omega_l^{(k)}} \right) \exp\left(-\frac{m_l^{(k)}}{\Omega_l^{(k)}} \beta^2 \right) \quad \forall \quad \beta \geq 0 \quad (4.11)$$

with

$$\Omega_l^{(k)} = E\{\beta_l^{(k)^2}\}. \tag{4.12}$$

where $\Omega_l^{(k)}$ denotes the average strength of path l from user k. In order to determine $\Omega_l^{(k)}$, an exponential multipath intensity profile is assumed. Specifically, the strength of path l is given by

$$\Omega_l^{(k)} = \Omega_1^{(k)} e^{\delta l} \quad 1 < l \leq L \tag{4.13}$$

where δ is the rate of decay of the multipath components. A typical value for δ in a micro cellular scenario is 1. Furthermore, $\Omega_l^{(k)}$ may be subject to log-normal shadowing effects [15]. However, in our analysis this effect is not taken into account and it is assumed that $\Omega_1^{(k)}$ is fixed at -10 dB.

In order to simplify the analysis of the cellular system, it is assumed that the network employs a limited power control algorithm capable of eliminating the effect of propagation loss for users in the reference cell. Therefore, the mean received signal power from users in the reference cell can be written as

$$\Omega_0^{(k)} = A_k^2/T \quad k \leq K \tag{4.14}$$

where $A_k = A_1 \delta_{pc}$ with δ_{pc} a Gaussian random variable with mean equal to 1 and variance equal to the error variance of the power control scheme. Furthermore, interfering users in neighboring cells have a mean received signal power denoted by

$$\Omega_0^{(k)} = A_1^2/T \int_R^{3R} (R/r)^n p_R(r) dr \tag{4.15}$$

where r denotes the distance from the base station to the interfering user, n denotes the path loss coefficient (see Tables 4.1 and 4.2) and $p_R(r)$ denotes the pdf of the distance distribution of the interfering users. Note that the power control error is excluded from the mean received signal as this error is much smaller than the propagation loss that is now taken into account. Clearly, a non-uniform pdf for the distance distribution of users in the neighboring cells will influence the total interference. For instance if the users are clustered near the cell border the interference level will be higher. However, in this analysis it is assumed that all interfering users in the neighboring cells have a uniform distance distribution between R and $3R$.

Next we consider $E\{A_k^2\}$. Clearly, A_k is constant for the duration of the analysis and therefore $E\{A_k^2\} = A_k^2$.

Assuming next that the steering vector for each subscriber is updated at a rate equal to the power control information and that it is constant over the period in which the bit error rate is evaluated, it is possible to calculate $E\{\|\bar{w}_k\|^2\}$. Each base station employs a M element antenna array, with each element considered to be an isotropic

SPATIAL FILTERING AND CDMA

Table 4.1 Propagation parameters for micro cellular environments.

Parameter	Symbol	Value
Path loss exponent (no LOS)	n	2.5
Log-normal fading variation	σ	6-12 dB
Short-term fading model		Rayleigh

Table 4.2 Propagation parameters for macro cellular environments.

Parameter	Symbol	Value
Path loss exponent (no LOS)	n	4
Log-normal fading variation	σ	6-12 dB
Short-term fading model		Rayleigh

radiator with complex gain factors I_m and spacing between elements d_x. Since we are considering a ULA, we have

$$\|\bar{w}_k\|^2 = \bar{w}_k^H \bar{w}_k = \sum_{m=0}^{M-1} I_m^{(k)^H} I_m^{(k)} \tag{4.16}$$

where

$$I_m^{(k)} = \begin{bmatrix} I_0^{(k)} & I_1^{(k)} e^{j\gamma d_x \cos\phi^{(k)}} & \cdots & I_{(M-1)}^{(k)} e^{j\gamma(M-1)d_x \cos\phi^{(k)}} \end{bmatrix}^T \tag{4.17}$$

and mobile k's signal incidents on the array from direction $\phi^{(k)}$. The parameter $\gamma = 2\pi/\lambda$, where λ is the carrier wavelength. Therefore

$$E\{\|\bar{w}_k\|^2\} = \|\bar{w}_k\|^2 = M \tag{4.18}$$

From (4.6) it follows readily that

$$E\{\bar{\mathcal{R}}_{k1}^2\} = \frac{E\left\{\left(\text{Re}[\bar{w}_1^H \bar{w}_k]\right)^2\right\}}{E\left\{(\|\bar{w}_1\|)^2\right\} E\left\{(\|\bar{w}_k\|)^2\right\}} \tag{4.19}$$

$$= \frac{1}{M^2} E\left\{\left(\text{Re}[\bar{w}_1^H \bar{w}_k]\right)^2\right\}.$$

In order to calculate $E\left\{\left(\text{Re}[\bar{w}_1^H \bar{w}_k]\right)^2\right\}$, it is noted that

$$\left(\text{Re}\left[\bar{w}_1^H \bar{w}_k\right]\right)^2 = \left(\text{Re}\left[\sum_{m=0}^{M-1} I_m^{(1)*} e^{-j\gamma m d_x \cos\phi^{(1)}} I_m^{(k)} e^{j\gamma m d_x \cos\phi^{(k)}}\right]\right)^2. \quad (4.20)$$

In our analysis we will assume our reference user to be located at $\phi_1 = 90°$, and all other mobile users' angles distributed according to the mobile location pdf that will be derived in section 4.5. Therefore, for our performance calculations

$$E\left\{\left(\text{Re}\left[\bar{w}_1^H \bar{w}_k\right]\right)^2\right\} = \left(\text{Re}\left[\bar{w}_1^H \bar{w}_k\right]\right)^2 \quad (4.21)$$

which is a constant.

To determine the variance of the multiple access interference caused by the interaction of spreading sequences, it is required to calculate

$$E\left\{\left(b_{-1}^{(1)} R_{k1}(\tau_l^{(k)}) + b_0^{(1)} \hat{R}_{k1}(\tau_l^{(k)}) \cos(\theta_l^{(k)})\right)^2\right\} \quad (4.22)$$

There have been many papers evaluating this term, many utilizing very computationally intensive algorithms. In [16] an efficient approximation based on a modified Gaussian approximation for random spreading sequences is presented, where

$$\sigma_{MA}^2 = \sum_{k=2}^{K} E\left\{\left(b_{-1}^{(1)} R_{k1}(\tau_l^{(k)}) + b_0^{(1)} \hat{R}_{k1}(\tau_l^{(k)}) \cos(\theta_l^{(k)})\right)^2\right\} \quad (4.23)$$

$$= \frac{2}{3}\frac{K-1}{3N} + \frac{1}{6}\frac{((K-1)N/3 + \sqrt{3}\sigma)}{N^2} + \frac{1}{6}\frac{((K-1)N/3 - \sqrt{3}\sigma)}{N^2}$$

and

$$\sigma = (K-1)\left[N^2 \frac{23}{360} + N\left(\frac{1}{20} + \frac{K-2}{36}\right) - \frac{1}{20} - \frac{K-2}{36}\right] \quad (4.24)$$

Equation (4.22) with $k = 1$ represents the self-noise caused by the interaction of delayed versions of the same spreading sequence. This term is somewhat more difficult to evaluate than the multiple access interference term, as it depends on the autocorrelation properties of the specific code used. Following the rationale presented in [16], the self-noise is modeled as interference caused by $L - 1$ paths of the $K + 1$'th user. Therefore, assuming that the autocorrelation properties at non-zero shifts are similar to cross correlation properties between sequences, the contribution of one delayed path of the reference user's sequence will be

$$E\left\{\left(b_{-1}^{(1)} R_{11}(\tau_l^{(1)}) + b_0^{(1)} \hat{R}_{11}(\tau_l^{(1)}) \cos(\theta_l^{(1)})\right)^2\right\} = \sigma_{MA}^2/(K-1) \quad (4.25)$$

To calculate the system BER, we follow a similar procedure as in [17], and the result to calculate the average error rate in Rayleigh fading is stated here as

$$P_e = \frac{1}{2}\left(1 - \sqrt{\frac{\Lambda \Omega_l^{(k)}}{1 + \Lambda \Omega_l^{(k)}}}\right) \qquad (4.26)$$

where

$$\frac{1}{\Lambda} = 1 + 2\frac{E_b}{N_0}E\{\alpha^2\} \qquad (4.27)$$

and $\frac{E_b}{N_0}$ is the energy per bit to noise ratio.

4.4 OPTIMUM ANTENNA SPACING CRITERIA

In this section we derive a parameter for the optimum antenna spacing from (4.10) in a cellular CDMA environment. It is clear that if $E\{\alpha^2\} = 0$ the system will have optimal performance, i.e. no multiple access interference. This can only happen when there is no multiple access interference or when $\left(\text{Re}\left[\bar{w}_1^H \bar{w}_k\right]\right)^2 = 0$, which is essentially the spatial correlation between $\phi^{(1)}$ and $\phi^{(k)}$. This is only possible when the interferers, relative to the reference user, are in the antenna nulls. Since we are considering a multiple access system, we would like to optimism $E\{\alpha^2\}$ in the presence of multiple access interference. The only parameter which can be optimized for this purpose when a ULA is assumed is the antenna spacing. Therefore it is required to determine an optimal antenna spacing, that is an optimum d_x in (4.20), that would minimize $E\{\alpha^2\}$. In order to achieve this, we define the following.

It is well known that for a broadside ULA, there is always a maximum gain at 90° [14]. We therefore assume that the desired user is at a spatially optimal position, $\phi^{(1)} = 90°$. This is arbitrary since it is assumed that, in the case of an adaptive antenna, the main beam of the antenna will always be on the desired user, i.e. maximum gain towards the desired user. The other users can be distributed anywhere in the cell. For our derivation of the optimal antenna spacing, we assume the mobile users to be uniformly distributed, that is, the pdf of $\phi^{(k)}$ can be written as

$$p(\phi^{(k)}) = \frac{1}{2\pi} \quad \forall \quad 0 \leq \phi^{(k)} \leq 2\pi. \qquad (4.28)$$

We can therefore determine

$$\begin{aligned}
\Omega &= E\left\{\left(\operatorname{Re}\left[\bar{w}_1^H \bar{w}_k\right]\right)^2\right\} \\
&= E\left\{\left(\operatorname{Re}\left[\sum_{m=0}^{M-1} e^{-j\gamma m d_x \cos(\pi/2)} e^{j\gamma m d_x \cos\phi^{(k)}}\right]\right)^2\right\} \\
&= \frac{1}{2\pi}\int_0^{2\pi}\left(\operatorname{Re}\left[\sum_{m=0}^{M-1} e^{j\gamma m d_x \cos\phi^{(k)}}\right]\right)^2 d\phi^{(k)} \\
&= \frac{1}{2}\sum_{x1=0}^{M-1}\sum_{x2=0}^{M-1}[J_0(\epsilon(x_1+x_2))+J_0(\epsilon(x_1-x_2))] \quad (4.29)
\end{aligned}$$

where J_0 is the zeroth order Bessel function and

$$\epsilon = \gamma d_x = \frac{2\pi}{\lambda}\frac{\lambda}{d} = \frac{2\pi}{d} \quad (4.30)$$

where $d \geq 1$ is an integer. Figure 4.5 shows Ω as a function of d for different values of M.

It is clear that for a given number of antenna elements M, the optimum value of $d = 2$. Therefore an optimum antenna element spacing for a ULA is $d_x = \frac{\lambda}{2}$ and will be used in the performance calculations of section 4.6.

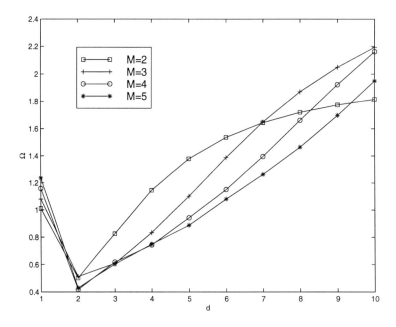

Figure 4.5 Optimum antenna spacing for uniformly distributed mobile users in a cell.

4.5 MOBILE LOCATION DISTRIBUTION

The position of mobiles in a cellular system has a significant influence on the system performance. Assuming that each cell, with radius R, has a base station located at its centre, with mobile users distributed throughout the cell, the position of a mobile user in the cellular structure is fully defined by its distance from the reference base station r, and its angle, ϕ_o, measured from some reference, both of which can be considered random variables. This is shown in Figure 4.6.

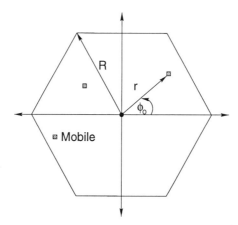

Figure 4.6 Modeling the location of mobiles in a cellular system.

As the performance of the cellular system depends on the spatial distribution of mobiles, true pdf's for both the angular and distance distributions of mobiles are required in order to calculate the performance of the system. These are, however, not available in the literature and various approximations such as uniform angular and distance distributions [18] and modified Gaussian distributions [19] are used. These assumptions are either not a good representation of the real world scenario or are somewhat inflexible as it is clear that these distributions will vary from situation to situation. Specifically, a single pdf applicable to many scenarios would be extremely useful. Thus, in terms of the angular distribution of mobiles the authors propose a pdf of the form

$$p_{\Phi_o}(\phi_o) = \frac{1}{A_{norm}} \left[1 + \sum_{l=1}^{N_{peak}} \gamma_l \left[\mathrm{rect}\left(\frac{w_l \phi_v}{\pi} - \alpha_l\right) \right. \right.$$
$$\left. \left. + \mathrm{rect}\left(\frac{w_l \phi_o}{\pi} - \alpha_l - 2\pi\right)\right] \right.$$
$$\left. \cos^2(w_l \phi_o - \alpha_l) \right] \qquad 0 \leq \phi_o \leq 2\pi \qquad (4.31)$$

where A_{norm} is a normalizing factor to ensure that $\int_0^{2\pi} p_{\Phi_o}(\phi_o) d\phi_o = 1$ and N_{peak} is the number of peaks in the pdf. This factor (N_{peak}) is a measure of the angular clustering of mobiles in a cell. Clearly, if $N_{peak} = 0$, (4.31) denotes a uniform angular subscriber distribution. Furthermore, rect(x) is defined as

$$\text{rect}(x) = \begin{cases} 1 & |x| < \frac{1}{2} \\ 0 & |x| > \frac{1}{2} \end{cases} \quad (4.32)$$

with w_l an integer controlling the width of peak l. Typically, values for w_l will be chosen to yield angular peaks of different maximum and minimum widths. For instance, if $w_l = 10$ the width of peak l in the pdf of the angular distribution of mobiles will be $\pi/10$ rad. The angular location of peak l is given by α_l, while the relative size of each peak is determined by γ_l (typically values between 0 and 1).

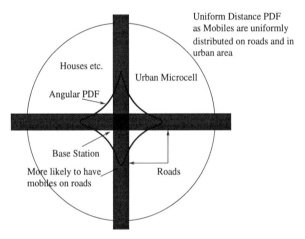

Figure 4.7 Qualitative description of the PDF of the mobile distributions in a typical urban micro cellular scenario.

As an example, consider the scenario depicted in Figure 4.7. This figure depicts a typical urban or suburban scenario where two main roads intersect with a base station situated at the crossing. As these roads carry a much larger portion of the total mobile traffic than other areas within the coverage region of the cell, the probability of receiving transmission with angles of incidence of 0, $\pi/2$, π and $3\pi/2$ rad is substantially higher, as the concentration of mobiles in these specific directions are substantially higher. The pdf of the location of mobiles for this scenario is shown in Figure 4.8, where the angular width of a street is taken to be $10°$

By using the variable transformation

$$r = \phi_o \frac{3R}{2\pi} \quad (4.33)$$

(4.31) can also describe the distance distribution of mobiles. However, a uniform distance distribution is often a good approximation to make (see for instance Figure

4.7), and will therefore be assumed in the analysis to follow. Using the above equations, different scenarios, such as clustering of mobiles, can be represented. In our BER calculations, the pdf of (4.31) is used to generate the $\phi^{(k)}$ values of (4.20), using a procedure in [20] to generate variables of arbitrary distributions.

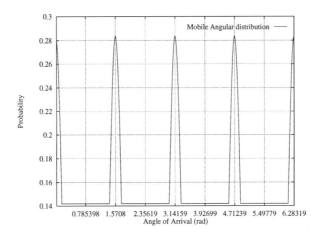

Figure 4.8 Example pdf of the angular distribution of mobiles.

In addition to the users being non-uniformly distributed throughout the cell, the multipath echoes received from each user are assumed to originate from a region around the mobile. This scattering region, shown in Figure 4.9 and futher described in [2], is assumed to have a Gaussian distribution. This means that more multipath echoes will be created by scattering points situated close to the mobile user than by scattering points far away from the user. Based on this assumption, it is shown in [2] that the pdf of the Angle of Arrival (AOA) of signals from a specific user, situated at 0° is

$$p_{\Phi_o}(\phi_o) = \frac{A}{2\sqrt{2\pi}\sigma_s} e^{\frac{D^2(\cos^2\phi_o - 1)}{2\sigma_s^2}} \operatorname{erfc}\left(\frac{-D\cos\phi_o}{\sqrt{2}\sigma_s}\right) \quad -180° \leq \phi_o \leq 180° \quad (4.34)$$

where ϕ_o is as shown in Figure 4.6, A is a normalising constant, D is the base station-mobile separation and σ_s^2 is the variance of the Gaussian distribution describing the scattering region surrounding the mobile. This variance is dependent on the mobile environment. As an example, the pdf of the angle of arrival (AOA) of signals from a specific user, situated at 0°, is shown in Figure 4.10, for the cases of macro, and N-LOS micro cells. In [2] a rule of thumb for this variance is presented which is used in the performance analysis presented in this chapter. Specifically, the variance of the scattering elements surrounding the mobile in the case of a macro cell with radius 10 000 m is 1 000, and for the case of a N-LOS micro cell with cell radius 1 000m it is 340. These numerical values have been used in all the performance evaluations presented in the next section.

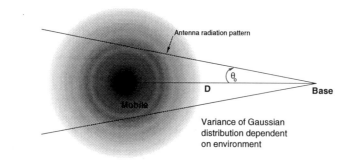

Figure 4.9 Modeling of scattering elements using a Gaussian approach.

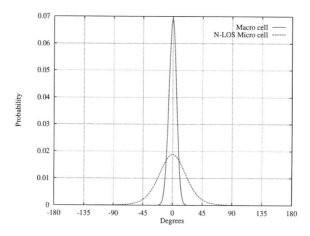

Figure 4.10 PDF of the AOA of a single user as seen at the base station, caused by scattering elements surrounding the mobile.

4.6 RESULTS

Using the results from the previous sections, numerical performance results obtained through Monte Carlo simulations are presented in this section. For convenience, the parameters affecting the average error rate are recapped in Table 4.3, indicating typical values in the calculations.

The basic system concept of a HSR/CDMA system using antenna arrays to limit the co-channel interference present in the cellular system is depicted in Figures 4.11 and 4.12. Figure 4.11 shows the radiating pattern of a 5 element ULA with a null located at approximately 150°. Positioning all users in the regio of this null (except for the reference user which is located at 90°), yields better BER performance compared to the cases where all interfering users are located at 180° or at 90°. As would be expected, co-locating the interfering users with the reference user in the main beam of the antenna results in the worst BER performance.

Table 4.3 Summary of most important simulation parameters and their vales.

Parameter	Value	Description
N	1023	Processing gain; number of chips per data bit
K_l	20	Number of users in reference cell
K	140	Total number of users in all cells
L	1	Number of multipaths
M	5	Number of elements in antenna array
δ_{pc}	0 dB	Power control error random variable
$\Omega_1^{(k)}$	-10 dB	Average path strength of path 1 from user k
γ_l	100	Relative size of mobile location peak
α_l		Angular location of user cluster
ω_l	10°	Width of cluster

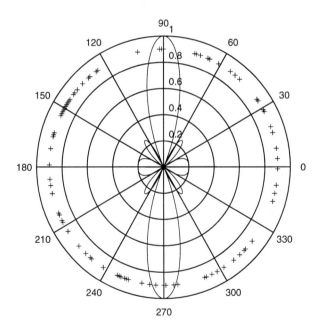

Figure 4.11 Basic concept of adaptive antenna system to limit interference and increase system capacity - users in antenna nulls do not contribute to overall interference levels.

However, in a cellular system all users are clearly not co-located, but are distributed according to (4.31). Furthermore, each user is surrounded by a scattering region with characteristics determined by the environment as discussed in section 4.5. Figure 4.13 summarizes the performance of the adaptive antenna system under various deployment

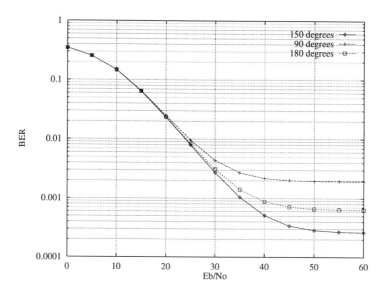

Figure 4.12 BER performance of cellular system with reference user located at $90°$, and interfering users at $90°$, $150°$ and $180°$ respectively.

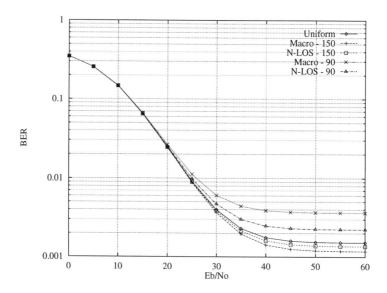

Figure 4.13 BER performance of macro, N-LOS micro cells with uniform and non-uniform user distributions

conditions. Specifically, the case of a macro cell is considered with the cell radius equal to 10 000 m and the variance of the scatterer distribution determined by the rule of thumb presented in [2] and equal to 1000. Furthermore, the antenna system shown in Figure 4.11 is used to limit interference from unwanted users. These users are distributed according to (4.31) where a single cluster of users of width 5° and height of 100 is placed at 90° and 150° respectively. This user pdf represents the distribution of users in both the reference cell, as well as in the first tier of surrounding cells. In all simulations (except where specifically noted otherwise) a user population consisting of 20 users in the reference cell and 120 users in the first tier of cells (yielding a total user population of 140) is assumed. Also, for reference purposes, the impact of the non-uniform distribution of users can be compared to the case where users are uniformly distributed. Concentrating all interfering users in the main radiation lobe results in the worst system performance in the case of macro as well as Non-Line of Sight (N-LOS) micro cells. Macro cells are more affected by these interfering users than N-LOS micro cells as the latter type of cells have received signals from a wider range of angles (see Figure 4.10) resulting in an increased limiting of received interfering signal power.

With the interfering users clustered in the vicinity of the null at 150°, the performance of the system is increased by approximately 14 dB. This figure is slightly higher than first order predictions presented in [6] of antenna gains in the order of $10 \log M$ for interference rejection systems used in the uplink of cellular networks. This increased gain is due to the clustering of users near the null in the antenna pattern.

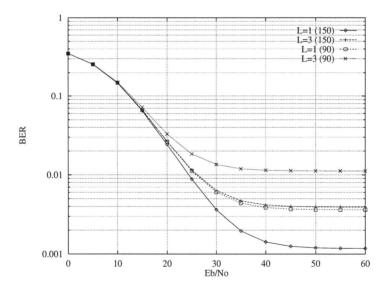

Figure 4.14 BER performance of a macro cellular system with $L = 1$ and $L = 3$ multipath components

In the case of the above mentioned BER curves, only one received signal was considered to have arrived from each user. Increasing the number of multipath echoes received from both the reference and interfering users decreases the system performance in the absence of a RAKE receiver and diversity schemes [21, 22, 23]. In

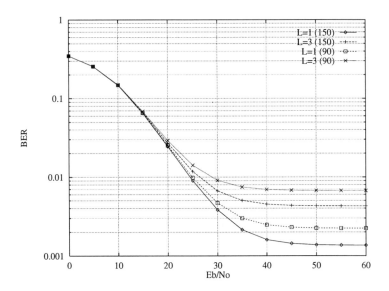

Figure 4.15 BER performance of a N-LOS micro cellular system with $L = 1$ and $L = 3$ multipath components

all deployment cases (macro and micro), the system's performance decreases as the number of multipath echoes increases.

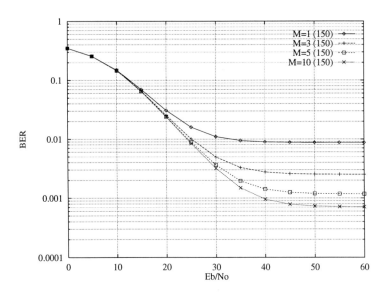

Figure 4.16 BER performance of a macro cellular system with $M = 1, 3, 5$ and 10 elements with users clustered at $150°$.

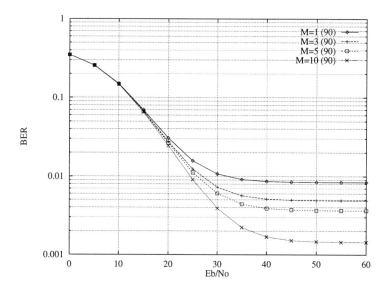

Figure 4.17   BER performance of a macro cellular system with $M = 1, 3, 5$ and 10 elements with users clustered at $90°$.

One of the key parameters determining the system performance is the number of antenna elements in the array. Whether the users are clustered at $90°$ or at $150°$, with subscriber peak width (from (4.31) equal to $5°$), increasing the number of antenna elements in the array leads to significant improvements in the required E_b/N_o ratio. For the case $M = 10$, this improvement is equal to 14 dB when compared to the system performance when $M = 1$. This value is also slightly higher than predicted by the first order approximation presented in [6], again due to the non-uniform user distribution. As presented above, it is clear from the figures that the absolute BER values are lower when the the users are clustered at $150°$.

The interference canceling effect that is evident from the graphs presented above would be expected to influence the sensitivity of the cellular system to other parameters, such as the power control error. Limiting the received signal power from interferers which may have substantially higher transmitted signal powers than the reference user will clearly increase the overall system performance. Figure 4.18 depicts the performance of macro and N-LOS micro cells when a normally distributed power control error with variance of 0 dB (perfect power control) and 5 dB respectively is present. As would be expected, the system performance decreases as the power control error increases.

Up to this point, only cluster sizes of $5°$ have been considered. This is a very narrow peak and the number of users in this peak is not extremely high. If this peak is widened without decreasing its height, more users are clustered around the antenna null, even though they have a slightly larger angular spread. In this case the system performance is better, indicating that it is not necessary to form sharp nulls in the direction of the interfering users, but that increased system performance can be obtained by ensuring

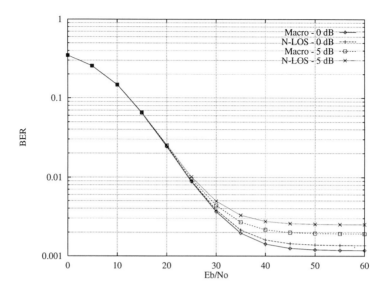

Figure 4.18 BER performance of a macro cellular system with $M = 5$ in the presence of power control errors.

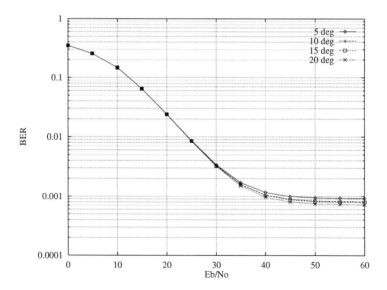

Figure 4.19 BER performance of a macro cellular system with $M = 5$ and various user location peak widths.

that clusters of interfering signals are only illuminated by antenna side lobes. The increase in BER performance that can be achieved is shown in Figure 4.19.

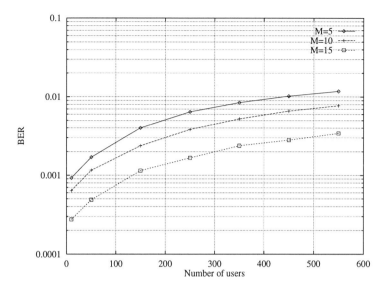

Figure 4.20 System capacity as a function of the number of antenna element in the array.

The final system parameter that is considered is the system capacity as a function of the number of antenna elements. It was already shown in Figures 4.16 and 4.17 that an increase in the number of antenna elements leads to an increase in the BER performance of the system. This fact is underlined in Figure 4.20 where the capacity of a macro cellular system is depicted as a function of the number of antenna elements. As can be seen from the Figure, increasing the number of antenna elements from $M = 5$ to $M = 15$, increases the capacity of the system from around 10 users to 150 users assuming that a BER of 10^{-3} is required for reliable system operation. In the simulation, the values of 10 and 150 refer to users in the reference cell and exclude an additional 120 users in the first tier of surrounding cells.

4.7 CONCLUSIONS

This chapter reviewed the basic operating principles of smart antenna techniques, and focussed on the analysis of a HSR/CDMA system. The BER performance of a multipath Rayleigh faded multiple cell HSR/CDMA system was derived, and a measure to determine the optimum antenna spacing in a cellular CDMA system employing a ULA was derived based on the BER analysis. Specifically, the derivation is based on reducing the multiple access interference by using the spatial correlation as criterion. This criterion can also be used to determine the optimum antenna geometry in a cellular CDMA system. Also, a mobile user location probability density function was derived to more realistically model the location of users in a cellular network. This pdf was used to analyse the performance of the HSR/CDMA system using Monte Carlo analysis.

From the results it is clear that antenna arrays are an effective way to reduce interference and to increase system capacity. What is of essential importance is to have an algorithm to determine the angle of arrival of the reference user and to be able to determine where the strongest interferers are located. Using this information, the antenna radiation pattern can be formed to ensure that clusters of interfering users are served only by antenna side lobes. From the results it is also clear that a HSR system will provide higher system gains in the case of macro cells as opposed to N-LOS micro cells. This is due to the fact that the angular spread of the arriving signals at the base station is narrower for macro cells as opposed to N-LOS micro cells, making it easier to filter out interference through the use of antenna nulls. In general, it is clear that, in addition to an antenna array, a RAKE receiver in a multipath environment is essential and that a good power control scheme is needed for acceptable system performance.

References

[1] J. S. da Silva, B. Arroyo-Fernandez, B.Barani, J. Pereira, and D. Ikonomou, "Mobile and personal communications:ACTS and beyond," in *Wireless Communications: TDMA versus CDMA*, pp. 379–414, Kluwer, 1997.

[2] M. Lötter and P. van Rooyen, "A probability density function for the angle of arrival of signals in cellular CDMA/SDMA systems," in *Proc. ICT'98*, (Chalkidiki, Greece), Aristotle University of Thessaloniki, 1998.

[3] E. Buracchini, F. Muratore, V. Palestini, and M. Sinibaldi, "Performance analysis of a mobile system based on combined SDMA/CDMA access techniques," in *Proc. ISSSTA'96*, (Mainz, Germany), pp. 370–374, University of Kaiserslautern, 1996.

[4] O. Norklit, P. Eggers, P. Zetterberg, and J. B. Andersen, "The angular aspects of wideband modelling and measurements," in *Proc. ISSSTA'96*, (Mainz, Germany), pp. 73–78, University of Kaiserslautern, 1996.

[5] C. Farsakh and J. Nossek, "Comparison of symmetric antenna array configurations by their spatial separation potential," in *Proc. ICUPC'96*, 1996.

[6] M. Tangeman, "Smart antenna technology fo GSM/DCS1800," in *Proc. PWC'96*, 1996.

[7] A. Viterbi, *CDMA - Principles of spread spectrum communications*, ch. 1. Addison-Wesley, 1995.

[8] D. Cox, "Wireless network access for personal communications," *IEEE Communications Magazine*, pp. 96–115, December 1992.

[9] A. Steil and J. Blanz, "Spectral efficiency of JD-CDMA mobile radio systems applying coherent receiver antenna diversity with directional antennas," in *Proc. ISSSTA'96*, pp. 313–319, 1996.

[10] P. Jung, B. Steiner, and B. Stilling, "Exploitation of intracell macrodiversity in mobile radio systems by deployment of remote antennas," in *Proc. ISSSTA'96*, (Mainz,Germany), pp. 302–307, University of Kaiserslautern, 1996.

[11] A. Klein, *Multi-user detection of CDMA signals - algorithms and their application to cellular mobile radio*. PhD thesis, University of Kaiserslautern, 1996.

[12] J. Cheung, M. Beach, and J. McGeehan, "Network planning for third-generation mobile radio systems," *IEEE Communications Magazine*, vol. 32, pp. 54–59, November 1994.

[13] W. Lee, "Smaller cells for greater performance," *IEEE Communications Magazine*, vol. 29, pp. 19–23, November 1991.

[14] C. Balanis, *Antenna Theory - Analysis and Design*, ch. 6. Wiley, 1997.

[15] J. Proakis, *Digital Communications*, ch. 14. McGraw-Hill, 1995.

[16] J. Holtzman, "A simple, accurate method to calculate spread-spectrum multiple access error probabilities," *IEEE Transactions on Communications*, vol. 40, pp. 461–464, March 1992.

[17] P. van Rooyen and R. Kohno, "DS-CDMA performance with maximum ratio combining and antenna arrays in Nakagami multipath fading," in *Proc. ISSSTA'96*, (Mainz, Germany), pp. 292–296, University of Kaiserslautern, 1996.

[18] C. Kchao and G. Stüber, "Analysis of a direct-sequence spread-spectrum cellular radio system," *IEEE Trans. Communications*, vol. 41, pp. 1507–1516, October 1993.

[19] R. Prasad, *CDMA for wireless personal communications*. Norwood:Ma: Artech House, 1996.

[20] W. Press, S. Teukolsky, W. Vetterling, and B. Flannery, *Numerical Recipes in C*. Cambridge University Press, second edition ed., 1992.

[21] P. Balaban and J. Salz, "Optimum diversity combining and equalization in digital data transmission with applications to cellular mobile radio - part I: Theoretical considerations," *IEEE Trans. Communications*, 1992.

[22] P. Balaban and J. Salz, "Optimum diversity combining and equalization in digital data transmission with applications to cellular mobile radio - part II: Numerical results," *IEEE Trans, Communications*, 1992.

[23] R. Kohno, "Spatially and temporally joint optimum transmitter - receiver based on adaptive array antenna for multi-user detection in DS/CDMA," in *Proc. ISSSTA'96*, (Mainz, Germany), pp. 365–369, University of Kaiserslautern, 1996.

5 TOPICS IN CDMA MULTIUSER SIGNAL SEPARATION[1]

(FROM RESEARCH RESULTS AT CCSPR, NJIT)

Y. Bar-Ness

Center for Communications and Signal Processing Research (CCSPR)
Department of Electrical and Computer Engineering
New Jersey Institute of Technology
University Heights, Newark, NJ 07102

barness@megahertz.njit.edu

Abstract: CDMA is predicted to become an important method for the third generation wireless communication systems. However, for CDMA to satisfy the need for an increased number of simultaneous users, multiuser interference must be effectively combatted. Many methods for handling this problem were suggested – of particular interest are those which are adaptive. In this paper, we summarize some of the recent results of research performed at CCSPR, NJIT, on multiuser signal separation and detection in CDMA systems.

5.1 INTRODUCTION

The recent and rapidly increasing interest in vehicular communication, the need for an increased number of simultaneous users in radio-based communications, and the limited availability of channel bandwidth has forced researchers to look into multiple-access schemes with frequency re-use, as in Code Division Multiplexing (CDMA) or time slot sharing, as in time domain multiplexing (TDMA).

The choice of CDMA is attractive because of its possible potential capacity increase and other technical factors, such as anti-multipath fading capability. For satisfactory performance, however, one must consider the effect of the "near-far" problem resulting from excessive Multiple Access Interference energy (MAI) from nearby users, com-

---

[1] This work was partially supported by NSF grant award NCR-9523954

pared with the desired user's signal energy. Power control [1-3], that is, adjustment of transmitter power of the desired user, depending on its location and the signal energies of the other users, has been suggested as a solution to this problem. Particularly with conventional matched filter receivers, this requires a significant reduction of the signal energies of the strong users in order for the weaker users to achieve reliable communication. This results in an overall reduction in communication ranges. Commercial digital cellular systems based on CDMA and which use stringent power control are described in [4]. It is noted in [4] that CDMA offers capacity increases over TDMA.

A different approach for combating the near-far problem was analyzed by Verdú[5]. There, a receiver that is optimum with respect to probability of error in the multiuser interference environment was proposed and shown to eliminate the near-far problem and provide much improved performance. The improvement comes at the expense of high computational complexity. A class of suboptimum receivers called decorrelating detectors are based on a linear transformation of the sampled matched filters' outputs were considered in [6] and [7]. The decorrelating decision-feedback detector presented in [8] utilizes the difference in received users' energies where the decisions of the stronger users are used to eliminate interference on weaker users. These decorrelating detectors do not require the knowledge of the receiver amplitude, but need the full information on cross-correlation between all user codes. In the asynchronous case, one must know accurately the relative delays between users' sequences so that adequate matched filtering can be implemented.

Another approach for suboptimum multiuser detection with low complexity was proposed in [9] and [10]. In these works two stage multiuser separation was considered, where in order to perform detection of the desired users, tentative decisions on the information bit of all other users are made after the first stage. The estimate of the multiple access interference is then obtained and is subtracted at the second stage from the desired signal. The performance of some of the suboptimum schemes is close to the performance of the optimum detector, particularly when the power of the interference increases, the two performances become indistinguishable. However, these schemes have to perform an estimation of the received signal energies: knowledge of which is required for the detectors' proper operation.

All these detectors, including the two stage arrangement, assumed knowledge of signal parameters, such as signal energies, channel response and relative delay between user's received signals. Sensitivity of detector's performance to estimation accuracy of these parameters varies. In some cases, without perfect knowledge, the detector is rendered useless. CDMA is a strong candidate for the third-generation wireless communication systems. Larger and larger capacity is a must. Loss of performance due to inaccurate estimation should be combatted. Because of these reasons adaptive versions of the aforementioned detectors were proposed.

Many authors have suggested different methods of implementing adaptive multiuser signal separators [11-15]. An overview of a paper on adaptive schemes can also be found in [16].

This paper will concentrate on reviewing recent results of work performed at CC-SPR. Other authors work will be referenced where it is relevant. Because of limit in space, this manuscript concentrates on non-faded channels, however, extensions to frequency selective and multipath environment results are referred to in different places. Furthermore, other relevant recent results on smart antenna cancelers, space-time processing, and other multiuser detection issues are left, hopefully, for other occasions to be reviewed.

5.2 MULTIUSER CDMA SIGNAL MODEL

For general multi-user CDMA, the equivalent low-pass signal at the input of the matched filter is given by

$$r(t) = \sum_{k=1}^{K} \sum_i b_k(i) \sqrt{a_k} s_k(t + iT - \tau_k) + n(t) \qquad (5.1)$$

where K is the number of users, and a_k, b_k, s_k and τ_k are the signal amplitude, user bit, signature waveform and relative delay (with respect to a reference time τ_0) of the kth user, respectively, and $n(t)$ is a zero-mean AWGN, with a variance of σ^2.

Depending on the structure of the matched filter bank, we will distinguish between (1) the one-shot matched filter bank and (2) the multi-shot matched filter bank. With the first structure, one filter is matched to one of the users code, called the desired user, while other filters are matched to the left (0 to τ_k, $k = 2 \cdots K$) and right (τ_k to T) parts of the other users codes, respectively. With the multi-shot structure, each filter is matched to the codes of the respective users and sampled at times (τ_k to $\tau_k + T$) corresponding to bit timing of these users.

5.2.1 One-Shot Matched Filter

For the asynchronous unlink channel, the τ_k are not necessarily equal to zero, we may write (5.1) in a form showing the different user's bit overlap [17] (see Figure 5.1).

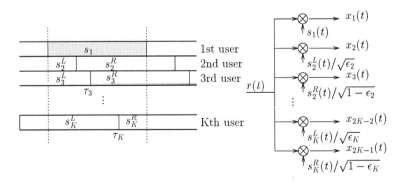

Figure 5.1 One-Shot matched filter,   (a) timing,   (b) structure.

$$r(t) = \sqrt{a_1}s_1(t)b_1(0) + \sum_{k=2}^{K}\sqrt{a_k\epsilon_k}\frac{1}{\sqrt{\epsilon_k}}s_k^L(t)b_k(-1)$$

$$+ \sum_{k=2}^{K}\sqrt{a_k(1-\epsilon_k)}\frac{1}{\sqrt{1-\epsilon_k}}s_k^R(t)b_k(0) + n(t), \quad (5.2)$$

where $0 \leq \tau_1 \leq \tau_2 \leq \cdots < \tau_K \leq T$

$$s_k^L(t) = \begin{cases} s_k(t+T-\tau_k) & \text{if } 0 \leq t \leq \tau_k \\ 0 & \text{if } \tau_k < t \leq T \end{cases}$$

$$s_k^R(t) = \begin{cases} 0 & \text{if } 0 \leq t \leq \tau_k \\ s_k(t-\tau_k) & \text{if } \tau_k < t \leq T \end{cases}$$

$$\epsilon_k = \int_0^{\tau_k} s_k^2(t+T-\tau_k)dt.$$

Note from (5.2) that $r(t)$ contains a linear combination of $b_k(i)$, $b_j(i)$ and $b_j(i-1)$ for $j = 1 \cdots K$ $j \neq k$. Therefore in matrix notation, the output of one-shot matched filter bank is given by

$$x(i) = PAb(i) + n(i) \quad (5.3)$$

where P is $(2K-1) \times (2K-1)$ matrix that will be termed "partial cross correlation" matrix (PCC), is given by

$$P = \begin{bmatrix} 1 & \rho_{12}^L & \rho_{12}^R & \cdots & \rho_{1K}^L & \rho_{1K}^R \\ \rho_{21}^L & 1 & 0 & & \rho_{2K}^{LL} & \rho_{2K}^{LR} \\ \rho_{21}^R & 0 & 1 & & \rho_{2K}^{RL} & \rho_{2K}^{RR} \\ \vdots & & & \ddots & & \vdots \\ \rho_{K1}^L & \rho_{K2}^{LL} & \rho_{K2}^{LR} & & 1 & 0 \\ \rho_{K1}^R & \rho_{K2}^{RL} & \rho_{K2}^{RR} & \cdots & 0 & 1 \end{bmatrix}, \quad (5.4)$$

$$A = \text{diag}[\alpha_1, \alpha_2, \cdots, \alpha_{K-1}]$$
$$= \text{diag}[\sqrt{a_1}, \sqrt{a_2\epsilon_2}, \sqrt{a_2(1-\epsilon_2)}, \cdots, \sqrt{a_K\epsilon_K}, \sqrt{a_K(1-\epsilon_K)}]$$

and $b(0) = [b_1(0), b_2(-1), b_2(0), \cdots, b_K(-1), b_K(0)]^T$. $n(t)$ is a zero-mean Gaussian vector with covariance $PN_0/2$, $x(0) = [x_1(0), x_2^L(0), x_2^R(0), \cdots, x_K^L(0), x_K^R(0)]^T$, which corresponds to the output of the filters matched to $s_1(t)$, $s_k^L(t)$ and $s_k^R(t)$, where $k = 2 \cdots K$ respectively. Note that the entries of P are the cross correlation between left and right parts of the code sequence $s_k(t)$, $k = 2, \cdots, K$ as well as with the code sequence $s_1(t)$ as shown in Figure 5.1.

For synchronous channels, $\tau_k = 0$ [2] for all k, the matched filters bank will contain only k filters whose output will be sampled at the same time. The output of the kth

---
[2]This might be and approximation for down-link channel where the transmitted antenna is unidirectional, single path propagation is assumed and inter-symbol interference and out of cell interference are ignored.

matched filter will contain a composite of bit b_k and all interfering bits given by the linear combination in (5.3). Here, however, P is the $(K \times K)$ cross-correlation matrix of the signature sequence $s_k(t)$, $\boldsymbol{A} = \text{diag}[\alpha_1, \cdots, \alpha_K] = \text{diag}[\sqrt{(a_1)}, \cdots, \sqrt{(a_K)}]$, $\boldsymbol{x}(i) = [x_1(i), \cdots, x_K(i)]^T$. This scenario has been intensively used by researchers.

Conventional single-user detectors make decision on the single output (without loss of generality called user 1).

$$x_1(0) = \sqrt{a_1}b_1(0) + \sum_{k=2}^{K} \sqrt{a_k\epsilon_k}\rho_{1_k}^L b_k(-1)$$
$$+ \sum_{k=2}^{K} \sqrt{a_k(1-\epsilon_k)}\rho_{1_k}^R b_k(0) + n(t),$$

Clearly, the probability of error will increase when a_k, $\rho_{1_k}^R$, and $\rho_{1_k}^L$ increase.

5.2.2 Multi-Shot Matched Filter

Particularly for asynchronous transmission τ_k's are not the same. It was noted by Ruxaudra Lupas [18], that for this channel the partial correlation matrix can be singular and hence processing the matched filter outputs with linear decorrelating detector [6,7] based on the Pseudo-inverse of the PCC matrix will not be effective. The ill-conditioned behavior of the PCC was also studied in [19].

The multi-shot matched filter structure [20] (Figure 5.2), however, does not suffer from such drawbacks. In fact, each filter of this bank of filters is matched to a different user bit time, the same as in applying conventional single-user detectors for multi-users system. Clearly with such arrangement, output samples are interfered by the current, previous and future bits of all other users. If only three bits are staked (minimum needed) to perform multi-user detection, then performance deteriorates quickly when interferences are increased. Increasing number of staked bits which are simultaneously detected, improves performance but requires higher complexity. However, with the associated signal processing technique proposed in [21] (and which would be detailed in section 5.3.3), it is possible to get a near-far resistant multiuser detector, through matrix decomposition, approximation and debiasing processing using only the minimum number of staked bits.

Consider the output of the kth matched filter; sampled at $t = \tau_k + iT$ $(i \in \mathbb{N}, k = 1, 2, \cdots, K)$ instants, we get

$$x_k(i) = \int_{iT+\tau_k}^{(i+1)T+\tau_k} s_k(t - iT - \tau_k)dt, (k = 1, \cdots K) \qquad (5.5)$$
$$= \sum_{l=1}^{K} \sqrt{a_l}b_l(i)\rho_{kl} + \sum_{l=k+1}^{K} \sqrt{a_l}b_l(i-1)\rho_{kl}^{(+)} + \sum_{l=1}^{k-1} \sqrt{a_l}b_l(i)\rho_{kl}^{(-)} + n_k(i),$$

116   CDMA TECHNIQUES FOR THIRD GENERATION MOBILE SYSTEMS

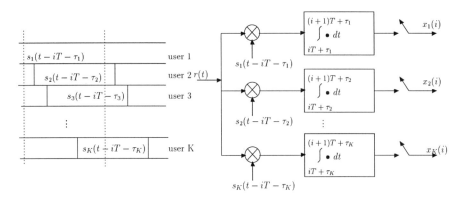

Figure 5.2   Multi-shot transformation tutorial.

where $\{n_k(i)\}_{k=1}^{K}$ is a sequence of colored Gaussian noise due to the multi-shot matched filtering; and

$$\rho_{kl} \triangleq \int_0^T s_k(t) s_l(t - \tau_l + \tau_k) dt$$

$$\rho_{lk}^{(+)} \triangleq \begin{cases} \int_0^T s_l(t + T - \tau_l + \tau_k) s_k(t) dt, & l > k \\ 0 & l \leq k \end{cases}$$

$$\rho_{lk}^{(-)} \triangleq \begin{cases} \int_0^T s_l(t) s_k(t - T - \tau_l + \tau_k) dt, & k > l \\ 0 & k \leq l \end{cases}$$

The outputs of the K multi-shot matched filters can be put into a $K \times 1$ vector

$$x(i) = [x_1(i), \cdots x_K(i)]^T.$$

Then in a matrix decomposition notation

$$x(i) = [P_U \vdots P \vdots P_L] \begin{bmatrix} z(i-1) \\ z(i) \\ \mathbf{b} z(i+1) \end{bmatrix} + n(i), \quad (5.6)$$

where $z(i) \triangleq Ab(i)$; $n(i) \sim \mathcal{N}(0, \sigma^2 P)$ is a colored Gaussian noise vector; P is a symmetric and positive definite matrix defined by $P(k,l) = \rho_{kl}$, P_U is an upper triangular matrix defined by $P_U(k,l) = \rho_{lk}^{(+)}$; P_L is a lower triangular matrix defined by $P_L(k,l) = \rho_{lk}^{(-)}$, and $P_U = P_L^T$.

Note from (5.6) that for the synchronous case, the matrices P_U and P_L vanish, and only multiuser interferences appear in $x(i)$. In the more general asynchronous case, besides the multiuser interferences, intersymbol interferences (ISI) also exist due to the additional terms in (5.6).

5.3 MULTIUSER DECORRELATING DETECTOR

5.3.1 Conventional Decorrelator

First in Lupas [18] and later in Lupas and Verdú[6] the inverse of the matrix P was used to separate the signals. That is:

$$z = P^{-1}x = Ab + P^{-1}n = Ab + \xi \quad (5.7)$$

where $z = [z_1, \cdots z_K]$, and without loss of generality we dropped the dependency on i. It follows that $E[\xi\xi^T] = \sigma^2 P$. Apart from noise, note that z contains data from only one user. Using a decision on the elements of z_i we get as output $\hat{b} = \text{sgn}(z)$. Clearly this detector is near-far resistant. The probability of error for $\hat{b}_k$ depends on the signal-to-noise ratio

$$SNR_i = \frac{a_i}{E[\xi_i^2]} = \frac{a_i |P|}{\sigma^2 |P_i|} \quad (5.8)$$

where $|P|$ is the determinant of P and $|P_i|$ is the ii-th co-factor of P.

5.3.2 The Adaptive Bootstrap Multiuser Detector

To motivate the bootstrap algorithm, we first proposed to use the linear transformation V instead of P^{-1}, that is

$$z = Vx = VPAb + Vn = VPAb + \xi, \quad (5.9)$$

and we take $V = I - W$, where

$$W^T = \begin{bmatrix} 0 & w_{12} & \cdots & w_{1K} \\ w_{21} & 0 & & w_{2K} \\ \vdots & & \ddots & \vdots \\ w_{K1} & w_{K2} & \cdots & 0 \end{bmatrix}. \quad (5.10)$$

as shown in Figure 5.3 and VP is diagonal.

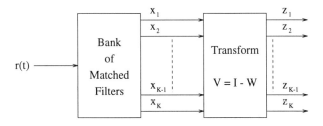

Figure 5.3 Transformation of matched filter outputs.

Note that W is not necessarily a symmetric matrix. The output of the detector is

$$z_k = x_k - w_k^T x_k \quad (5.11)$$

where $\boldsymbol{w}_k$ is the kth column of $\boldsymbol{W}$ without kkth element and $\boldsymbol{x}_k$ is the vector $\boldsymbol{x}$ with x_k taken out. To satisfy the condition for VP being diagonal, it is sufficient to have

$$\text{E}[z_k \boldsymbol{b}_k] = 0 \qquad k = 1, \cdots K \tag{5.12}$$

when $\boldsymbol{b}_k$ is the data vector $\boldsymbol{b}$ without b_k. Solving (5.12), leads to

$$\boldsymbol{w}_k = \boldsymbol{P}_k^{-1} \boldsymbol{\rho}_k \tag{5.13}$$

where ρ is the kth column of $\boldsymbol{P}$ without ρ_{kk}, $\boldsymbol{P}_k$ is the matrix $\boldsymbol{P}$ without the kth row and column, and $\boldsymbol{w}_k$ is the kth column of $\boldsymbol{W}$ without w_{kk}.

The corresponding SNR_k

$$\text{SNR}_k = \frac{a_k}{\sigma^2}(1 - \boldsymbol{\rho}_k^T \boldsymbol{P}_k \boldsymbol{\rho}_k) \tag{5.14}$$

It can be shown that

$$0 \leq (1 - \boldsymbol{\rho}_k^T \boldsymbol{P}_k \boldsymbol{\rho}_k) = \frac{|\boldsymbol{P}|}{|\boldsymbol{P}_k|} \leq 1 \tag{5.15}$$

so, as expected, this decorrelator gives the same performance as the conventional decorrelator of Lupas and Verdú, given in section 5.3.1.

The bootstrap algorithm, is based on choosing the elements of $\boldsymbol{w}_k$ to satisfy,

$$\text{E}[\boldsymbol{z}_k \text{sgn}(z_k)] = 0 \tag{5.16}$$

where $\boldsymbol{z}_k$ is the vector $\boldsymbol{z}$ without the kth element. That is the bootstrap adaptive algorithm uses the following updating formula (see Figure 5.4)

$$\boldsymbol{w}_k(i+1) = \boldsymbol{w}_k(i) + \mu \boldsymbol{z}_k \text{sgn}(z_k) \qquad k = 1, \cdots K \quad ^3 \tag{5.17}$$

Note that (in the mean) $\boldsymbol{w}_k$ will reach a steady state if (5.16) is satisfied. Also it is obvious that for high SNR_i at $z_i, i = 1, 2, \cdots, K, i \neq k$, $\text{sgn}(z_k) \simeq b_k$, (5.12) will be satisfied and $\boldsymbol{w}_k$ will be given (in the mean) by (5.13). Under this condition, which will be called "in the limit", the performance of the bootstrap will be the same as that of the conventional decorrelator.

Furthermore, except in these limiting conditions, the steady state linear transformation obtained with (5.17) will not separate the signal totally, but will have some interference residue at the output. The additive noise, however, will be smaller so that the output signal-to-noise-plus-interference ratio will be better than (or at least as good as) the signal-to-noise ratio with total separation. This gives an improved performance

---

[3] Although not stated explicitly, one may deduce from (5.17) that the cost function for this algorithm is the square of correlation between any output of the linear transformation with $\boldsymbol{V}$ and the sgn of all other outputs. Hence, in the limit, the algorithm minimizes contributions of other bits at outputs of the decorrelator.

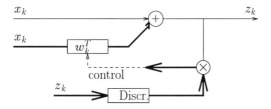

Figure 5.4 Adaptive Bootstrap Multiuser Separator.

in the low interference region as compared to that of the conventional decorrelating detector.

This fact was demonstrated with simulations as well as analytically in [22]. To show these analytical results, we use the simple case of the two user bootstrap detector shown in Figure 5.5. Let ρ be the cross-correlation between the two codes, then

$$x_1 = \sqrt{a_1}b_1 + \sqrt{a_2}\rho b_2 \qquad x_2 = \sqrt{a_2}b_2 + \sqrt{a_1}\rho b_1$$

$$\begin{aligned} z_1 &= x_1 - w_{12}x_2 \\ &= \sqrt{a_1}\left(1 - w_{12}\rho\right)b_1 + \sqrt{a_2}\left(\rho - w_{12}\right)b_2 + n_1 - w_{12}n_2, \end{aligned} \quad (5.18)$$

and

$$z_2 = \sqrt{a_2}\left(1 - w_{21}\rho\right)b_2 + \sqrt{a_1}\left(\rho - w_{21}\right)b_1 + n_2 - w_{21}n_1$$

with $E\left[n_1^2\right] = E\left[n_2^2\right] = \sigma^2$, $E\left[n_1 n_2\right] = \rho\sigma^2$, $\sigma^2_{noise\ z_1} = \sigma^2\left(1 - 2w_{12}\rho + w_{12}^2\right)$ and the signal-to-noise plus interference at this output is:

$$\text{SNIR}_{z_1} = \frac{a_1\left(1 - w_{12}\rho\right)^2}{\sigma^2\left(1 - 2w_{12}\rho + w_{12}^2\right) + a_2\left(\rho - w_{12}\right)^2} \quad (5.19)$$

Assuming $\text{SNIR}_{z_1} \gg 1 \Rightarrow \text{sgn}(z_1) \approx b_1$ z_2 can't contain $b_1 \Rightarrow w_{12} = \rho$; On the other hand, let $\text{SNIR}_{z_2} \gg 1 \Rightarrow w_{21} \ne \rho$ Let $w_{21} = \rho + \delta$. Using $E[z_1 \text{sgn}(z_2)] = 0$, we find (see [22])

$$\delta = \frac{-\rho}{1 + \sqrt{\frac{\pi}{2}\text{LSNR}}\frac{1}{1-\rho^2}(1 - 2Q(\sqrt{\text{LSNR}}))e^{\frac{\text{LSNR}}{2}}}, \quad (5.20)$$

where $\text{LSNR} = \frac{a_2(1-\rho^2)}{\sigma^2}$.

The probability of error is given by

$$P_{e_1} = \frac{1}{2}\left(Q\left(\frac{\sqrt{a_1}(1 - \rho^2 + \sqrt{a_2}\delta)}{\sigma\sqrt{1 - \rho^2 + \delta^2}}\right) + Q\left(\frac{\sqrt{a_1}(1 - \rho^2 - \sqrt{a_2}\delta)}{\sigma\sqrt{1 - \rho^2 + \delta^2}}\right)\right) \quad (5.21)$$

For small a_2, LSNR is small, $\delta \to -\rho$ and

$$P_{e_1} \to Q(\frac{\sqrt{a_1}}{\sigma}) \qquad \text{single user BPSK error}$$

For large a_2, $\delta \to 0$ and

$$P_{e_1} \to Q(\frac{\sqrt{a_1}\sqrt{1-\rho^2}}{\sigma}) \quad \text{The same as the conventional decorrelating detector}$$

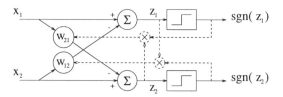

Figure 5.5 2 user synchronous bootstrap detector.

These results are depicted in Figure 5.6, which shows P_{e_1} as a function of E_{b_2}/N_0 for a fixed E_{b_1}/N_0. The limits of (5.21) are also emphasized with the asymptotes, which correspond to different ρ. It might not be as surprising that this detector should outperform the conventional decorrelator in the low SNR region. The decorrelator is completely independent of the amplitudes of the active users and the additive noise variance, while the bootstrap, by using the recursion (5.17) is indirectly implementing information about these parameters. Therefore, it seems logical that the bootstrap receiver would perform better in this region.

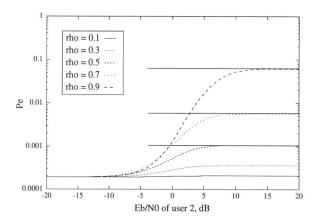

Figure 5.6 Theoretical error probability of user 1 as a function of the energy of user 2. $E_b/N_{0_1} = 8\text{dB}$. The asymptotes show Verdú's decorrelating detector performance.

It is of particular interest to compare the steady state performance of the bootstrap algorithm to that of the LMMSE of Madhow and Honig [23]. In this work, they propose to use a linear transformation, which minimizes
$E\{b - Wx\}^2$, the mean square error between the actual data b and a mapping W of

the matched filter outputs. Such W is given by:

$$W_{\text{MMSE}} = [P + \sigma^2 A^{-2}]^{-1} \quad (5.22)$$

In [24], a new approach for analyzing the performance of the LMMSE from statistical signal processing and parameter estimation point of view is shown. Furthermore, it was shown in [25], by theoretical analysis and computer simulation, that LMMSE and the bootstrap algorithm (steady state) perform quite similarly. To summarize these results from [24] and [25], we first rewrite the matched filter output

$$x = PAb + n = P\theta + n \quad (5.23)$$

The random variables θ and n are statistically independent. Defining a new random variable vector

$$y \triangleq \begin{bmatrix} \theta \\ x \end{bmatrix} = \begin{bmatrix} Ab \\ x \end{bmatrix} = \underbrace{\begin{bmatrix} A & 0 \\ PA & I \end{bmatrix}}_{H} \cdot \begin{bmatrix} b \\ n \end{bmatrix} \quad (5.24)$$

where $E[y] = 0$ and

$$\text{cov}(y) = H \text{cov}\left(\begin{bmatrix} b \\ n \end{bmatrix}\right) H^T = \begin{bmatrix} A^2 & A^2 P \\ PA^2 & PA^2 P + \sigma^2 P \end{bmatrix} \triangleq \begin{bmatrix} \Sigma_{\theta\theta} & \Sigma_{\theta x} \\ \Sigma_{x\theta} & \Sigma_{xx} \end{bmatrix}$$

The LMMSE estimate of θ is given by;

$$\begin{aligned}\hat{\theta}_{\text{LMMSE}} &= E[\theta] + \Sigma_{\theta x} \Sigma_{xx}^{-1} (x - E[x]) = \Sigma_{\theta x} \Sigma_{xx}^{-1} \cdot x \\ &= \underbrace{(P + \sigma^2 A^{-2})^{-1}}_{W} \cdot x \end{aligned} \quad (5.25)$$

and the decision is made on $\hat{\theta}_{\text{LMMSE}}$

$$\hat{b} = \text{sign}(\hat{\theta}_{\text{LMMSE}}) \quad (5.26)$$

One can easily recognize that $\sigma^2 A^{-2}$ is a diagonal matrix whose elements are the inverse of the SNR. Therefore, if the interference of other users is small compared to the noise level then $\sigma^2 A^{-2}$ is large.

$$W = (P + \sigma^2 A^{-2})^{-1} \simeq (I + \sigma^2 A^{-2})^{-1} \quad \text{as in single user detector (BPSK)}$$

If the interference level is large then $\sigma^2 A^{-2}$ is small

$$W = (P + \sigma^2 A^{-2})^{-1} \simeq P^{-1} \qquad \text{as in conventional decorrelating detector}$$

Using (5.25) we have

$$\begin{aligned}\hat{\theta}_{\text{LMMSE}} &= Wx = (P + \sigma A^{-2})^{-1} x \\ &= (P + \sigma A^{-2})^{-1} \cdot (PAb + n)\end{aligned}$$

and applying the inversion lemma, we get

$$\hat{\boldsymbol{\theta}}_{\text{LMMSE}} = \boldsymbol{\theta} - \sigma^2 \boldsymbol{L}\boldsymbol{A}^{-1}\boldsymbol{b} + \boldsymbol{L}\boldsymbol{n} = \boldsymbol{\theta} + \boldsymbol{e}$$

The error term contains both bias and noise. The improved performance of the LMMSE detector is achieved as by trading off a little bias for less noise variance. Finally, the probability of error of the kth user is given by

$$\begin{aligned}\text{P}_{\text{e}}(k) &= \text{P}_{\text{e}}(k|b_k, \boldsymbol{b}_k)\text{P}(b_k)\text{P}(\boldsymbol{b}_k)\\ &= 2^{-K}\sum_{\boldsymbol{b}_k}\left[Q\left(\frac{\sqrt{a_k} - R_k(b_k = -1, \boldsymbol{b}_k)}{\sigma_k}\right) + \right.\\ &\qquad\left. Q\left(\frac{\sqrt{a_k} + R_k(b_k = +1, \boldsymbol{b}_k)}{\sigma_k}\right)\right]\end{aligned} \quad (5.27)$$

where

$$R_x(\cdot) = -\sigma^2 \boldsymbol{w}_k^T \boldsymbol{A}^{-1} \boldsymbol{b} \quad \sigma_k^2 = \sigma^2 \boldsymbol{w}_k^T \boldsymbol{P} \boldsymbol{w}_k$$

and $\boldsymbol{w}_k^T$ is the kth row of $\boldsymbol{W}$

$$\boldsymbol{w}_k^T = \frac{\left[1 - \boldsymbol{\rho}_k^T\left(\boldsymbol{P}_k + \sigma^2 \boldsymbol{A}_k^{-2}\right)^{-1}\right]}{1 + \sigma^2/a_k - \boldsymbol{\rho}_k\left(\boldsymbol{P}_k + \sigma^2 \boldsymbol{A}_k^{-2}\right)^{-1}\boldsymbol{\rho}_k}$$

It can be verified that in the limit the above probability of error expression will reduce either into that of decorrelating detector or the single user BPSK

$$\text{P}_{\text{e}}(k) = \begin{cases} Q\left(\frac{\sqrt{a_k}}{\sigma}\right) & \text{small interferences} \\ Q\left(\frac{\sqrt{a_k(1 - \boldsymbol{\rho}_k^T \boldsymbol{P}_k^{-1} \boldsymbol{\rho}_k)}}{\sigma}\right) & \text{strong interferences} \end{cases}$$

similar to the bootstrap separator. Figure 5.7 shows a sample of analytical and simulation results which clearly compare the performance of LMMSE and the bootstrap.

Although the "Bootstrap" algorithm is easily implemented and experimentally shown to provide satisfactory performance, its convergence property is not easy to predict. The difficulty arises from the usage of hard limiters that makes the algorithm control nonlinear. Nevertheless convergence and stability condition of the bootstrap algorithm as it is applied to CDMA multiuser detector, has recently been analyzed [26].

First we rewrite (5.17) as

$$\boldsymbol{w}_k(i+1) = (\boldsymbol{I} - \mu \boldsymbol{T}_k)\boldsymbol{w}_k(i) + \mu \boldsymbol{t}_k \quad (5.28)$$

where

$$\boldsymbol{T}_k = \text{E}[\hat{\boldsymbol{b}}_k \boldsymbol{x}_k^T], \boldsymbol{t}_k = \text{E}[\hat{\boldsymbol{b}}_k x_k] \quad (5.29)$$

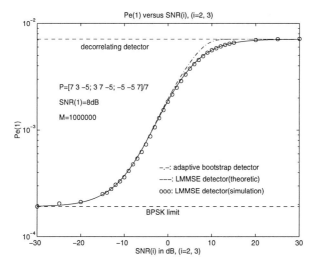

Figure 5.7 Simulation results which compare performance of LMMSE and bootstrap algorithm: $K = 3$, $\text{SNR}_1 = 8\text{dB}$.

and the steady state weight vector is $\mathbf{w}_k^o = \boldsymbol{T}_k^{-1} \mathbf{t}_k$. First, it is easy to conclude from (5.28) that the sufficient condition for $\boldsymbol{w}_k(i)$ to converge to $\mathbf{w}_k^o$ is

$$|1 - \mu\lambda\{\boldsymbol{T}_k\}| < 1,$$

where $\lambda\{\boldsymbol{T}_k\}$ denotes the eigenvalues of $\boldsymbol{T}_k$. Next, if we take μ positive then by using the Gershgorin's theorem one can show (see [26]) that a sufficient condition for convergence:

1)

$$[\boldsymbol{T}_k]_{jj} > \sum_{\substack{i=1 \\ i\neq j}}^{K-1} |[\boldsymbol{T}_k]_{ji}| \quad \forall j = 1, 2, \cdots, K-1 \qquad (5.30)$$

That is, $\boldsymbol{T}_k$ is diagonally dominant with diagonal entries greater than zero.

2)

$$\mu < 2/(\max_j\{[\boldsymbol{T}_k]_{jj} + \sum_{\substack{i=1 \\ i\neq j}}^{K-1} |[\boldsymbol{T}_k]_{ji}|\}) \qquad (5.31)$$

or with further simplification, it is sufficient to take

$$\mu < 1/(\max_j\{[\boldsymbol{T}_k]_{jj}\}) \qquad (5.32)$$

Therefore, the sufficient conditions for the convergence (in the mean) of the "Bootstrap" algorithm in the mean are (5.30) and (5.32). From (5.29), however, the matrix T_k is not easy to explore due to the hard limiter used to determine $\hat{b}_k$. Nevertheless for special cases applicable to the CDMA systems, condition could be found which related to the cross-correlation matrix entries and the amplitude of the incoming signal.

Case 1. Consider $K = 1$ and $\hat{b}_1 = b_1$ when $b_1 = [b_2, \cdots, b_K]^T$, i.e. the bits of all other outputs are perfectly estimated, the sufficient conditions are (see[26])

$$\sum_{\substack{i=2\\i\neq j}}^{K} |\rho_{ji}| < 1, \quad \forall j = 2, \ldots, K, \tag{5.33}$$

$$\mu < 1/(\max_{j\neq 1} \sqrt{a_j}) \tag{5.34}$$

That is P_1 (P without the first column and row) should be diagonally dominant and the step size μ should be smaller than the inverse of the maximal signal amplitude.

Case 2. Assume that $\mathbf{w}_1(i)$ has converged to $\mathbf{w}_1^o$ ($= P_1^{-1}\rho_1$) and the estimated bit of the first output is given by

$$\hat{b}_1(i) = \text{sgn}(z_1(i)) = \text{sgn}(x_1(i) - (\mathbf{w}_1^o)^T \mathbf{x}_1(i)). \tag{5.35}$$

However, due to noise in the output $z_1(i)$, $\hat{b}_1(i)$ is not a perfect estimate of $b_1(i)$. Therefore, we next consider the convergence of "Bootstrap" algorithm for the kth weight vector

$$\mathbf{w}_k(i+1) = \mathbf{w}_k(i) + \mu z_k(i)\hat{\mathbf{b}}_k(i), \tag{5.36}$$

where $\hat{\mathbf{b}}_k(i) = [\hat{b}_1(i), b_2(i), \cdots, b_{k-1}(i), b_{k+1}(i), \cdots, b_K(i)]^T$, and $\hat{b}_1(i)$ is given by (5.35). It was shown in [26] that sufficient conditions for (5.30) are P_K is dominant and

$$\sqrt{a_1}\gamma_1 (1 - \sum_{\substack{j=2\\j\neq K}}^{K} |\rho_{1j}|) > \sum_{\substack{j=1\\j\neq K}}^{K} |c_{1j}| \tag{5.37}$$

where c_{1j} denotes the jth element of c_1, $c_1 = E\{\hat{b}_1(i)n_K(i)\}$ and $\gamma_1 = E\{\hat{b}_1 b_1\}$. By using (5.37), it can be shown that the second sufficient condition of (5.32) is reduced to

$$\mu < \frac{1}{\max_{j\neq k} \sqrt{a_j}} \tag{5.38}$$

Although these two cases do not show the totality of possible cases, they demonstrate together with the many simulations performed, that the bootstrap is stable and converge to its steady state if one uses step size adequately defined in relation to input signal amplitudes. Worth noting that for the 2 users case the algorithm is unconditionally stable provided that the step size is taken as $\mu < 2/\sqrt{a_j}(j \neq k, j = 1, 2, \cdots)$. Besides its ability to perform better to or equal than the conventional decorrelating detector and almost similar to MMSE detector, it is suitable for cases where codes have been altered by the channel and hence the exact cross-correlation matrix is not available [27-28].

Furthermore it performs adequately in non-frequency selective fading [29] and multipath fading channel [30]. The concept of bootstrap algorithm was also extended for handling complex data as in QAM [31].

5.3.3 Multiuser Separator for the Multi-Shot Matched Filter [21]

In order to obtain a near-far resistant receiver free from interference, and have a minimum ISI, it customary to combine N-bit data sequences into a $NK \times 1$ vector (see [21])[4].

$$\underbrace{\begin{bmatrix} x(i) \\ x(i+1) \\ x(i+2) \\ \cdots \\ x(i+N-1) \end{bmatrix}}_{X_{(t,i)}} = \tag{5.39}$$

$$\underbrace{\begin{bmatrix} P & P_L & 0 & \cdots & 0 \\ P_U & P & P_L & \cdots & 0 \\ 0 & P_U & P & \ddots & \cdots \\ \cdots & \cdots & \ddots & \ddots & P_L \\ 0 & \cdots & 0 & P_U & P_L \end{bmatrix}}_{\mathcal{P}} \underbrace{\begin{bmatrix} z(i) \\ z(i+1) \\ z(i+2) \\ \cdots \\ x(i+N-1) \end{bmatrix}}_{Z_{(t,i)}}$$

$$+ \underbrace{\begin{bmatrix} P_U z(i-1) \\ 0 \\ \cdots \\ 0 \\ P_L x(i+N) \end{bmatrix}}_{\text{bias}} + \underbrace{\begin{bmatrix} n(i) \\ n(i+1) \\ n(i+2) \\ \cdots \\ n(i+N-1) \end{bmatrix}}_{\text{noise} N_{(t,i)}}$$

$\mathcal{P}$ is a $NK \times NK$ matrix which is symmetric, positive definite, block tridiagonal and full rank. The noise vector is colored Gaussian with probability density function

---

[4]The multi-shot decorrelator also mentioned in the paper by S. Verdú "Multiuser Detection". In H. V. Poor and J. B. Thomas, editors, Advances in Statistical Signal Processing, Vol. 2: Signal Detection, pp. 139-199, JAI Press, Greenwich, CT, 1993.

(PDF) of $\mathcal{N}(0, \mathcal{P}\sigma^2)$. We can use $\mathcal{P}^{-1}$ on the data vector $\boldsymbol{X}(\cdot, i)$ (5.39) to obtain the multiuser information. One can see from (5.40) that ISI due to asynchronous transmission come from the neighboring two bits. Hence in order to detect and separate multiuser information bits, we need jointly process at least 3 bits. We also notice that by arranging estimates of the vectors $\{\hat{\boldsymbol{Z}}(:, i)\}^N$ in $KN \times N$ matrix; $\hat{\boldsymbol{Z}} = [\hat{\boldsymbol{Z}}(:, 1), \cdots, \hat{\boldsymbol{Z}}(:, N)]$, then $\hat{\boldsymbol{Z}}$ can be used to improve the estimate of the user bit by anti-diagonal processing. Increasing N further improves this estimate, but not without added complexity.

However, we can also notice that $\boldsymbol{P}_L$ is lower triangle matrix, $\boldsymbol{P}_U$ is an upper triangle matrix, and $\boldsymbol{P}_U = \boldsymbol{P}_L^T$, hence, $\mathcal{P}$ is a diagonal dominant matrix and so its inverse. Therefore by using $\mathcal{P}^{-1}$ only the first $\hat{\boldsymbol{Z}}(i)$ and the last $\hat{\boldsymbol{Z}}(i + N - 1)$ bit multiuser information will be effected by the truncation bias term. Thus to minimize the effect of bias (debiasing), we only use the middle sub-block of $\hat{\boldsymbol{Z}}(:, i)$. That is we operate on $\boldsymbol{x}(:, i)$ by the central $K \times NK$ row block matrix $(\boldsymbol{W})$ of $\mathcal{P}^{-1}$. For $N = 3$, $\boldsymbol{W}$ is given by

$$\boldsymbol{W} = -[\boldsymbol{P} - \boldsymbol{P}_L \boldsymbol{P}^{-1} \boldsymbol{P}_U - \boldsymbol{P}_U \boldsymbol{P}^{-1} \boldsymbol{P}_L]^{-1} \cdot$$
$$[\boldsymbol{P}_U \boldsymbol{P}^{-1} \vdots \boldsymbol{I} \vdots \boldsymbol{P}_L \boldsymbol{P}^{-1}] \qquad (5.40)$$

and

$$\hat{\boldsymbol{z}}(i+1) = \boldsymbol{W} \begin{bmatrix} \boldsymbol{x}(i) \\ \boldsymbol{x}(i+1) \\ \boldsymbol{x}(i+2) \end{bmatrix} = \boldsymbol{W} \begin{bmatrix} A & & \\ & A & \\ & & A \end{bmatrix} \begin{bmatrix} \boldsymbol{b}(i) \\ \boldsymbol{b}(i+1) \\ \boldsymbol{b}(i+2) \end{bmatrix} \qquad (5.41)$$

It can be seen that when the transmission delay changes, instead of re-inverting the whole $(NK \times NK)$ matrix $\mathcal{P}$, we only need to update the sub-blocks matrices $\boldsymbol{P}$, $\boldsymbol{P}_U$ and $\boldsymbol{P}_L$ which are of less dimension and thus requires less computation than a direct inversion of $\mathcal{P}$. Simulation results in Figure 5.8 show the performance achieved with anti-diagonal averaging and by the proposed computationally simple multiuser detector with debiasing technique of (5.41).

To capitalize on the simplicity and robustness of the bootstrap algorithm, we recently [32] modified this algorithm such that it is suitable for the multi-shot case and can easily handle multipath fading channels as well.

5.3.4 One-Shot Detector with Singular PCC Matrix

As mentioned earlier when a one-shot matched filter bank is used, the partial correlation matrix might become singular depending on the relative delay τ_k, $k = 2, \cdots K$. In such cases the conventional decorrelator, using a generalized inverse, fails to separate the multiuser signal, and the bootstrap algorithm becomes non-convergent. Two methods were used to remedy this problem. In the first [33], we eliminated the rows of the PCC matrix which caused reduced rank and decorrelated only the independent outputs. To determine these rows, we used an adaptive algorithm that minimized the decorrelator output powers, and show that in fact, these rows correspond to zero output power. In

TOPICS IN CDMA MULTIUSER SIGNAL SEPARATION   127

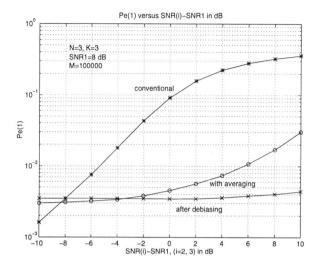

Figure 5.8   Performance comparison of various detectors in asynchronous multi-user CDMA communication systems. The performance improvement obtained by debiasing processing is shown.

the second scheme, for controlling the bootstrap weights, we proposed a soft limiter (see Figure 5.9) instead of a hard limiter [34]. We show (see Figure 5.10) that with a wide range of the threshold, this modified bootstrap decorrelating detector's weights converge regardless of the PCC matrix singularity. Some sketch of convergence is given in [34], but further work is needed.

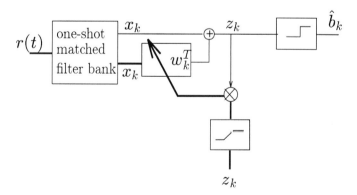

Figure 5.9   Bootstrap algorithm with soft limiter control.

128 CDMA TECHNIQUES FOR THIRD GENERATION MOBILE SYSTEMS

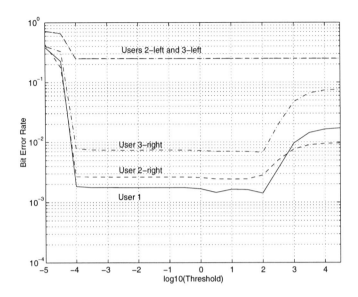

Figure 5.10 Adaptive weights convergence for various threshold levels: $\text{SNR}_1 = \text{SNR}_i = 8$ dB, $i = 2, 3$.

5.4 ADAPTIVE MULTISTAGE RECEIVER FOR MULTIUSER CDMA

The multiuser decorrelating detector discussed in the previous section (except the bootstrap algorithm) results in a total separation of the active user's signal and hence in near-far resistant performance. However, even the adaptive bootstrap detector suffers from relatively high probability of error (at high INR) as a result of the zero forcing nature of the transformation used. To further improve the probability of error (POE), multistage receivers were proposed. Sometimes, these are also called "subtractive interference canceler". They can be grouped into successive interference cancelers (SIC) [35, 8], and parallel interference canceler (PIC) [9, 36-38]. In comparing the performance of these two groups, it was concluded in [39] that although the performance of the parallel IC scheme was found to be superior under equal received powers, the successive IC scheme outperformed it under realistic fading environment. We will concentrate on the PIC, in which the decorrelating detectors of the previous section are used as a first stage. Functionally, such a receiver is shown in Figure 5.11. The matched filter bank uses the user's code (synchronous), partial codes (asynchronous, one-shot), or delayed full codes (asynchronous, multi-shot).

The linear separator could be any of the decorrelator detectors mentioned earlier. A tentative decision is made on the data vector z and used as an estimate for interference fed to the canceler stage.

A synchronous multistage multi-user CDMA receiver that employs a combination of a decorrelator and an interference canceler, and whose weights are adaptively

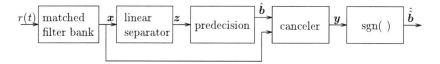

Figure 5.11   Multistage receiver for multiuser CDMA.

controlled to minimize output energies, was presented in [36]. This scheme which did not require the knowledge of the received signal energies, was shown both analytically and through simulation to perform as well as the fixed weight counterpart, described in [10]. The convergence of the adaptive scheme was studied in [40]. Both schemes with adaptive and nonadaptive canceler used a hard decision on the decorrelator output to obtain a tentative data bit estimate. As it is expected, with most multi-stage receivers which employ some kind of interference cancellation and use some form of tentative decision, the performance of this two-stage decorrelator-canceler, performs very well when the interference's signal-to-noise ratios are high in comparison to that of the desired signal. The performance degrades when the former becomes of the same level or lower than the latter (see Figure 5.12, curve with hard limiter).

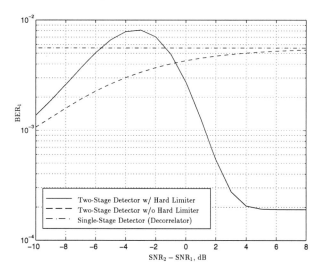

Figure 5.12   Performance comparison of multistage detector with hard limiter and without hard limiter.

5.4.1   Predecision

For the canceler to operate well, the (auxiliary) inputs to the controlled weights must represent a high quality estimate of the interference. A hard limiter improves estimator's quality if interference-to-desired signal plus noise ratio (ISNR) at these inputs are high, resulting in better interference cancellation. If the ISNR are not sufficiently high, then it is preferred not to add a limiter. This motivates the use of a soft limiter in which tentative decision is made only at high ISNR (see performance curves with and without limiter in Figure 5.12). Such a decorrelator-canceler with a soft limiter was proposed in [41] for the synchronous case. In [42] the idea was extended to the asynchronous case, wherein the one-shot decorrelator was used as a first stage. Zhang and Brady [43] also used soft decision tentative statistics in their multiuser asynchronous detector.

The question that remained is where the threshold point of the soft limiter should be located. In all the aforementioned references the threshold value was determined heuristically from the observed values of the decorrelator outputs. For example we define

$$f(z) = \begin{cases} z/t & if \ |z| < t \\ \text{sgn}(z) & \text{otherwise} \end{cases} \quad (5.42)$$

where t is the threshold point of the soft tentative decision. In [41], t was chosen such that

$$t_{lk} = \rho_{lk} \left\{ \frac{(E[|z_k|])^2}{E[|z_l|]} \right\} \quad (5.43)$$

where z_k is the output of the decorrelator corresponding to the desired user and z_l corresponds to the interfering lth user. ρ_{kl} is the cross-correlation between the code sequences of user l and k. Note, from (5.43), that if the interference power is larger comparable to the desired signal power ($E|z_1| \gg E|z_k|$), then the corresponding threshold at the kth output is small as required. A similar choice was used in [42], for the asynchronous case. Zhang and Brady [43], chose the soft decision nonlinearity for the kth interferer of user 1, for example so as to maximize the asymptotic multiuser efficiency. Notice that for setting the threshold a continual measure of energy at the output of the decorrelator is needed.

In [44], adaptive setting of the threshold, which minimizes the detector output energy, was proposed. It obviously requires no measurements. Performance comparison of multistage receiver with hard limiter, soft limiter with heuristic threshold and optimal threshold setting is depicted in Figure 5.13.

5.4.2 Adaptive Canceler

In contrast to the scheme proposed in [10], which uses a constant weight canceler, we proposed and analyzed in [36] an adaptive canceler stage. It uses minimization of output energies as its criterion. It was shown (analytically as well as by simulation) that after it converges, it performs as well as the fixed weight canceler that assumes knowledge of signals amplitude. The structure of the multi-stage receiver with adaptive canceler that used in [36] is depicted in Figure 5.14. It considered the synchronous case, used the conventional decorrelator of section 5.3.1 as its first stage, followed by hard limiter, and finally the adaptive canceler. With the notation of Figure 5.14,

$$\hat{b} = \text{sgn}(z) \quad (5.44)$$

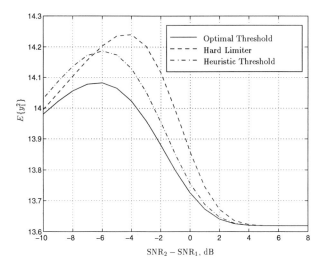

Figure 5.13 Performance comparison of multistage detector with hard limiter, soft limiter with heuristic threshold and optimal threshold setting.

$$y = x - U\hat{b} \qquad (5.45)$$

and U has a structure similar to that of W in (5.10). We use the updating algorithm

$$u_k \leftarrow u_k - \mu \frac{\partial}{\partial u_k} \mathrm{E}(y^2) \qquad (5.46)$$

where u_k is the kth row of U without the kkth element. This, in fact the steepest decent algorithm which minimizes the energy at the canceler output y_k with respect to the elements of u_k. In practice, a stochastic algorithm is used, wherein an estimate y_k^2 of $\mathrm{E}(y_k^2)$ is implemented. This leads to steady state in the mean of the canceler weights

$$u_k = \mathrm{E}\{\hat{b}_k\hat{b}_k^T\}^{-1} A_k \mathrm{E}\{b_k\hat{b}_k\}\rho_k \qquad (5.47)$$

where $\hat{b}_k$ is $\hat{b}$ without the kth element, A_k is the diagonal matrix A without the kth element and ρ_k is the kth column of P without ρ_{kk}, all the same as in definitions made in previous sections. The probability of error at the kth output of such a canceler is calculated in [36].

$$\mathrm{P}_{ek} = \frac{1}{2^{K-1}} \sum_{\hat{b}_k, b_k} Q[\chi] \cdot \mathrm{Pr}\{\hat{b}_k | b_k\} \qquad (5.48)$$

with

$$\chi = \frac{\sqrt{a_k} - [b_k^T - \hat{b}_k^T (\mathrm{E}\{\hat{b}_k\hat{b}_k^T\})^{-1} (\mathrm{E}\{b_k\hat{b}_k^T\}) A_k \rho_k}{\sqrt{\frac{N_0}{2}}}$$

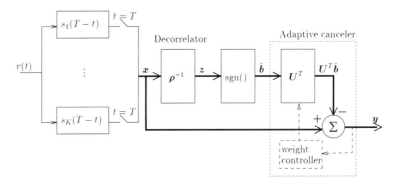

Figure 5.14 Multi-stage receiver with adaptive canceler.

where $P_r\{\hat{\boldsymbol{b}}_k|\boldsymbol{b}_k\}$ is defined by $(K-1)$ variant Gaussian density function.

The probability of error for five users system was evaluated, using gold sequences of length seven, whose cross-correlation matrix $\boldsymbol{P}$ is given by

$$\boldsymbol{P} = \frac{1}{7}\begin{bmatrix} 1 & -1 & 3 & 3 & 3 \\ -1 & 1 & -1 & 3 & -1 \\ 3 & -1 & 1 & -1 & -1 \\ 3 & 3 & -1 & 1 & -1 \\ 3 & -1 & -1 & -1 & 1 \end{bmatrix}, \tag{5.49}$$

Figure 5.15 shows the steady state error probability of user 1, whose interferences are determined by the first column of $\boldsymbol{P}$ in (5.49) as a function of $\text{SNR}_i - \text{SNR}_1$, $i = 2, 3, 4, 5$ (near-far resistance curve and $\text{SNR}_1 = 8\text{dB}$).

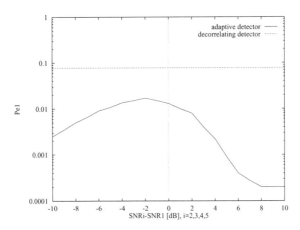

Figure 5.15 Error probability of user 1: $K = 5$ and $\text{SNR}_1 = 8\text{dB}$.

Figure 5.16 depicts the probability of error of user 1 as a function of other users' SNR (taken to be of equal powers) when total number of these active users are 2, 3, 4 and 5. For comparison, the corresponding error probability curves with conventional decorrelators are added to these figures.

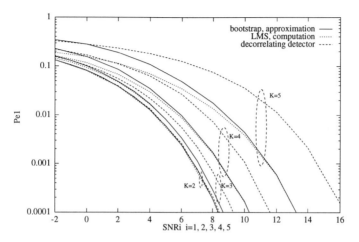

Figure 5.16 Error Probability of User 1, for $K = 2$ to $K = 5$.

It is important to note that for high level of interference, the probability of error reached that obtained with a single user (no CCI). When the number of users increases, the performance degrades, but always stays better than a conventional decorrelator.

As mentioned earlier, the adaptive multistage receiver presented in this section, was for the synchronous case and with the conventional decorrelator of section 5.3.1. Extension to one-shot asynchronous case is done by using a partial cross-correlation matrix, with the bank of filters matched to the one-shot user and to the right and left partial codes of all other users.

With this:

→ P becomes $(2K − 1) \times (2K − 1)$ matrix

→ The user amplitudes bA are modified as in (5.4), $A = \text{diag}[\alpha_1, \alpha_2, \cdots, \alpha_{K-1}]$ with α_i depending on the relative delays $\tau_i (i = 2, 3, \cdots, K − 1)$.

When using the bootstrap algorithm the decorrelator transform matrix V, will become, instead of P^{-1}, $V = I − W$ with $w_k = f(P_k, \text{snr}_k)$, where P_k is the matrix P with the kth column and row deleted, and snr_k is the vector of snr except for the kth user.

In the limit:

$$w_k = 0 \quad \text{if } \text{snr}_k = 0$$
$$w_k = P_k^{-1} \rho_k \quad \text{if } \text{snr}_k \to \infty$$

When using LMMSE for decorrelating detector, then the vector w_k is obtained from the minimum mean square estimator, derived earlier.

With a multi-shot approach, the transformation W was calculated directly from the correlation matrix $\mathcal{P}$.

Regardless of the kind of decorrelator used, the output of the decorrelator is given by

$$z = \Lambda Ab + n_D \tag{5.50}$$

where $\Lambda = \text{diag}(\lambda_1, \lambda_2, \cdots, \lambda_{2K-1})$ and $\lambda_i = 1 - w_i^T P_i^{-1} \rho_i, i = 1, 2, \cdots, 2K$.

Easy to show, that in the limit when snr_i is large

$$\text{SNR}_i = \frac{\alpha_i^2(1 - \rho_i^T P_i^{-1} \rho_i)}{N_o/2} \quad i = 1, 2, \cdots, 2K - 1 \tag{5.51}$$

Note that α_i except for $i = 1$ (the first user) is less than $\sqrt{a_i}$. Therefore the outputs corresponding to these users might have low SNR_i, unless the one shot matched filter is repeated for users other than the first. This motivate introducing, in the next section, the "decorrelator, combiner, canceler, combiner (DC^3)", as new multistage receiver structure.

5.4.3 DC^3 (decorrelator, combiner, canceler, combiner) Multistage Receiver Structure [45]

Instead of repeating the one-shot multistage receiver, once for every active user, we propose to add data combiners after the decorrelator and the canceler. Such structure is depicted in Figure 5.17. We will consider the case in which the conventional decorrelator (equivalently the adaptive bootstrap or the MMSE with high interference) is used, for which the SNR_i at the output is given by (5.51) with α_i defined in (5.4).

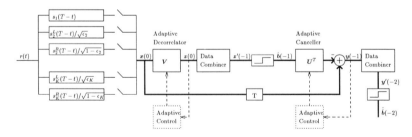

Figure 5.17 One-shot asynchronous multistage detector with DC^3 structure.

The Decorrelator Output Data Combiner. Following the first stage of the decorrelator, we use a data combiner shown in Figure 5.18, from which the relation between

$z'(i-1)$ and the data vector $b(i)$ is given by

$$z'(-1) = \begin{bmatrix} D & 0 & 0 & & 0 & 0 \\ 0 & D & \gamma_2 D^2 & . & 0 & 0 \\ 0 & 1 & \gamma_2 D & & 0 & 0 \\ & & & . & & \vdots \\ 0 & 0 & 0 & & D & \gamma_K D^2 \\ 0 & 0 & 0 & . & 1 & \gamma_K D \end{bmatrix} (\mathbf{\Lambda A b}(0) + \mathbf{n}_D(0)), \quad (5.52)$$

where A and $b(i)$ are defined in (5.4), $\Lambda = \text{diag}(\lambda_1, \lambda_2, \cdots, \lambda_{2K-1})$, $\lambda_i = (1-\rho_i^T P_i^{-1} \rho_i)$, $i = 1, 2, \cdots, 2K-1$, D stands for one bit delay, γ_k, $k = 2, \cdots, K$ is the combining gain, and the noise vector $\mathbf{n}_D(0) = [n_{D,1}(0), n_{D,2}(-1), n_{D,3}(0), \cdots, n_{D,2K-2}(-1), n_{D,2K-1}(0)]$, having independent elements of $n_{D,2i-2}(0)$ and $n_{D,2i-1}(-1)]$, $i = 2, 3, \cdots, K$.

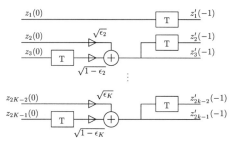

Figure 5.18 Decorrelator output data combiner.

The combined output SNR_i

$$\text{SNR}_{z'_{2k-2}} = \text{SNR}_{z'_{2k-1}} = \frac{a_k \epsilon_k \lambda_{2k-2}[1 + \gamma_k \sqrt{\frac{1-\epsilon_k}{\epsilon_k}} \frac{\lambda_{2k-1}}{\lambda_{2k-2}}]^2}{[1 + \gamma_k^2 \frac{\lambda_{2k-1}}{\lambda_{2k-2}}] \frac{N_0}{2}} \quad (5.53)$$

One can show that $\gamma_k = \sqrt{\frac{1-\epsilon_k}{\epsilon_k}}$ will result in maximum $\text{SNR}_{z'_{2k-2}} = \text{SNR}_{z'_{2k-1}}$,

$$\text{SNR}_{z'_{2k-2}} = \text{SNR}_{z'_{2k-1}} = \left[1 + \frac{1-\epsilon_k}{\epsilon_k} \frac{\lambda_{2k-1}}{\lambda_{2k-2}}\right] \text{SNR}_{z_{2k-2}}$$

$$= \left[1 + \frac{\epsilon_k}{1-\epsilon_k} \frac{\lambda_{2k-1}}{\lambda_{2k-2}}\right] \text{SNR}_{z_{2k-1}} \quad (5.54)$$

It can be shown that the combined $\text{SNR}_{z_{2k-2}}$ and $\text{SNR}_{z_{2k-1}}$ are greater than both uncombined SNR for any ϵ_k and any P. Therefore, the tentative decision will give better estimates of $b_k(-2)$ and $b_k(-1)$ needed for cancelling interference.

The output of the combiner in matrix notation

$$z'(-1) = \Lambda' A b(-1) + \xi(-1), \quad (5.55)$$

where

$$\begin{aligned}\mathbf{\Lambda}' &= \mathrm{diag}[\lambda'_1,\cdots,\lambda'_{2K-1}] \\ &= \mathrm{diag}[\lambda_1,\lambda_2+\gamma_2\lambda_3,\lambda_2+\gamma_2\lambda_3,\cdots,\lambda_{2K-2}+\gamma_K\lambda_{2K-1}]\end{aligned}$$

and

$$\begin{aligned}\boldsymbol{\xi}(-1) &= [\xi_1(-1),\xi_2(-1),\xi_3(-1),\cdots,\xi_{2K-1}(-1)]^T \\ &= [n_{D1}(-1),n_{D2}(-1)+\gamma_2 n_{D3}(-2),n_{D2}(0)+\gamma_2 n_{D3}(-1),\\ &\quad \cdots,n_{D,2K-2}(-1)+\gamma_K n_{D,2K-1}(-2),n_{D,2K-2}(0)+\\ &\quad \gamma_K n_{D,2K-1}(-1)]^T\end{aligned}$$

Adaptive Canceler Stage. Referring to Figure 5.17, we have

$$\boldsymbol{y}(-1) = \boldsymbol{x}(-1) - \boldsymbol{U}^T\hat{\boldsymbol{b}}(-1), \tag{5.56}$$

where

$$\hat{\boldsymbol{b}} = \mathrm{sgn}(\boldsymbol{z}')$$

and $\boldsymbol{U}$ is $(2K-1)\times(2K-1)$ matrix given by

$$\boldsymbol{U} = \begin{bmatrix} 0 & u_{12} & \cdots & u_{1,2K-1} \\ u_{21} & 0 & \cdots & u_{2,2K-1} \\ \vdots & \vdots & \ddots & \vdots \\ u_{2K-1,1} & u_{2K-1,2} & \cdots & 0 \end{bmatrix}. \tag{5.57}$$

Note, even though we combine data at the output of the decorrelator, we still need to use both versions of data for cancellation. This is the reason for regenerating all the $2K-1$ outputs in Figure 5.18.

The ith output of (5.56) is given

$$y_i = x_i - \boldsymbol{u}_i^T\hat{\boldsymbol{b}}_i, \quad i=1,\cdots,2K-1 \tag{5.58}$$

where $\boldsymbol{u}_i$ is the ith column of $\boldsymbol{U}$ without the element u_{ii} and $\hat{\boldsymbol{b}}_i$ is $\hat{\boldsymbol{b}}$ without $\hat{b}_i$.

Except for the fact that we are dealing with $(2K-1)$ dimensional system, we use the steepest descent the same as (5.46) of section 5.4.2

$$\boldsymbol{u}_i \leftarrow \boldsymbol{u}_i - \frac{\mu}{2}\frac{\partial}{\partial \boldsymbol{u}_i}\mathrm{E}\{y_i^2\}$$

which leads to

$$\boldsymbol{u}_i = \left[\mathrm{E}\{\hat{\boldsymbol{b}}_i\hat{\boldsymbol{b}}_i^T\}\right]^T \boldsymbol{A}_i \mathrm{E}\{\boldsymbol{b}_i\hat{\boldsymbol{b}}_i^T\}\boldsymbol{\rho}_i$$

where $\boldsymbol{A}_i$ is a diagonal $(2K-2)\times(2K-2)$ sub-matrix of $\boldsymbol{A}$ without the ith entry and $\boldsymbol{\rho}_i$ is the ith column of the PCC matrix $\boldsymbol{P}$ without the ith element.

The matrices $\mathrm{E}\{\hat{\boldsymbol{b}}_i\hat{\boldsymbol{b}}_i^T\}$ and $\mathrm{E}\{\boldsymbol{b}_i\hat{\boldsymbol{b}}_i^T\}$ will depend on the random vector of the data $\boldsymbol{b}$ as well as on the colored Gaussian vector $\boldsymbol{\xi}$ through the tentative decision $\hat{\boldsymbol{b}}$. Derivation of these matrices are detailed in [45].

Canceler Output Data Combining. The canceler output $y(-1)$ corresponds to the data vector

$$b(-1) = [b_1(-1), b_2(-2), b_2(-1), \cdots, b_K(-2), b_K(-1)]^T$$

These outputs are combined using the transformation (see Figure 5.19)

$$y'(-2) = \begin{bmatrix} D & 0 & 0 & \cdots & 0 & 0 \\ 0 & 1 & \gamma'_2 D & & 0 & 0 \\ \vdots & & & \ddots & & \vdots \\ 0 & 0 & 0 & \cdots & 1 & \gamma'_K D \end{bmatrix} y(-1) \quad (5.59)$$

$$\hat{b} = \text{sgn}(y')$$

It can be shown that if the tentative decision $\hat{b}_k = b_k$ with high probability, then $u_k = A_k \rho_k$ and the interference term is canceled (with high probability) leaving

$$y'_k(-2) = (\alpha_{2k-2} + \gamma'_k \alpha_{2k-1}) b_k(-2) + n_{2k-2}(-1) + \gamma'_k n_{2k-1}(-2)$$

Hence

$$\text{SNR}_{y'_k} = \frac{\alpha^2_{2k-2}(1 + \gamma'_k \frac{\alpha_{2k-1}}{\alpha_{2k-2}})^2}{(1 + \gamma'^2_k) N_0/2} \quad (5.60)$$

Maximum is obtained with $\gamma'_k = \frac{\alpha_{2k-1}}{\alpha_{2k-2}} = \sqrt{\frac{1-\epsilon_k}{\epsilon_k}}$, making

$$\text{SNR}_{y'_k} = \frac{\alpha^2_{2k-2}}{N_0/2}(1 + \frac{1-\epsilon_k}{\epsilon_k}) = \frac{a_k}{N_0/2}, \quad k = 2, \cdots K$$

means that the output attains the BPSK single user limit, as in the synchronous case.

For $b \neq \hat{b}$, the value of combining gain is not necessarily equal to the one found earlier, so that the performance will be somewhat worse.

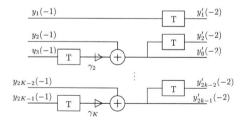

Figure 5.19 Canceler output data combiner.

Output Probability of Error. One might first consider the performance of multistage receiver at the output of the canceler before combining, which can be deduced from (5.49) with the one-shot parameter. In fact for user No. 1, the probability of error is exactly that of (5.48). For the other combined outputs we have

$$P_{ekC} = 2^{-2(2K-1)} \sum_{b_{2k-2},\hat{b}_{2k-2},b_{2k-1},\hat{b}_{2k-1}} Q\{\sqrt{2}(\sqrt{SNR_k}$$
$$-\sqrt{\epsilon_k}(b_{2k-2}^T - \hat{b}_{2k-2}^T E\{\hat{b}_{2k-2}\hat{b}_{2k-2}^T\}^{-1} E\{b_{2k-2}\hat{b}_{2k-2}^T\}^{-1}) \cdot$$
$$R_{2k-2}\rho_{2k-2}$$
$$-\sqrt{1-\epsilon_k}(b_{2k-1}^T - \hat{b}_{2k-1}^T E\{\hat{b}_{2k-1}\hat{b}_{2k-1}^T\}^{-1} E\{b_{2k-1}\hat{b}_{2k-1}^T\}^{-1}) \cdot$$
$$R_{2k-1}\rho_{2k-1}\}$$
$$\cdot P_r(\hat{b}_{2k-2}|b_{2k-2}) P_r(\hat{b}_{2k-1}|b_{2k-1}) \qquad (5.61)$$

where

$$R = \text{diag}\left[\sqrt{SNR_1'}, \cdots, \sqrt{SNR_{2K-1}'}\right]$$

$\sqrt{SNR_1'} = \sqrt{SNR_1}$, $\sqrt{SNR_{2k-2}'} = \epsilon_k \sqrt{SNR_{2k-2}}$, $\sqrt{SNR_{2k-1}'} = (1-\epsilon_k)\sqrt{SNR_{2k-1}}$,
$SNR_i = \frac{a_i}{N_0}, i = 1 \cdots 2K-1$
and

$$P_r(\hat{b}_i|b_i) = \frac{\prod_{j=1, j\neq i}^{2K-1} \sqrt{SNR_i'}}{\sqrt{(\pi)^{2(K-1)}|\Gamma_i|}} \int_{\substack{b_j\xi_i = \alpha_j b_j \hat{b}_j \\ j=1,\ldots,2K-1, j\neq i}}^{\infty} \cdots \int \exp(-\lambda_i^T R_i \Gamma_i^{-1} R_i \lambda_i) d\lambda_i$$

R_i is R without the ith element.

$$\lambda_i = A_i^{-1}\xi_i \qquad \Gamma_i = \left(\frac{N_0}{2}\right)^{-1} Z_i$$

and Z_i is the $2(K-1) \times 2(K-1)$ covariance matrix of ξ_i.

Numerical results. In this section we present a set of numerical results which depicts the error probability at different points of the proposed multiuser detector. Some simulations performed showed similar results in steady state.

The emphasis of these calculations is to show in particular that, although the one-shot decorrelator is matched to one particular code (termed user 1), the performance of all other users is high, irrespective of the fact that its bit is split due to lack of synchronization with user 1:

- Figure 5.20 shows the effect of data combining on the second user predecision performance (2 users).

- Figure 5.21 show the effect of data combining following the decorrelator, on performance of user 1 at the output of DC^3 (2 users).

- Figure 5.22 compares the performance of user 1 and user 2 (combined) as a function of other user energy.

- Figure 5.23 is the same as Figure 5.6 for 3 users.

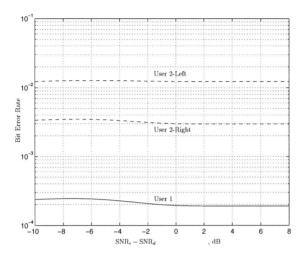

Figure 5.20 The effect of data combining on the 2nd user's predecision performance (2 user asynchronous CDMA system).

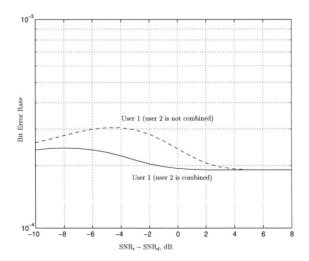

Figure 5.21 The effect of data combining following the decorrelator on the performance of user 1 at the output of DC^3 (2-user asynchronous CDMA system).

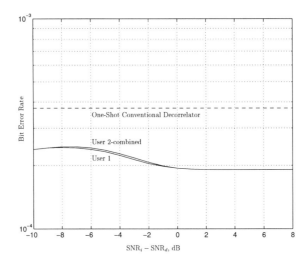

Figure 5.22 BER Performance of DC^3 multistage detection structure for 2-user asynchronous CDMA system, SNR_d = 8 dB. (d=1,2) while other user's SNR is variable.

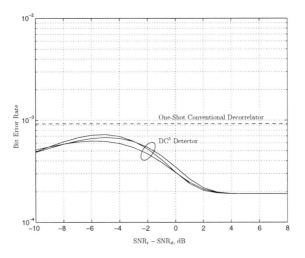

Figure 5.23 BER Performance of DC^3 multistage detection structure for 3-user asynchronous CDMA system, SNR_d = 8 dB. (d=1,2,3) while other user's SNR are variable.

5.5 CONCLUSION

We presented in this overview paper some results of research on multiuser signal separation and detection in CDMA system. It contains different decorrelating detectors for synchronous and asynchronous channel, some of which are of adaptive nature. Multistage receiver with adaptive parallel interference canceler (PIC) is presented.

Novel structure of this PIC receiver suitable for asynchronous channel is also discussed. To limit this paper in length, only the results of analysis are given, with reference to where the reader can find its detail. Some performance curves are also included. Although nonfaded environments are considered, reference to work extension that handled these scenarios are mentioned. As the title emphasizes, the paper deals with research results at CCSPR, NJIT. Nevertheless, some references (though not very complete) to work of others was added.

Acknowledgments

The author would like to acknowledge the contribution of his colleagues and students from CCSPR, who contributed to the work reviewed in this paper. Among these are Prof. Z. Siveski, Prof. H. Ge. Dr. D.W. Chen, Dr. N. Sezgin, J.B. Punt, N. van Waes and X. Li and others.

References

[1] D.Everitt, "Performance and Power Control in CDMA Cellular Mobile Communications," *Journal of Electrical and Electronics Engineering,* Australia, Vol.15, June 1995, pp.151-160.

[2] M.Zorzi, "Simplified Forward-Link Power Control Law in Cellular CDMA," *IEEE Trans. on Vehicular Technology,* Vol.43, No.4, Nov.1994, pp.1088-1093.

[3] R.D.Yates, "Integrated Power Control and Base Station Assignment," *IEEE Trans. on Vehicular Technology,* Vol.44, No.3, Aug.1995, pp.638-644.

[4] K.S.Gilhouson, L.M.Jacobs, R.Padovani, A.J.Viterbi. L.A.Weaver and C.Wheatley II, "On the Capacity of Cellular CDMA Systems," *IEEE Trans. on Vehicular Technology,* vol.VT-40, No.2, pp.303-311, may 91.

[5] S.Verdú, "Minimum Probability of Error for Asynchronous Gaussian Multiple Access Channels," *IEEE Trans. Inform. Theory,* vol.IT-32, No.1, pp.85-96, 1986.

[6] R.Lupas and S.Verdú, "Linear Multiuser Detectors for Synchronous Code Division Multiple Access Channels," *IEEE Trans. Inform. Theory,* vol.IT-35, No.1, pp.123-136. Jan.1989.

[7] R.Lupas and S.Verdú, "Near-Far Resistance of Multiuser Detectors for Synchronous Channels,"*IEEE Trans. Comm,* vol.COM-38, No.4, pp.496-508. April 1990.

[8] A.Duel-hallen, "Decorrelating Decision-Feedback Multiuser Detector for Synchronous Code Division Multiple Access Channels," *IEEE Trans. Comm,* vol.COMM-41, pp.285-290, Feb.1993.

[9] M.K.Varanasi and B.Aazhang, "Multistage Detector in Asynchronous Code Division Multiple Access Communications," *IEEE Trans. Comm.,* vol.COM-38, No.4, pp.509-519, April 1990.

[10] M.K.Varanasi and B.Aazhang, "Near-optimum Detector in Synchronous Code Division Multiple Access System," *IEEE Trans. Comm.,* vol.COM-39, No.5, pp.725-736, May 1991.

[11] M.Honig, U.Madhow and S.Verdú, "Blind Adaptive Multiuser Detection," *IEEE Trans. on Information Theory,* Vol.41, No.4, pp.944-960, July 1995.

[12] D.S.Chen and S.Roy, "An Adaptive Multi-user Receiver for CDMA Systems," *IEEE Journal on Selected Areas of Comm.,* Vol.12, No.5, pp.808-816, June 1994. Issue on Code-Division Multiple-Access Networks II.

[13] N.B.Narayan and B.Aazhang, "Gradient Estimation for Sensitivity Analysis and Adaptive Multiuser Interference Rejection in Code Division Multiple Access Systems," *IEEE Trans. on Communications,* Vol.45, No.7, pp.848-858, July 1997.

[14] S.N.Batalama and D.A.Pados, "On Wiener Signal Reconstruction Filters and DS/CDMA Multiuser Detectors," In *Proceedings ITC'96-International Conference on Telecommunications,* Istanbul, Turkey, April 1996.

[15] S.N.Batalama and D.A.Pados, "Blind Real-Time Low-Complexity Receivers for DS/CDMA Mobile Users," In *Proceedings of the Communication Theory Miniconference,* Globecom, Sigapore, pp. 122-125, November 1995.

[16] S.Verdú, "Adaptive Multiuser Detection," *Proceedings IEEE International Symp. on Spread Spectrum Theory and Application,* Oulu, Finland, 1994.

[17] S.Verdú, "Recent Progress in Multiuser Detection," in Advances in Communications and Signal Processing, Springer Verlag, 1989.

[18] R.Lupas, "Near-Far Resistant Linear Multi-User Detector," PhD thesis, Princeton University, Dec.1988.

[19] N.van Waes and Y.Bar-Ness, "On the Ill Conditioning of the Correlation Matrix of the One-shot Asynchronous Multiuser CDMA Detector", Internal report.

[20] S.Verdú, "Multiuser Detection," in H.V.Poor and J.B.Thomas, editor, Advances in Statistical Signal Processing, vol.2, pp.139-199, JAI Press, Greenwich, CT, 1993.

[21] H.Ge and Y.Bar-Ness, "Multi-shot Approaches to Multiuser Separation and Interference Suppression in Asynchronous CDMA," *The 30th Annual Conference on information Science and Systems,* Princeton NJ, pp.590-595, March 1996.

[22] Y.Bar-Ness and J.B.Punt, "Adaptive Bootstrap CDMA Multiuser Detector," *Wireless Personal Comm., an international Journal, Special issue on Signal Separation and Interference Cancellation (PIMRC)* Vol.3, No.1-2 pp.55-71, 1996

[23] U.Madhow and M.L.Honig, "MMSE Interference Suppression for Direct-Sequence Spread-Spectrum CDMA," *IEEE Trans. Comm.,* Vol.COM-42, No.12. pp. 3178-3188, Dec.1994.

[24] H.Ge and Y.Bar-Ness, "Performance Analysis of Linear Minimum Mean Square Error (LMMSE) Estimator-Based Multiuser Detector," *The IEEE 8th Signal Processing Conference (SSAP-96),* Greece, June 1996.

[25] H.Ge and Y.Bar-Ness, "Comparative study of the Linear Minimum Mean Square Error (LMMSE) and the Adaptive Bootstrap Multiuser Detector for CDMA Communication," International Conference on Comm., Dallax Tx, pp 78-82, June 1996.

[26] C.C.Lee and Y.Bar-Ness, "Convergence and Stability Analysis of the Adaptive 'Bootstrap' CDMA Multi-User Detector," to be submitted to *IEEE Communications Letter*.

[27] Y.Bar-Ness and N.Sezgin, "Adaptive Multiuser Decorrelating CDMA Detector for One-Shot Asynchronous Unknown Channel," ICASSP'95, Detroit, MI, pp.1733-1736, May 1995.

[28] X.Li and Y.Bar-Ness, "The Bootstrap Algorithm: A Robust multiuser CDMA Detector with Time Delay Variation," PIMRC'97, pp. 68-72, Helsinki, Finland

[29] X.Li and Y.Bar-Ness, "Performance of the Bootstrap Multiuser CDMA Detector with Frequency Non-Selective Fading Channel," International Conference on Comm., Atlanta, GA, pp. 53-57, June 1998.

[30] N.van Waes and Y.Bar-Ness, "The Bootstrap Algorithm for One-Shot Matched Filtering, Multiuser Detector in a Multipath Environment," *48th IEEE Vehicle Technology Conference (VTC'98)*, Ottawa, Canada, pp.184-188, May 98.

[31] N.van Waes and Y.Bar-Ness, "The Complex Bootstrap Algorithm for Blind Separation of Co-channel QAM Signals," submitted to *Wireless Personal Commun, An International Journal*, Kluwer.

[32] Y.Bar-Ness and N.van Waes, "The Bootstrap Algorithm to Multi-shot Matched Filtering, Multiuser Detection in Multipath Fading Channels," *48th IEEE Vehicle Technology Conference (VTC'98)*, Ottawa, Canada, pp.179-183, May 98.

[33] N.Sezgin and Y.Bar-Ness, "Adaptive One-shot Asynchronous Signal Separator with Singular Partial Cross-Correlation Matrix," Submitted to *Wireless Personal Comm. An international Journal*, Kluwer Publishers

[34] N.Sezgin and Y.Bar-Ness, "Adaptive Softlimiter Bootstrap for One-Shot Asynchronous CDMA Channel with Singular Partial Cross-Correlation Matrix," *the International Conference on Comm*, Dallas, TX, pp.546-550, June 96.

[35] P.Patel and J.Holtzman, "Analysis of a Simple Successive Interference Cancellation Scheme in DS/CDMA System," *IEEE Journal on Selected Area in Communications*, Vol.12, No.5, pp.796-807, June 1996.

[36] Z.Siveski, Y.Bar-Ness and D.W.Chen, "Error Performance of a Synchronous Multiuser Code Division Multiple Access Detector with a Multidimensional Adaptive Canceler," *The European Trans. on Telecomm. and Related Tech.*, Vol.5, No.6, pp. 719-724, Nov.-Dec.1994.

[37] Y.C.Yoon, R.Kohono and H.Imai, "Cascaded Co-channel Interference Cancellation and Diversity Combining for Spread Spectrum Multiaccess over Multipath Fading Channels", *ICICE Trans. on Comm.*, No.2, pp.163-, Feb.1993.

[38] L.B.Nelson and H.V.Poor, "Iterative Multiuser Receivers for CDMA Channels: An EM-Based Approach," *IEEE Trans on Comm.*, Vol.44, No.12, pp.1700- , Dec.1996.

[39] P.Patel and J.Holtzman, "Performance Comparison of DS/CDMA System Using Successive Interference Cancellation (IC) Scheme and Parallel IC Scheme Under Fading," *The International Conference on Communications (ICC)*, New Orleans, LA, pp.510-514, May 1994.

[40] B.Zhu, N.Ansari, Z.Siveski and Y.Bar-Ness, "Convergence and Stability Analysis of Synchronous Adaptive CDMA-Based PCS Receiver," *MILCOM 94*, Fort Monmouth, NJ, pp.923-927, Oct.1994.

[41] D.W.Chen, Z.Siveski and Y.Bar-Ness, "Synchronous Multiuser CDMA Detector with a Soft Decision Adaptive Canceler," The 28th Annual CISS, Princeton, NJ, pp.139-143, March.1994.

[42] F.Viehofer, Z.Siveski and Y.Bar-Ness, "Soft Decision Adaptive Multiuser CDMA Detector for Asynchronous AWGN Channel," *The 32nd Annual Allerton Conference on Comm., Computing and Control,* Monticello, IL, pp. 362-371, Sept.28, 1994.

[43] X.Zhang and D.Brady, "Soft-Decision Multistage Detection for Asynchronous AWGN Channels," *31st Annual Allerton Conference on Comm., Computing and Control,* Monticello, IL, Sept.1993.

[44] Y.Bar-Ness and N.Sezgin, "Adaptive Threshold Setting for Multiuser CDMA Signal Separator with Soft Tentative Decision," *The 29th annual CISS,* John Hopkins University, March 1995.

[45] Y.Bar-Ness, "Asynchronous Multiuser CDMA Detector Made Simpler: Novel decorrelator, combiner, canceler, combiner (DC^3) Structure," to appear in *IEEE Trans. on Comm.*.

6 LMMSE RECEIVERS FOR DS-CDMA SYSTEMS IN FREQUENCY-SELECTIVE FADING CHANNELS

Matti Latva-aho

University of Oulu
Centre for Wireless Communications (CWC)
Oulu, Finland

Abstract: The linear minimum mean squared error (LMMSE) criterion can be used to obtain near-far resistant single-user receivers in direct-sequence code-division multiple-access (DS-CDMA) systems. The standard version (postcombining) of the LMMSE receiver minimizes the mean squared error (MSE) between the filter output and the actual transmitted data sequence. Since the detector depends on the channel coefficients of all users, it cannot be implemented adaptively in fading channels due to severe convergence problems. A modified criterion for deriving LMMSE receivers in fading channels is presented. The modified (precombining) LMMSE receiver is independent from users' complex channel coefficients, and it effectively converts the pathological Rayleigh fading channel to an equivalent fixed AWGN channel from the detector updating perspective. The precombining LMMSE receiver leads to blind adaptive implementations. Some alternative blind adaptive receivers are reviewed and their performance in frequency-selective fading channels is addressed. Since the blind adaptive implementations require explicit information on multipath delays, delay estimation issue will also be considered. The blind adaptive interference suppression methods are applied furthermore to the residual interference suppression in parallel interference cancellation receivers.

6.1 INTRODUCTION

The third generation system for cellular mobile communications will be based on code-division multiple-access (CDMA) techniques. The system capacity of CDMA systems depends among other things on the receiver methods applied. The conventional approach to reception in CDMA systems is to neglect multiple-access interference (MAI) and the near-far problem. This imposes tight limits on the system capacity due to interference even if strict power control is used. A more efficient way to detect different users in CDMA systems is based on multiuser receivers. An optimal multiuser receiver requires joint estimation of channel parameters and data symbols [1, 2]. Optimal receivers are far too complex for practical implementations and hence several suboptimal receivers [3, 4, 5] have been proposed.

Most near-far resistant receivers are centralized, i.e., all user signals are processed jointly in the receiver. When considering downlink receivers, only the desired user signal should be received while suppressing the interference due to other users. The linear minimum mean squared error (LMMSE) single-user receivers [6, 7, 8] are an option for the downlink receivers. The adaptive versions of the LMMSE receivers are usually designed in such a way that only one user is demodulated, as desired in the downlink. The so-called *postcombining LMMSE receiver* [6, 7, 8] minimizes the mean squared error (MSE) between the receiver output and the training sequence. The LMMSE receivers are capable of handling both multipath induced and multiple-access interference under severe near-far situations. The postcombining LMMSE receiver [6] depends on the channel complex coefficients of all users, and hence it must be adapted as the channel changes. If the fade rate of the channel is fast enough, the adaptive postcombining LMMSE receivers need to be updated continuously. Thus, the postcombining LMMSE receivers will have severe convergence problems in relatively fast fading channels. Nevertheless, they can be applied if the rate of fading is low enough with respect to the data rate [9].

The LMMSE optimization criterion can be modified [10, 11] to overcome the convergence problems of the postcombining LMMSE receiver. The modified criterion leads to the *precombining LMMSE receiver* which no longer depends on the instantaneous channel coefficient values. The precombining LMMSE receiver coefficients depend only on the normalized signature sequence cross-correlations and the average channel profiles of the users. Since the delays and the average channel profiles change rather slowly, the adaptation requirements of the precombining LMMSE receiver are significantly less stringent than those of the adaptive postcombining LMMSE receivers [8]. The adaptive implementations of the precombining LMMSE receiver do not necessarily require training sequences, since the decisions made by the conventional RAKE receiver can often be used to train the adaptive receiver. Thus, the precombining LMMSE receiver can be viewed as an add-on feature in the conventional coherent RAKE receivers.

The adaptive version of the precombining LMMSE receiver with decision directed training is blind in the sense that no training sequences are required. The minimum output energy (MOE) criterion was used in [12] to derive blind adaptive receivers. It is well known that the MSE and MOE criterion lead to the same receivers and have the same performance under ideal conditions. However, their performance depends on the adaptive implementations which may differ significantly. There are also other blind adaptive algorithms which are suitable for single-user CDMA receivers, such as the unconstrained [13] and constrained constant modulus algorithms [13, 14, 15], Griffiths' algorithm [14] and generalized sidelobe canceller [13, 14, 16]. Also the blind adaptive algorithms based on higher order statistics [17] have been proposed [18], but they are not considered here. Blind adaptive receivers have been studied for AWGN channels in [12, 14, 15, 16, 19, 20], for flat fading in [21], for non-fading multipath channels in [22] and for frequency-selective fading channels in [13, 11]. Blind adaptive least squares receivers have been studied in [21, 23, 24, 25]. A blind receiver performing both the MOE filtering and timing estimation has been studied in [26].

The rest of the chapter is organized as follows. In Section 6.2, the CDMA system model is presented. The principles of the pre and the postcombining LMMSE receivers are presented in Section 6.3. Bit error probability analysis for the precombining LMMSE receiver will be developed in Section 6.4. Blind adaptive LMMSE receivers are derived in Section 6.5. Delay acquisition in the precombining LMMSE receivers is discussed in Section 6.6 and the delay tracking is considered in Section 6.7. A combination of the blind adaptive and parallel interference cancellation receivers is presented in Section 6.8. Numerical results are presented in Section 6.9, followed by concluding remarks in Section 6.10.

6.2 SYSTEM MODEL

A standard model for an asynchronous DS-CDMA system with K users and L propagation paths will be considered. BPSK modulated data of the kth user is spread by multiplying the data modulated signal by a binary pseudorandom noise sequence given by $s_k(t) = \sum_{j=0}^{G-1} s_k(j) p(t - jT_c)$, where G is the number of chips per symbol, $s_k(j) \in \{-1, 1\}$ is the jth chip of kth user, and $p(t)$ is the chip waveform. It will be assumed that the length of the PN sequence equals one symbol interval. Now, the complex envelope of the received signal can be expressed as

$$r(t) = \sum_{n=0}^{N_b-1} \sum_{k=1}^{K} A_k b_k^{(n)} s_k(t - nT) * c_k(t) + n(t), \qquad (6.1)$$

where N_b is the number of received symbols, K is the number of users, $A_k = \sqrt{E_k}$, E_k is the energy per symbol, $b_k^{(n)}$ is the nth transmitted data symbol, $s_k(t)$ is the kth user's signature signal (in the sequel $\int_0^T |s_k(t)|^2 dt = 1$), T denotes the symbol interval, $n(t)$ is complex zero mean additive white Gaussian noise with variance σ^2, $*$ denotes convolution and $c_k(t) = \sum_{l=1}^{L_k} c_{k,l}^{(n)} \delta(t - \tau_{k,l})$ is the impulse response of the kth user's radio channel, where L_k is the number of propagation paths (here $L_k = L$, $\forall k$, for notational simplicity), $c_{k,l}$ is the complex attenuation factor of the kth user's lth path and $\tau_{k,l}$ is the propagation delay. The received signal has the form

148   CDMA TECHNIQUES FOR THIRD GENERATION MOBILE SYSTEMS

$$r(t) = \sum_{n=0}^{N_b-1} \sum_{k=1}^{K} \sum_{l=1}^{L} A_k b_k^{(n)} c_{k,l}^{(n)} s_k(t - nT - \tau_{k,l}) + n(t). \quad (6.2)$$

The received signal is time-discretized by anti-alias filtering and sampling $r(t)$ at the rate $T_s^{-1} = \frac{SG}{T}$, where S is the number of samples per chip. The received discrete-time signal over a data block of N_b symbols is

$$\mathbf{r} = \mathbf{SCAb} + \mathbf{n} \in \mathbb{C}^{SGN_b}, \quad (6.3)$$

where $\mathbf{r} = \left[\mathbf{r}^{\mathrm{T}}(0), \ldots, \mathbf{r}^{\mathrm{T}}(N_b-1)\right]^{\mathrm{T}}$ is the input sample vector with

$$\mathbf{r}^{\mathrm{T}}(n) = [r(T_s(nSG+1)), \ldots, r(T_s(n+1)SG)],$$

$$\mathbf{S} = [\mathbf{S}^{(0)}, \mathbf{S}^{(1)}, \ldots, \mathbf{S}^{(N_b-1)}] \in \mathbb{R}^{SGN_b \times KLN_b}$$

is the sampled spreading sequence matrix with $\mathbf{S}^{(n)} = [\mathbf{s}_{1,1}^{(n)}, \ldots, \mathbf{s}_{1,L}^{(n)}, \ldots, \mathbf{s}_{K,L}^{(n)}]$, where $\mathbf{s}_{k,l}^{(n)} = [\mathbf{0}_{(nSG+\tau_{k,l})\times 1}^{\mathrm{T}}, \mathbf{s}_k^{\mathrm{T}}, \mathbf{0}_{((N_b-n-1)SG-\tau_{k,l})\times 1}^{\mathrm{T}}]$, where $\tau_{k,l}$ is time-discretized delay in sample intervals ($\tau_{1,1} = 0$) and $\mathbf{s}_k = [s_k(T_s), \ldots, s_k(T_s SG)]^{\mathrm{T}}$ is the sampled signature sequence of the kth user, $\mathbf{C} = \mathrm{diag}[\mathbf{C}^{(0)}, \ldots, \mathbf{C}^{(N_b-1)}] \in \mathbb{C}^{KLN_b \times KN_b}$ is the channel coefficient matrix with $\mathbf{C}^{(n)} = \mathrm{diag}[\mathbf{c}_1^{(n)}, \ldots, \mathbf{c}_K^{(n)}] \in \mathbb{C}^{KL \times K}$, $\mathbf{c}_k^{(n)} = [c_{k,1}^{(n)}, \ldots, c_{k,L}^{(n)}]^{\mathrm{T}} \in \mathbb{C}^L$, $\mathbf{A} = \mathrm{diag}[\mathbf{A}^{\mathrm{T}}(0), \ldots, \mathbf{A}^{\mathrm{T}}(N_b-1)] = \mathbf{I}_{N_b} \otimes \mathbf{A}^{(n)} \in \mathbb{R}^{KN_b \times KN_b}$ is the matrix of total received energies with $\mathbf{A}^{(n)} = \mathrm{diag}[A_1, \ldots, A_K]$, $\mathbf{b} = [\mathbf{b}^{\mathrm{T}}(0), \ldots, \mathbf{b}^{\mathrm{T}}(N_b-1)]^{\mathrm{T}} \in \Xi^{KN_b}$ is the data vector with modulation symbol alphabet Ξ and $\mathbf{b}^{(n)} = [b_1^{(n)}, \ldots, b_K^{(n)}]$, and $\mathbf{n} \in \mathbb{C}^{SGN_b}$ is the channel noise vector.

6.3 LMMSE RECEIVERS IN FADING CHANNELS

There are two methods for employing linear multiuser detection in multipath channels. The multiuser filtering can take place either after the multipath combining or prior to it. In other words, the multiuser receiver can be either a *postcombining interference suppression* type of receiver (Figure 6.1(a)), or a *precombining interference suppression* type of receiver (Figure 6.1(b)). Performance differences of the two structures for the decorrelating detector in known fixed channels have been compared in [27]. The results show that the order of multipath combining and interference suppression does not have a significant impact on the bit error probability of the decorrelator when the product of number of users and multipath components (KL) is relatively low. As the product KL becomes large the cross-correlation matrix of users' signature sequences becomes ill-conditioned. In such a case, multipath combining prior to interference suppression usually results in stable matrix inversion and robust performance. However, multipath combining prior to the interference suppression makes the channel estimation more difficult since the multiuser detector depends on the channel estimates, which cannot be estimated at the output of the detector in that case (see Figure 6.1 (a)). Therefore,

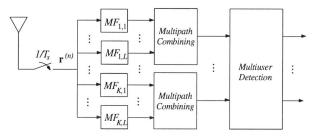

(a) Postcombining interference suppression receiver.

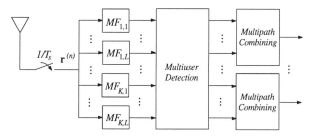

(b) Precombining interference suppression receiver.

Figure 6.1 Multiuser receiver structures.

practical versions of the multiuser receivers first perform interference suppression and after that channel estimation [27, 2, 28, 29]. Such receivers also have the advantage that the detector does not depend on the fading channel state.

The postcombining LMMSE receiver minimizes the cost function $E\{|\mathbf{b} - \hat{\mathbf{b}}|^2\}$ elementwise, where $\hat{\mathbf{b}} = \mathbf{L}^H \mathbf{r}$, and $|\mathbf{x}|$ is the vector of absolute values of the vector $\mathbf{x}$. It is easy to show [30, pp. 391] that the postcombining LMMSE receiver is

$$\mathbf{L} = \mathbf{SCA} \left(\mathbf{AC}^H \mathbf{RCA} + \sigma^2 \mathbf{I}\right)^{-1} \in \mathbb{C}^{SGN_b \times KN_b}, \tag{6.4}$$

where $\mathbf{R} = \mathbf{S}^T \mathbf{S} \in \mathbb{R}^{KLN_b \times KLN_b}$ is the signature sequence cross-correlation matrix. The output of the postcombining LMMSE receiver is

$$\mathbf{y}_{[\mathbf{L}]} = \left(\mathbf{AC}^H \mathbf{RCA} + \sigma^2 \mathbf{I}\right)^{-1} (\mathbf{SCA})^H \mathbf{r} \in \mathbb{C}^K, \tag{6.5}$$

where $(\mathbf{SCA})^H \mathbf{r}$ is the multipath combined matched filter bank output. Thus, the postcombining LMMSE detector clearly leads to a multipath-combining-interference-suppression type of receiver (Figure 6.1(a)). If the channel is a non-fading AWGN channel, i.e., all non-zero elements in $\mathbf{C}$ are equal to 1, the detector can be written in the standard form [6]: $\mathbf{L} = \mathbf{S}\left(\mathbf{R} + \sigma^2(\mathbf{A}^H\mathbf{A})^{-1}\right)^{-1}$. It can be seen from (6.4) that the postcombining LMMSE receiver in fading channels depends on the channel complex coefficients of all users and paths. If the channel is changing rapidly, the optimal LMMSE receiver changes continuously. Thus, the adaptive versions of the LMMSE receivers have increasing convergence problems as the fading rate increases.

The dependence on the fading channel state can be removed by applying an interference-suppression-multipath-combining type of receiver (Figure 6.1(b)). This can be accomplished by minimizing each element of $E\{|\mathbf{h} - \hat{\mathbf{h}}|^2\}$, where $\mathbf{h} = \mathbf{CAb}$ is the data-amplitude product vector and $\hat{\mathbf{h}} = \mathbf{M}^T\mathbf{r}$ its estimate. This leads to the *precombining LMMSE receiver*:

$$\mathbf{M} = \mathbf{S}\left(\mathbf{R} + \sigma^2 \mathbf{\Sigma_h}^{-1}\right)^{-1} \in \mathbb{R}^{SGN_b \times KLN_b}, \quad (6.6)$$

where $\mathbf{\Sigma_h} = \text{diag}\left[A_1^2 \mathbf{\Sigma_{c_1}}, \ldots, A_K^2 \mathbf{\Sigma_{c_K}}\right] \in \mathbb{R}^{KLN_b \times KLN_b}$ is the covariance matrix of $\mathbf{h}$ which consists of transmitted user powers and the average channel tap powers, with $\mathbf{\Sigma_{c_k}} = \text{diag}\left[E[|c_{k,1}|^2], \ldots, E[|c_{k,L}|^2]\right] \in \mathbb{R}^{L \times L}$ where $E[|c_{k,l}|^2]$ is the average power of the kth user's lth propagation path. The output of the precombining LMMSE receiver is

$$\mathbf{y}_{[M]} = \left(\mathbf{R} + \sigma^2 \mathbf{\Sigma_h}^{-1}\right)^{-1} \mathbf{S}^T \mathbf{r} \in \mathbb{C}^{KL}, \quad (6.7)$$

where $\mathbf{S}^T\mathbf{r}$ is the matched filter bank output vector without multipath combining. Thus, the precombining LMMSE receiver leads to a receiver of the type shown in Figure 6.1(b). As can be seen from (6.6), the precombining LMMSE receiver no longer depends on the instantaneous values of the channel complex coefficients but on the average power profiles of the channels. The receiver is of exactly the same form as the postcombining LMMSE receiver in a non-fading AWGN channel. The adaptation requirements are now significantly milder and the receiver can be made adaptive even in relatively fast fading channels. Since interference is suppressed in each multipath component separately before multipath combining, the precombining LMMSE receiver has the same structure as the conventional RAKE receiver. Hence, the precombining LMMSE receiver is called the *LMMSE-RAKE receiver*.

6.4 BIT ERROR PROBABILITY ANALYSIS FOR THE PRECOMBINING LMMSE RECEIVER IN FADING CHANNELS

The performance of the precombining LMMSE receiver is analyzed in a known channel to obtain the expression for the average bit error probability. The analysis is based on the characteristic function method presented in [31]. The characteristic function is solved via eigen-analysis for the matrix formed from the decision variable. The analysis uses the same principles as for the decorrelating receiver presented in [2, 32].

The decision variable of the precombining LMMSE receiver for user k after maximal ratio combining can be expressed in the form

$$y_{[M,MRC]k}^{(n)} = \mathbf{c}_k^{H(n)} \mathbf{y}_{[M]k}^{(n)}, \quad (6.8)$$

where $\mathbf{c}_k^{(n)} = \left[c_{k,1}^{(n)}, c_{k,2}^{(n)}, \ldots, c_{k,L}^{(n)}\right] \in \mathbb{C}^L$ is the combining vector, and $\mathbf{y}_{[M]k}^{(n)} = \left[y_{[M]k,1}^{(n)}, y_{[M]k,2}^{(n)}, \ldots, y_{[M]k,L}^{(n)}\right] \in \mathbb{C}^L$ includes the LMMSE filter output for kth user $\left(\mathbf{y}_{[M]}^{(n)} = \mathbf{M}^T \bar{\mathbf{r}}^{(n)} = \left[(\mathbf{y}_{[M]1}^{T(n)}, \ldots, (\mathbf{y}_{[M]K}^{T(n)}\right]^T\right)$. Let

$$\mathbf{Q} = \frac{1}{2} \begin{pmatrix} \mathbf{0}_L & \mathbf{I}_L \\ \mathbf{I}_L & \mathbf{0}_L \end{pmatrix} \in \{0, \tfrac{1}{2}\}^{2L \times 2L}, \quad (6.9)$$

and

$$\boldsymbol{\nu} = \left[\mathbf{c}_k^{T(n)}, \mathbf{y}_{[M]k}^{T(n)}\right]^T \in \mathbb{C}^{2L}. \quad (6.10)$$

By rewriting (6.8), the decision variable can be expressed in the form

$$y_{[M,MRC]k}^{(n)} = \boldsymbol{\nu}^H \mathbf{Q} \boldsymbol{\nu}. \quad (6.11)$$

The LMMSE receiver output vector $\mathbf{y}_{[M]k}^{(n)}$ conditioned on the data vector $\bar{\mathbf{b}}^{(n)}$ over the processing window of size M for all users $\left(\bar{\mathbf{b}}^{(n)} = [\bar{\mathbf{b}}_1^{T(n)}, \ldots, \bar{\mathbf{b}}_K^{T(n)}]^T, \bar{\mathbf{b}}_k^{T(n)} = [b_k^{(n-D)}, \ldots, b_k^{(n)}, \ldots, b_k^{(n+D)}]^T\right)$ is a complex Gaussian random vector. Since the weight vector $\mathbf{c}_k^{(n)}$ is also Gaussian, the probability of bit error for user k conditioned on $\bar{\mathbf{b}}^{(n)}$ can be expressed in the case of BPSK modulation as [31]

$$Pr\{\text{error}|\bar{\mathbf{b}}^{(n)}\} = \sum_{\substack{i=1 \\ \lambda_i < 0}}^{2L} \prod_{\substack{j=1 \\ j \neq i}}^{2L} \frac{1}{1 - \frac{\lambda_j}{\lambda_i}}, \quad (6.12)$$

where $\lambda_i, i = 1, 2, \ldots, 2L$ are the eigenvalues of the matrix $\Sigma_{\boldsymbol{\nu}|\bar{\mathbf{b}}^{(n)}} \mathbf{Q}$, and

$$\Sigma_{\boldsymbol{\nu}|\bar{\mathbf{b}}^{(n)}} = \begin{pmatrix} \Sigma_{\mathbf{c}_k^{(n)}} & \Sigma_{\mathbf{c}_k^{(n)}, \mathbf{y}_{[M]k}^{(n)}|\bar{\mathbf{b}}^{(n)}} \\ \Sigma_{\mathbf{c}_k^{(n)}, \mathbf{y}_{[M]k}^{(n)}|\bar{\mathbf{b}}^{(n)}}^H & \Sigma_{\mathbf{y}_{[M]k}^{(n)}|\bar{\mathbf{b}}^{(n)}} \end{pmatrix} \quad (6.13)$$

is the covariance matrix of the vector $\boldsymbol{\nu}$. Finally, the bit error probability for the kth user is expressed as

$$P_k = \frac{1}{2^{MK-1}} \sum_{\substack{\bar{\mathbf{b}}^{(n)} \in \{-1,1\}^{MK-1} \\ b_k = 1}} Pr\{\text{error}|\bar{\mathbf{b}}^{(n)}\}. \quad (6.14)$$

The analysis can be simplified by assuming that the other user data bits are random variables of the discrete uniform distribution, i.e., using the Gaussian approximation presented in [33].

6.5 ADAPTIVE IMPLEMENTATIONS OF THE PRECOMBINING LMMSE RECEIVERS

The modified MSE criterion $\mathrm{E}\{|\mathbf{h}-\hat{\mathbf{h}}|^2\}$ requires that the reference signal $\mathbf{h} = \mathbf{CAb}$ is known in adaptive implementations. Hence, the adaptive versions of the precombining LMMSE receiver need to know spreading sequence timing, data bits, and channel complex coefficients of each desired multipath component or estimates thereof. This side information of the channel parameters may not be available in all applications. However, in the conventional coherent RAKE receiver, this information is available.

Another approach to derive adaptive algorithms for the precombining LMMSE receivers is based on the constrained minimum output energy (MOE) optimization [12]. The MOE ($\mathrm{E}[|y_{k,l}|^2]$) and the MSE criteria lead to the receivers $\mathbf{w}_{[MOE]k,l} = \Sigma_{\bar{\mathbf{r}}}^{-1}\bar{\mathbf{s}}_{k,l}/(\bar{\mathbf{s}}_{k,l}^T\Sigma_{\bar{\mathbf{r}}}^{-1}\bar{\mathbf{s}}_{k,l})$ [12] and $\mathbf{w}_{[MSE]k,l} = \Sigma_{\bar{\mathbf{r}}}^{-1}\Sigma_{\bar{\mathbf{r}}d_{k,l}} = \Sigma_{\bar{\mathbf{r}}}^{-1}\bar{\mathbf{s}}_{k,l}\mathrm{E}[|c_{k,l}|^2]$. Hence the receivers are equal up to a scalar, which does not influence on the BEP. In the following subsections, different adaptive implementations of the precombining LMMSE receivers are reviewed.

6.5.1 Adaptive LMMSE-RAKE Receiver

The modified cost function results in separate LMMSE receivers for each multipath component, as can be seen in Figure 6.1. The adaptive LMMSE receiver is actually an adaptive RAKE receiver, where each receiver branch is adapted independently to suppress MAI (see Figure 6.2).

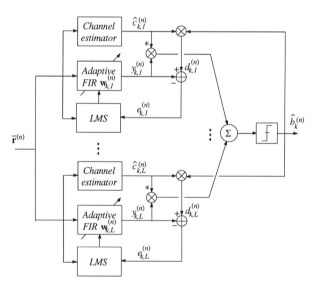

Figure 6.2 General block diagram of the adaptive LMMSE-RAKE receiver.

The received signal is processed in blocks of M symbols ($M < N_b$). Let the input sample vector during the nth symbol be $\bar{\mathbf{r}}^{(n)} = \left[\mathbf{r}^{\mathrm{T}}(n-D), \ldots, \mathbf{r}^{\mathrm{T}}(n), \ldots, \mathbf{r}^{\mathrm{T}}(n+D)\right]^{\mathrm{T}} \in$

$\mathbb{C}^{MSG}$, where $M = (2D + 1)$ is the sample vector length in symbol intervals. The received signal vectors are fed into linear filters with impulse response $\mathbf{w}_{k,l}^{(n)} = [w_{k,l}^{(n)}(0), \ldots, w_{k,l}^{(n)}(MSG - 1)]^T \in \mathbb{C}^{MSG}$. The output of the lth receiver branch can be written as $y_{k,l}^{(n)} = \mathbf{w}_{k,l}^{H\,(n)} \bar{\mathbf{r}}^{(n)}$. The decisions in an LMMSE-RAKE receiver are made according to

$$\hat{b}_k^{(n)} = \text{sign}\left(\sum_{l=1}^{L} \hat{c}_{k,l}^{*\,(n)} y_{k,l}^{(n)}\right). \qquad (6.15)$$

The optimal filter coefficients are derived using the MSE criterion ($E[|e_{k,l}^{(n)}|^2]$), which leads to the optimal filter coefficients $\mathbf{w}_{[MSE]k,l} = \mathbf{\Sigma}_{\bar{\mathbf{r}}}^{-1} \mathbf{\Sigma}_{\bar{\mathbf{r}} d_{k,l}}$ [34], where $\mathbf{\Sigma}_{\bar{\mathbf{r}} d_{k,l}}$ is the cross-correlation vector between the input vector $\bar{\mathbf{r}}$ and the desired response $d_{k,l}$. Adaptive filtering is based on iteratively solving the optimization problem at hand. The most widely used methods are based on estimating the gradient of the error function and aiming at the negative gradient directions, which provide steepest descent on the error surface. The filter weights are updated according to

$$\mathbf{w}_{k,l}^{(n+1)} = \mathbf{w}_{k,l}^{(n)} - \mu \nabla_{k,l}, \qquad (6.16)$$

where μ is a step-size parameter and $\nabla_{k,l}$ is the gradient of the MSE with respect to the filter weights[1]. Let us first assume that the receiver processing window M is only one symbol interval, i.e., $\bar{\mathbf{r}}^{(n)} = \mathbf{r}^{(n)} \hat{=} \mathbf{r}$. The gradient of the MSE ($J_{k,l} = E[|(\mathbf{CAb})_{k,l} - \mathbf{w}_{k,l}^H \mathbf{r}|^2] = E\{|c_{k,l} A_k b_k - \mathbf{w}_{k,l}^H \mathbf{r}|^2\}$) with respect to the filter weight vector $\mathbf{w}_{k,l}$ is

$$\begin{aligned}\nabla_{k,l} &= -2E\left[\mathbf{r}(c_{k,l} A_k b_k)^*\right] + 2E\left[\mathbf{r}\mathbf{r}^H\right] \mathbf{w}_{k,l} \\ &= -2\mathbf{\Sigma}_{\mathbf{r} d_{k,l}} + 2\mathbf{\Sigma}_{\mathbf{r}} \mathbf{w}_{k,l}, \end{aligned} \qquad (6.17)$$

where $d_{k,l} = c_{k,l} A_k b_k$ (In the rest of this chapter it will be assumed that $A_k = 1$, $\forall k$.). The resulting iterative algorithm includes recursions

$$\mathbf{w}_{k,l}^{(n+1)} = \mathbf{w}_{k,l}^{(n)} + 2\mu\left(\mathbf{\Sigma}_{\mathbf{r} d_{k,l}} - \mathbf{\Sigma}_{\mathbf{r}} \mathbf{w}_{k,l}^{(n)}\right) \qquad (6.18)$$

and is called the steepest descent algorithm [34]. The widely used approximation of the gradient is the so-called stochastic approximation: $\nabla_{k,l} \approx -2\mathbf{r}(c_{k,l} b_k)^* + 2\mathbf{r}\mathbf{r}^H \mathbf{w}_{k,l}^{(n)} = -2\mathbf{r}(c_{k,l} b_k)^* + 2\mathbf{r} r y_{k,l}^*$. Using this as a gradient estimate and assuming now that $M \geq 1$, the least mean squares (LMS) algorithm has the form

$$\mathbf{w}_{k,l}^{(n+1)} = \mathbf{w}_{k,l}^{(n)} + 2\mu \bar{\mathbf{r}}^{(n)} \left(c_{k,l}^{(n)} b_k^{(n)} - y_{k,l}^{(n)}\right)^* \in \mathbb{C}^{MSG}. \qquad (6.19)$$

The receiver vector can be decomposed into adaptive and fixed components such that

$$\mathbf{w}_{k,l}^{(n)} = \bar{\mathbf{s}}_{k,l} + \mathbf{x}_{k,l}^{(n)} \in \mathbb{C}^{MSG}, \qquad (6.20)$$

---

[1] $\nabla_{k,l} = \frac{\partial J_{k,l}}{\partial \text{Re}\{\mathbf{w}_{k,l}\}} + j \frac{\partial J_{k,l}}{\partial \text{Im}\{\mathbf{w}_{k,l}\}} = 2 \frac{\partial J_{k,l}}{\partial \mathbf{w}_{k,l}^*}$ [34, pp. 197,894].

where $\mathbf{x}_{k,l}^{(n)}$ is the adaptive filter component and

$$\bar{\mathbf{s}}_{k,l} = \begin{bmatrix} \mathbf{0}_{(DSG+\tau_{k,l})\times 1}^{\mathrm{T}}, \mathbf{s}_k^{\mathrm{T}}, \mathbf{0}_{(DSG-\tau_{k,l})\times 1}^{\mathrm{T}} \end{bmatrix}^{\mathrm{T}} \qquad (6.21)$$

is the fixed spreading sequence[2] of the kth user with the delay $\tau_{k,l}$. Hence, the adaptive component can be laid on top of the conventional RAKE receiver, as is shown in Figure 6.3. The updates for the adaptive component can be written as

$$\begin{aligned} \mathbf{x}_{k,l}^{(n+1)} &= \mathbf{x}_{k,l}^{(n)} + 2\mu_{k,l}^{(n)} \left(c_{k,l}^{(n)} b_k^{(n)} - y_{k,l}^{(n)} \right)^* \bar{\mathbf{r}}^{(n)} \\ &= \mathbf{x}_{k,l}^{(n)} + 2\mu_{k,l}^{(n)} e_{k,l}^{*(n)} \bar{\mathbf{r}}^{(n)}, \end{aligned} \qquad (6.22)$$

where $\mu_{k,l}^{(n)}$ is a time-variant step-size parameter, which controls the rate of convergence of the algorithm.

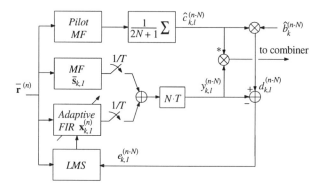

Figure 6.3 Block diagram of one receiver branch in the adaptive LMMSE-RAKE receiver.

The optimal step-size depends on the eigenvalues of the input vector covariance matrix $\Sigma_{\bar{\mathbf{r}}}$ [34]. Quite often the step-size is chosen as $\mu_{k,l}^{(n)} = \mu/(\bar{\mathbf{r}}^{\mathrm{H}(n)} \bar{\mathbf{r}}^{(n)}); 0 < \mu < 1$. An LMS algorithm with this step-size is called the normalized LMS (NLMS) algorithm [34]. The actual value of the step-size is crucial for the adaptive postcombining LMMSE receiver when used in fading channels [9]. Since the precombining LMMSE receiver does not need to track the fading channel coefficients, the step-size can be set more freely.

---

[2]With this definition interference is not suppressed symmetrically from past and future samples. With the definition $\bar{\mathbf{s}}_{k,l} = [\mathbf{0}_{DSG\times 1}^{\mathrm{T}}, \mathbf{s}_k^{\mathrm{T}}, \mathbf{0}_{DSG\times 1}^{\mathrm{T}}]^{\mathrm{T}}$ symmetric interference suppression is achieved. In such a case, however, the received signal must be delayed separately for each RAKE branch. It should be noted that in one-shot receivers ($M = 1$), only the latter definition results in satisfactory performance with large delay spreads. In the numerical examples (Section 6.9.3), the "one-shot" receiver in a two-path fading channel used sequences $\bar{\mathbf{s}}_{k,1} = [\mathbf{s}_k^{\mathrm{T}}, \mathbf{0}_{\tau_{[max]}\times 1}^{\mathrm{T}}]^{\mathrm{T}}$ and $\bar{\mathbf{s}}_{k,2} = [\mathbf{0}_{\tau_{[max]}\times 1}^{\mathrm{T}}, \mathbf{s}_k^{\mathrm{T}}]^{\mathrm{T}}$, where $\tau_{[max]}$ is the maximum delay difference between the multipath components.

The error signals, $e_{k,l}^{(n)} = d_{k,l}^{(n)} - y_{k,l}^{(n)}$, produced by the difference between the filter outputs and the reference signals, are used to update the filter weights. Either known or estimated data symbols are used as reference signals in the adaptive postcombining LMMSE receivers. The product of the estimated channel coefficients and data symbols is the refence signal in the adaptive precombining LMMSE receiver ($d_{k,l}^{(n)} = \hat{c}_{k,l}^{(n)} b_k^{(n)}$ or $d_{k,l}^{(n)} = \hat{c}_{k,l}^{(n)} \hat{b}_k^{(n)}$, respectively). The data decisions produced initially by a conventional RAKE receiver are often sufficiently reliable for adapting the receiver as will be demonstrated by some numerical examples.

6.5.2 Adaptive MOE Receiver

The adaptive and non-adaptive components are constrained to satisfy $\bar{\mathbf{s}}_{k,l}^T \mathbf{x}_{k,l}^{(n)} = 0$ in the MOE receivers. The stochastic approximation of the gradient for the MOE criterion is $\nabla_{k,l} = \bar{\mathbf{r}}^{(n)} \bar{\mathbf{r}}^{H(n)} \mathbf{w}_{k,l}$. The orthogonality condition is maintained at each step of the algorithm by projecting the gradient onto the linear subspace orthogonal to $\bar{\mathbf{s}}_{k,l}$. In practice, this is accomplished by subtracting an estimate of the desired signal component from the received signal vector. Hence, the blind receiver based on the MOE criterion for the one-shot synchronous case [12] and for the three-shot asynchronous case [20] includes recursions

$$\mathbf{x}_{k,l}^{(n+1)} = \mathbf{x}_{k,l}^{(n)} - 2\mu_{k,l}^{(n)} \bar{\mathbf{r}}^{H(n)} \left(\bar{\mathbf{s}}_{k,l} + \mathbf{x}_{k,l}^{(n)}\right)\left(\bar{\mathbf{r}}^{(n)} - \mathbf{F}_{k,l}\left(\mathbf{F}_{k,l}^T \bar{\mathbf{r}}^{(n)}\right)\right), \quad (6.23)$$

where

$$\mathbf{F}_{k,l} = \begin{pmatrix} \mathbf{0}_{\tau_{k,l} \times 1}^T, & \mathbf{s}_k^T, & \mathbf{0}_{(2DSG-\tau_{k,l}) \times 1}^T \\ \mathbf{0}_{(SG+\tau_{k,l}) \times 1}^T, & \mathbf{s}_k^T, & \mathbf{0}_{((2D-1)SG-\tau_{k,l}) \times 1}^T \\ & \vdots & \\ \mathbf{0}_{(2DSG+\tau_{k,l}) \times 1}^T, & [s_k(T_s), \ldots, s_k(T_s(SG-\tau_{k,l}))] \end{pmatrix}^T \in \mathbb{R}^{MSG \times M}$$

(6.24)

is a block diagonal sampled spreading sequence. Effectively M separate filters are adapted in (6.23) as was proposed in [20].

6.5.3 Griffiths' Algorithm

In Griffiths' algorithm [35, pp. 85-86], it is assumed that the cross-correlation vector in (6.17) $\Sigma_{\mathbf{r}d_{k,l}}$ is known. Furthermore the instantaneous estimate for the covariance $\hat{\Sigma}_{\bar{\mathbf{r}}}$ is used, i.e., $\Sigma_{\bar{\mathbf{r}}} = \bar{\mathbf{r}}^{(n)} \bar{\mathbf{r}}^{H(n)}$. In this case, the cross-correlation is $\Sigma_{\bar{\mathbf{r}}d_{k,l}} = E[|c_{k,l}|^2] \bar{\mathbf{s}}_{k,l}$. Hence Griffiths' algorithm results in adaptation according to

$$\mathbf{x}_{k,l}^{(n+1)} = \mathbf{x}_{k,l}^{(n)} + 2\mu_{k,l}^{(n)} \left(E[|c_{k,l}|^2] \mathbf{F}_{k,l} \mathbf{1}_M - \bar{\mathbf{r}}_{k,l}^{*(n)} (\bar{\mathbf{s}}_{k,l} + \mathbf{x}_{k,l}^{(n)})^H \bar{\mathbf{r}}^{(n)}\right). \quad (6.25)$$

In practice the energies of multipath components ($E[|c_{k,l}|^2]$) are not known and must be estimated.

6.5.4 Constant Modulus Algorithm

The optimization criterion with the constant modulus algorithms [35, Chapt. 6] is $\mathrm{E}[(|y_{k,l}|^2 - \delta)^2]$, where δ is the so-called constant modulus (CM), which is set according to the received signal power, i.e, $\delta = \mathrm{E}[|c_{k,l}|^2]$ or $\delta^{(n)} = |c_{k,l}^{(n)}|^2$. Using the CM algorithm, it is possible to avoid the use of the data decisions in the reference signal in the adaptive LMMSE-RAKE receiver by taking the absolute value of the estimated channel coefficients ($|\hat{c}_{k,l}^{(n)}|$) in adapting the receiver [15]. In the precombining LMMSE receiver framework, the cost function for BPSK data modulation is $\mathrm{E}[||\hat{\mathbf{h}}|^2 - |\mathbf{h}|^2|^2]$. The stochastic approximation of the gradient for the CM criterion is $\nabla_{k,l} = (|y_{k,l}^{(n)}|^2 - |\hat{c}_{k,l}^{(n)}|^2)\bar{\mathbf{r}}^{(n)}\bar{\mathbf{r}}^H{}^{(n)}\mathbf{w}_{k,l}$. Hence, the constant modulus algorithm can be expressed as

$$\mathbf{x}_{k,l}^{(n+1)} = \mathbf{x}_{k,l}^{(n)} - 2\mu_{k,l}^{(n)} y_{k,l}^{*(n)}(|y_{k,l}^{(n)}|^2 - |\hat{c}_{k,l}^{(n)}|^2)\bar{\mathbf{r}}^{(n)}. \quad (6.26)$$

6.5.5 Constrained LMMSE-RAKE, Griffiths' and Constant Modulus Algorithms

The adaptive LMMSE-RAKE (6.22), the Griffiths' (6.25) and the constant modulus algorithm (6.26) contain no constraints. By applying the orthogonality constraint $\bar{\mathbf{s}}_{k,l}^T \mathbf{x}_{k,l}^{(n)} = 0$ to each of these algorithms, an additional term [15] $\bar{\mathbf{s}}_{k,l}^T \mathbf{x}_{k,l}^{(n)} \bar{\mathbf{s}}_{k,l}$ is subtracted from the new update $\mathbf{x}_{k,l}^{(n+1)}$ at every iteration. The constrained LMMSE-RAKE receiver becomes then

$$\mathbf{x}_{k,l}^{(n+1)} = \mathbf{x}_{k,l}^{(n)} + 2\mu_{k,l}^{(n)}\left(\hat{c}_{k,l}^{(n)}\hat{b}_k^{(n)} - y_{k,l}^{(n)}\right)^* \bar{\mathbf{r}}^{(n)} - \bar{\mathbf{s}}_{k,l}^T \mathbf{x}_{k,l}^{(n)} \bar{\mathbf{s}}_{k,l}. \quad (6.27)$$

The Griffiths' and the constant modulus algorithms can be defined in a similar fashion.

6.5.6 Blind Least Squares (LS) Receivers

All blind adaptive algorithms described in the previous section are based on the gradient of the cost function. In actual adaptive algorithms the gradient is estimated, i.e., the expectation in the optimization criterion is not taken but is replaced in most cases by some stochastic approximation. This means that the optimization criterion is only approximately satisfied. In fact, the stochastic approximation is accurate for only small stepsizes μ. This results in rather slow convergence, which may be intolerable in practical applications. Another drawback with the blind adaptive receivers presented above is the delay estimation. Those receiver structures as such support only conventional delay estimation based on matched filtering (MF). The MF based delay estimation is sufficient for the downlink receivers in systems with an unmodulated pilot channel since the zero-mean MAI can be averaged out if the rate of fading is sufficiently low. However, all emerging CDMA systems do not have the pilot channel. In such cases, it would be beneficial to use some near-far resistant delay estimators.

One possible solution to both the convergence and the synchronization problems is based on blind linear least squares receivers [23], which are based on the least squares cost function

$$J_{[LS]k,l} = \sum_{j=n-N+1}^{n} \left(c_{k,l}^{(j)} b_k^{(j)} - \mathbf{w}_{k,l}^{H(n)} \bar{\mathbf{r}}^{(j)} \right)^2, \qquad (6.28)$$

where N is the observation window in symbol intervals. The LS estimate for the filter weights can be written as

$$\mathbf{w}_{k,l}^{(n)} = \hat{\mathbf{\Sigma}}_{\bar{\mathbf{r}}}^{-1(n)} \bar{\mathbf{s}}_{k,l}. \qquad (6.29)$$

The estimate for the covariance matrix is called the sample-covariance matrix, which can be expressed as

$$\hat{\mathbf{\Sigma}}_{\bar{\mathbf{r}}}^{(n)} = \sum_{j=n-N+1}^{n} \bar{\mathbf{r}}^{(j)} \bar{\mathbf{r}}^{H(j)}. \qquad (6.30)$$

Analogous to the MOE criterion, the LS criterion can be modified as

$$J_{[LS']k,l} = \sum_{j=n-N+1}^{n} \left(\mathbf{w}_{k,l}^{H(n)} \bar{\mathbf{r}}^{(j)} \right)^2, \qquad (6.31)$$

which results in the receiver coefficients

$$\mathbf{w}_{k,l}^{(n)} = \frac{\hat{\mathbf{\Sigma}}_{\bar{\mathbf{r}}}^{-1(n)} \bar{\mathbf{s}}_{k,l}}{\bar{\mathbf{s}}_{k,l}^{T} \hat{\mathbf{\Sigma}}_{\bar{\mathbf{r}}}^{-1(n)} \bar{\mathbf{s}}_{k,l}}. \qquad (6.32)$$

The adaptation of the blind LS receiver means updating the inverse of the sample-covariance. It is clear that the blind adaptive LS receiver is significantly more complex than the stochastic gradient based blind adaptive receivers. Recursive methods such as the RLS algorithm [34] for updating the inverse and iteratively finding the filter weights are known. Also the methods based on eigen-decomposition of the covariance matrix have been proposed to avoid explicit matrix inversion [22]. Using the matrix inversion lemma [36], the inverse of the sample-covariance can be updated according to [37, p. 407]

$$\hat{\mathbf{\Sigma}}_{\bar{\mathbf{r}}}^{-1}(n) = \frac{1}{\gamma} \left(\hat{\mathbf{\Sigma}}_{\bar{\mathbf{r}}}^{-1}(n-1) - \frac{\hat{\mathbf{\Sigma}}_{\bar{\mathbf{r}}}^{-1}(n-1) \bar{\mathbf{r}}(n) \bar{\mathbf{r}}^{H}(n) \hat{\mathbf{\Sigma}}_{\bar{\mathbf{r}}}^{-1}(n-1)}{\gamma + \bar{\mathbf{r}}^{H}(n) \hat{\mathbf{\Sigma}}_{\bar{\mathbf{r}}}^{-1}(n-1) \bar{\mathbf{r}}(n)} \right). \qquad (6.33)$$

where $0 < \gamma < 1$ is the so-called forgetting factor.

6.6 DELAY ACQUISITION IN THE PRECOMBINING LMMSE RECEIVER

Adaptive precombining LMMSE receivers require explicit information on the delays of the desired user. Quite a many different approaches to delay estimation have been applied to multiuser CDMA communications since delay estimation has been found

158   CDMA TECHNIQUES FOR THIRD GENERATION MOBILE SYSTEMS

to be one of the most demanding problems in practical systems. Maximum-likelihood type delay estimators [38, 39, 40] for multiuser CDMA systems are very complex. One popular research topic in recent years has been subspace based delay estimators [41, 42] derived originally for spectrum estimation [43, 44]. However, it has been shown that the subspace based delay estimators are not efficient in highly loaded systems [45]. The blind least squares single-user receivers can be used for delay acquisition as was discussed in Section 6.5.6. By using the minimum variance principle in deriving the delay estimator, the resulting estimator is of the same form as the blind MOE receiver [26].

6.6.1 Minimum Variance Delay Estimator

Once the inverse for the sample-covariance matrix is obtained, the blind receiver can be solved if the delay for the signal component of interest is known. As was indicated earlier, the blind LS receiver structure can be used for delay estimation as well. Specifically, the minimum-variance delay estimator developed orginally for beamforming [37] gives the blind MOE receiver for the correct delay. In delay estimation, the power at the estimator output is maximized for the correct delay. Clearly, this property can be used in delay estimation.

The average output power of the minimum variance or minimum output energy receiver is given by

$$\mathcal{P}_{[MV]}(\bar{\mathbf{s}}_{k,l}) = \mathrm{E}\big[(\mathbf{w}_{k,l}^T \bar{\mathbf{r}})^2\big] = (\bar{\mathbf{s}}_{k,l}^T \Sigma_{\bar{\mathbf{r}}}^{-1} \bar{\mathbf{s}}_{k,l})^{-1}. \tag{6.34}$$

As can be seen, the output power is a function of the sampled spreading sequence. This can be utilized in single-user delay acquisition in the following fashion: given a user with the sampled spreading sequence $\bar{\mathbf{s}}_k$, find the delay for the the sequence that maximizes the receiver output power. More formally,

$$\hat{\tau}_{k,l} = \arg\max_{\tau_{k,l}} \mathcal{P}_{[MV]}\big(\bar{\mathbf{s}}_{k,l}(\tau_{k,l})\big), \tag{6.35}$$

where $\bar{\mathbf{s}}_{k,l}(\tilde{\tau}_{k,l}) = \bar{\mathbf{s}}_k = [\mathbf{0}_{DSG+\tau_{k,l}}^T, \mathbf{s}_k^T, \mathbf{0}_{DSG-\tau_{k,l}}^T]^T$ and $\tau_{k,l}$ is time-discretized delay in samples. In practice, the covariance matrix $\Sigma_{\bar{\mathbf{r}}}$ is not known and it must be estimated, as was discussed in Section 6.5.6.

The sample-covariance estimation interval must be quite long to obtain small enough steady-state error for the receiver processing windows larger than one symbol. It is therefore beneficial to use as small a receiver processing window size as possible. For a processing window size of one symbol interval, the minimum variance delay estimator must be defined in a different way in the case there is no pilot channel available, i.e., the signal for the desired user contains data modulation. In such a case, the signature sequence must be split into two parts corresponding to two different symbol intervals such that the upper portion of the sequence is $\mathbf{s}_{[u]k}(\tilde{\tau}_{k,l}) = \big[s_k(T_s), \ldots, s_k(T_s \tilde{\tau}_{k,l}), \mathbf{0}_{(SG-\tilde{\tau}_{k,l}) \times 1}^T\big]^T$, and the lower part is $\mathbf{s}_{[l]k}(\tilde{\tau}_{k,l}) = \big[\mathbf{0}_{(\tilde{\tau}_{k,l}) \times 1}^T, s_k(T_s(\tilde{\tau}_{k,l}+1)), \ldots, s_k(T_s SG)\big]^T$. The minimum-variance delay estimation problem will be then

$$\hat{\tau}_{k,l} = \arg\max_{\tilde{\tau}_{k,l}} \left(\mathbf{s}_{[u]k}(\tilde{\tau}_{k,l})^{\mathrm{T}} \hat{\boldsymbol{\Sigma}}_{\tilde{\mathrm{F}}}^{-1} \mathbf{s}_{[u]k}(\tilde{\tau}_{k,l}) + \mathbf{s}_{[l]k}(\tilde{\tau}_{k,l})^{\mathrm{T}} \hat{\boldsymbol{\Sigma}}_{\tilde{\mathrm{F}}}^{-1} \mathbf{s}_{[l]k}(\tilde{\tau}_{k,l}) \right)^{-1}. \quad (6.36)$$

If the receiver processing window is large enough the scheme (6.35) can be used even if there is no pilot channel available. In fact, a sufficient condition for using it in delay acquisition is that the receiver processing window is two symbol intervals long. This guarantees that at least one full symbol is inside the processing window. Another possibility would be to estimate the covariance matrix over one symbol interval with all trial delays while keeping the sampled spreading sequence fixed. This approach, however, is computationally demanding and not very practical.

6.6.2 Subspace Based Delay Estimators

The subspace delay estimators [41] are based on the eigenvalue decomposition of the covariance matrix [34, p. 166]: $\boldsymbol{\Sigma}_{\tilde{\mathrm{F}}} = \mathbf{U}\boldsymbol{\Lambda}\mathbf{U}^{\mathrm{H}}$, where $\mathbf{U} = [\mathbf{u}_1, \ldots, \mathbf{u}_{SGM}] \in \mathbb{C}^{SGM}$ is a matrix of the eigenvectors ($\mathbf{u}_i$, $i \in \{1, \ldots, SGM\}$) and $\boldsymbol{\Lambda} = \mathrm{diag}(\lambda_i) \in \mathbb{R}^{SGM}$ is a diagonal matrix of the eigenvalues ($\lambda_i, i \in \{1, \ldots, SGM\}$) of the covariance matrix. The connection between the minimum variance, MUSIC and the eigenvector based estimators can be explained by different estimators for the inverse of the sample-covariance matrix. The inverse of the covariance matrix can be presented by using the eigenvectors and the corresponding eigenvalues [34]. By doing so, the inverse of the covariance matrix for the minimum variance delay estimator can be written as

$$\boldsymbol{\Sigma}_{\tilde{\mathrm{F}}}^{-1} = \mathbf{U}\boldsymbol{\Lambda}^{-1}\mathbf{U}^{\mathrm{H}} = \sum_{i=1}^{SGM} \lambda_i^{-1} \mathbf{u}_i \mathbf{u}_i^{\mathrm{H}}. \quad (6.37)$$

The eigenvector based method [46, 37] uses only the $SGM - KL$ smallest eigenvalues and corresponding eigenvectors, which span the noise subspace. Hence, the inverse of the covariance matrix is estimated as

$$\hat{\boldsymbol{\Sigma}}_{\tilde{\mathrm{F}}[EV,n]}^{-1} = \sum_{i=KL+1}^{SGM} \lambda_i^{-1} \mathbf{u}_i \mathbf{u}_i^{\mathrm{H}}, \quad (6.38)$$

where $\lambda_i \geq \lambda_j, \forall i < j$. The MUSIC algorithm [44] neglects the weights for the eigenvectors, i.e., all eigenvalues are set to unity. The inverse of the covariance matrix with the MUSIC algorithm is approximated as

$$\hat{\boldsymbol{\Sigma}}_{\tilde{\mathrm{F}}[MUSIC,n]}^{-1} = \sum_{i=KL+1}^{SGM} \mathbf{u}_i \mathbf{u}_i^{\mathrm{H}}. \quad (6.39)$$

Is is also possible to use the signal-plus-noise subspace [37, p. 377] (referred to as the signal subspace in the sequel), which results in the covariance matrix inverse estimates as

$$\hat{\Sigma}_{\bar{\mathbf{r}}[MUSIC,s]}^{-1} = \sum_{i=1}^{KL} \mathbf{u}_i \mathbf{u}_i^H. \tag{6.40}$$

The output power must be minimized when using the signal subspace inseted of the noise subspace.

6.6.3 Conventional Delay Estimators

The conventional non-coherent delay estimator is simply obtained by setting $\mathbf{w}_{k,l} = \bar{\mathbf{s}}_{k,l}$. The output power of the conventional matched filter is then

$$\mathcal{P}_{[MF]}(\bar{\mathbf{s}}_{k,l}) = \mathrm{E}\big[(\bar{\mathbf{s}}_{k,l}^T \bar{\mathbf{r}})^2\big] = \bar{\mathbf{s}}_{k,l}^T \Sigma_{\bar{\mathbf{r}}} \bar{\mathbf{s}}_{k,l}. \tag{6.41}$$

The conventional delay estimator with an extended despreading interval is obtained by modifying the sampled signature sequence so that the sequence is repeated several times, i.e.,

$$\begin{aligned}\bar{\mathbf{s}}_{k,l} &= \Big[[s_k(T_s(SG - \tau_{k,l} + 1)), \ldots, s_k(T_sSG)], [\mathbf{1}_{(M'-1)} \otimes \mathbf{s}_k]^T, \\ &\quad [s_k(T_s), \ldots, s_k(T_s(SG - \tau_{k,l}))]\Big]^T \in \mathbb{R}^{M'SG},\end{aligned} \tag{6.42}$$

where M' is the so-called coherent (or extended) integration time in symbols.

6.7 DELAY TRACKING IN THE PRECOMBINING LMMSE RECEIVERS

The principles of the blind LS receiver of Section 6.5.6 will be extended to the delay tracking in this section. The motivation to use the minimum variance or MOE principle in delay tracking is, of course, to improve the tracking performance. In delay acquisition, the received signal is used to construct an estimate of the covariance matrix. Once the inverse of the covariance matrix exists, the delay can be acquired by finding the maximum output power for the MOE based receiver. In delay tracking, the inverse of the covariance matrix already exists and it is updated as the channel changes. The approach taken in delay tracking is to pre-process the received signal by multiplying it by the inverse of the sample-covariance matrix in order to suppress interference due to other users. After the pre-processing, the received signal is despread and a delay-locked loop (DLL) is used to track the best timing position [47]. Since the early and the late phased sequences do not match the best timing position (on-time) some of the near-far resistance will be lost[3]. However, if the early-late difference is not very large, the loss should be tolerable.

The signal at the output of the despreading device during the nth symbol interval is

$$y_{k,l}^{(n)}(\hat{\tau}_{k,l}^{(n)}) = \bar{\mathbf{s}}_{k,l}^T(\hat{\tau}_{k,l}^{(n)}) \hat{\Sigma}_{\bar{\mathbf{r}}}^{-1} \bar{\mathbf{r}}^{(n)}, \tag{6.43}$$

---

[3]The sample-covariance is estimated for the on-time code phase.

where $\hat{\tau}_{k,l}^{(n)}$ is the time-discretized delay estimate. The delay estimator with an observation interval $\mathcal{T}$ maximizes the function

$$\hat{\tau}_{k,l}^{(n)} = \arg\max_{\tau_{k,l}} \sum_{j=n-\mathcal{T}+1}^{n} \left(y_{k,l}^{(j)}(\tau_{k,l})\right)^2. \quad (6.44)$$

Instead of direct maximization, this can be solved by setting the derivative of (6.44) to zero. The DLLs are based on the approximation of the derivative by the early and late difference. In practice the received signal is despread with early and late phased sequences and the difference produced by them is driven towards zero in the feedback loop [47]. By applying the early-late approximation for the derivative results in the timing error signal

$$\begin{aligned} e_{k,l}^{(n)} &= \left(\bar{\mathbf{s}}_{k,l}^{T}(\hat{\tau}_{k,l}^{(n)}+\Delta)\hat{\boldsymbol{\Sigma}}_{\bar{\mathbf{r}}}^{-1}\bar{\mathbf{r}}^{(n)}\right)^{2} - \left(\bar{\mathbf{s}}_{k,l}^{T}(\hat{\tau}_{k,l}^{(n)}-\Delta)\hat{\boldsymbol{\Sigma}}_{\bar{\mathbf{r}}}^{-1}\bar{\mathbf{r}}^{(n)}\right)^{2} \\ &= \left(y_{k,l}^{(n)}(\hat{\tau}_{k,l}^{(n)}+\Delta)\right)^{2} - \left(y_{k,l}^{(n)}(\hat{\tau}_{k,l}^{(n)}-\Delta)\right)^{2}, \end{aligned} \quad (6.45)$$

where 2Δ is the early-late difference in chip intervals. Due to the squaring operation in (6.45), the resulting DLL is non-coherent. The block diagram of the proposed improved DLL is given in Figure 6.4.

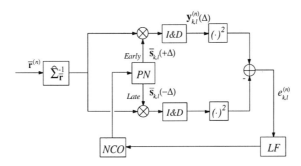

Figure 6.4 The block diagram of the MV-DLL suitable for the blind LS receivers.

6.7.1 Tracking Error Analysis

The exact analysis of the feedback loops leads to difficult mathematical problems [48]. In particular, the dependence of the loop noise power spectral density on the estimated parameter value causes problems in the analysis. To simplify the task, the so-called linear analysis methods are used [49], where the loop noise power spectrum is evaluated at the equilibrium point, i.e., $e_{k,l}=0$. Furthermore, if the loop bandwidth [50] is narrow, the loop noise power spectrum can be assumed to be constant within the loop band in many cases. With these assumptions, the code tracking error variance can be presented as [48, 49, 51, 52]

$$\mathrm{var}(\hat{\tau}) = \frac{1}{(s'(0))^2} \frac{\sigma^2}{E_b}\left(1+\zeta\frac{\sigma^2}{E_b}\right)2B_L T, \quad (6.46)$$

where $s'(0)$ is the slope of the s-curve at the origin, ζ is the squaring-loss obtained from the pulse shape after pre-processing and $2B_L T$ is the two-sided loop bandwidth.

Using this simplified analysis the aim is to show that the tracking error variance can be made smaller by pre-processing the received signal by multiplying the received signal vector by the inverse of the covariance matrix prior to squaring. The analysis will be carried out in a two path fading channel assuming that the multipath components are distinguishable and do not cause any distortion to the s-curves. This assumption means that the fading processes are uncorrelated and the loop filter is narrow enough to average the fading, i.e., the s-curves are completely stable. For more detailed analysis in fading multipath channels, one could use the methods presented in [53]. It is also assumed that the loop noise is white and Gaussian. Therefore, the multiple-access interference is modelled as additional white Gaussian noise. The noise variance in a multipath channel with equal energy paths for the improved DLL will be

$$\sigma^2 = \sum_{k'=1, k' \neq k}^{K} \frac{A_{k'}^2}{A_k^2} \sum_{l'=1}^{L} \bar{\mathbf{s}}_{k',l'}^T \Sigma_{\bar{\mathbf{F}}}^{-1} \bar{\mathbf{s}}_{k,l}(\Delta)$$
$$+ \sum_{l'=1, l' \neq l}^{L} \left(\bar{\mathbf{s}}_{k,l'}^T \Sigma_{\bar{\mathbf{F}}}^{-1} \bar{\mathbf{s}}_{k,l}(\Delta)\right)^2 + \sigma^2 \left(\Sigma_{\bar{\mathbf{F}}}^{-1} \bar{\mathbf{s}}_{k,l}(\Delta)\right)^2. \quad (6.47)$$

For the conventional DLL, $\Sigma_{\bar{\mathbf{F}}}^{-1} = \mathbf{I}$. The Gaussian approximation is degraded slightly due to the timing offset of the spreading sequence of the desired component $\bar{\mathbf{s}}_{k,l}(\Delta)$.

6.8 RESIDUAL BLIND INTERFERENCE SUPPRESSION IN PIC RECEIVERS

If LMMSE receivers are the most promising for the single-user downlink receivers, parallel interference cancellation receivers are for the multiuser uplink receivers. The capacity and the performance of the PIC receivers has been found to result in superior performance in comparison to other practical multiuser receivers in frequency-selective fading channels [2].

The parallel interference cancellation relies on the knowledge of the number of users and propagation paths. In practice, the exact number of users and paths is not known exactly, e.g., due to inter-cell interference, unknown propagation paths or new users trying to connect to the base station. As a result, there will be some residual interference even with perfect cancellation of the known signal components. In many cases, the inter-cell interference can be large enough to significantly degrade the performance of the PIC receivers. For that reason, residual interference suppression is crucial to guarantee that the PIC receivers can operate reliably. In this section, one possibility for residual interference suppression in PIC receivers is considered. The approach taken is to combine the precombining LMMSE and PIC receivers [11].

6.8.1 Parallel Interference Cancellation Receivers

The idea of interference cancellation (IC) receivers is to estimate the multiple-access and multipath induced interference and then subtract the interference estimate from the received signal. There are several approaches to estimating interference leading to different IC techniques. The interference can be canceled simultaneously from all users leading to parallel interference cancellation (PIC), or on a user-by-user basis leading to successive (or serial) interference cancellation (SIC). The interference cancellation utilizing tentative data decisions is called hard decision (HD) interference cancellation and it requires explicit channel estimation. The soft decision (SD) interference cancellation utilizes only the composite signal of the data and the channel coefficient and no explicit channel estimation is needed. Usually the interference is estimated iteratively in several receiver stages, which leads to the so-called multistage receivers. The multistage HD-PIC receiver has been proposed and analyzed for AWGN channels in [54, 55, 56, 57, 58, 59], for slowly fading channels in [60, 61, 62, 63, 64, 65, 66], and for relatively fast fading channels in [67, 68, 2, 69]. The HD-PIC receivers for transmissions with diversity encoding has been analyzed in [70], for systems with multiple data rates in [71], and with trellis-coded CDMA systems in AWGN channels in [72]. The application of the HD-PIC to multiuser delay estimation in relatively fast fading channels has been considered in [68], and the effect of delay estimation errors on the performance of the HD-PIC receiver in [73], and to the SD-PIC receiver in [74, 75]. The SD-PIC receivers with linear data-amplitude product estimation for slowly fading channels have been considered in [74], and for multi-cellular systems in [76]. The SD-PIC receivers with soft nonlinearity have been considered for AWGN channels in [77]. A PIC receiver cancelling interference partially in purpose has been proposed in [78]. In [79, 80, 81, 82] the expectation maximization (EM) [83] and the space alternating generalized EM (SAGE) algorithms [84] have been used in deriving different forms of IC receivers. The serial interference cancellation is performed on a user-by-user basis [85, 86]. SD-SIC has been considered in [85, 87], and HD-SIC in [88, 89, 90, 86]. The SIC for multirate CDMA communications has been studied in [91, 92]. The effect of delay estimation errors to the SD-SIC has been considered in [75] and to the HD-SIC in [93]. The combination of the PIC and the SIC receivers has been studied in [94]. In [95] it was shown that the SD-PIC receiver with an infinite number of stages is actually a decorrelating receiver. Interference cancellation based on user grouping has been considered for serial interference cancellation in [96], for parallel interference cancellation in [97, 98], groupwise SIC for multiple data rates in [99], and for generic multiuser receivers in [100, 101].

6.8.2 Hybrid LMMSE-PIC Receiver

There are two possibilities to implement PIC receivers in practice. Interference cancellation can take place either before matched filtering or after that. The receivers are mathematically equivalent and have the same performance. However, the receiver performing interference cancellation from the wideband signal can be used more easily together with blind adaptive schemes and will be considered in the reminder of this section.

The received signal vector after interference cancellation at the pth stage for the desired signal component is given by

$$\bar{\mathbf{r}}^{(n)}_{[PIC]k,l}(p) = \bar{\mathbf{r}}^{(n)} - \hat{\mathbf{\Psi}}^{(n)}_{[\mathbf{r}]k,l}, \qquad (6.48)$$

164   CDMA TECHNIQUES FOR THIRD GENERATION MOBILE SYSTEMS

where $\hat{\boldsymbol{\Psi}}_{[r]k,l}^{(n)}$ is the wideband interference estimate for the kth user's lth path. In the standard PIC receiver the signal is despread by matched filtering

$$y_{[PIC]k,l}^{(n)}(p) = \bar{\mathbf{s}}_{k,l}^{T} \bar{\mathbf{r}}_{[PIC]k,l}^{(n)}(p). \tag{6.49}$$

The matched filters can be replaced with the precombining LMMSE filters. The resulting hybrid LMMSE-PIC receiver processes each multipath component according to

$$y_{[L-PIC]k,l}^{(n)}(p) = \bar{\mathbf{s}}_{k,l}^{T} \boldsymbol{\Sigma}_{\bar{\mathbf{r}}_{[PIC]k,l}(p)}^{-1} \bar{\mathbf{r}}_{[PIC]k,l}^{(n)}(p) = \mathbf{w}_{[PIC]k,l}^{T}(p) \bar{\mathbf{r}}_{[PIC]k,l}^{(n)}(p). \tag{6.50}$$

The techniques presented in earlier sections can be used to iteratively calculate the residual interference suppression filters $\mathbf{w}_{[PIC]k,l}(p)$ for every cancellation stage.

6.9 NUMERICAL EXAMPLES

Some computer simulations and numerical analysis results are presented to demonstrate the usefulnes of the precombining LMMSE receivers in receivers operating over fading channels. The main parameters used are the following: carrier frequency 1.8 GHz, symbol rate 16 kbits/s, 31 chip Gold code, and rectangular chip shape. A synchronous downlink case with equal energy two-path ($L = 2$) Rayleigh fading channel with vehicle speeds of 40 km/h and maximum delay spreads of 10 chip intervals are considered. Channel estimation based on an unmodulated pilot channel and a moving average estimator of length 11 symbol intervals is used in the simulations. The number of users is 1 - 30 including the unmodulated pilot channel. The average energy is the same for the pilot channel and user data channels.

6.9.1 BEP Analysis

The validity of the Gaussian approximation [33] in frequency-selective fading channels was studied in order to simplify further the analysis. The results for both the conventional RAKE receiver and the precombining LMMSE receiver are presented in Figure 6.5. The approximation is very accurate for the precombining LMMSE receiver in multipath fading channels, whereas slightly optimistic for the conventional RAKE receiver, as was expected. Also the simulation results suggest that the Gaussian approximation is accurate for the precombining LMMSE receiver (Figure 6.6). According to the results, the precombining LMMSE receiver can significantly improve the performance of the conventional RAKE receiver. Furthermore, the conventional RAKE receiver is known to be very sensitive to the near-far problem, and the potential gains are greater.

The BER sensitivity to timing errors was studied by using both the Gaussian approximation and the exact analysis. It should be emphasized that the rectangular chip waveform was used and hence, the results are slightly pessimistic. Both multipath components had the same fixed absolute delay error. The results are presented in Figure 6.7. Firstly, the Gaussian approximation is quite accurate for the precombining LMMSE receivers with delay errors at least in the case of no near-far problem. The results clearly indicate that the LMMSE receivers are not necessarily more sensitive

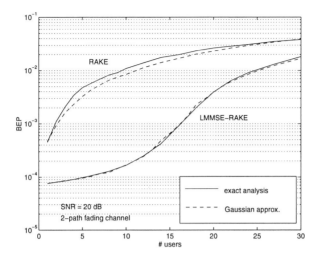

Figure 6.5 Bit error probabilities (BEP) as a function of the number of users for the conventional RAKE and the precombining LMMSE (LMMSE-RAKE) receivers with exact analysis and Gaussian approximation in a two-path fading channel at vehicle speeds of 40 km/h and average SNR of 20 dB.

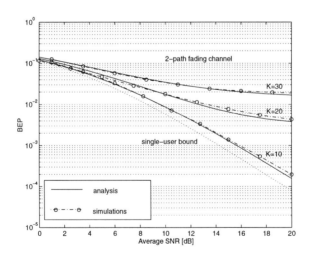

Figure 6.6 Simulated and analytical bit error probabilities as a function of the average signal-to-noise ratio for the precombining LMMSE receiver in two-path fading channels at vehicle speed 40 km/h with different numbers of users.

to delay estimation errors than other receivers, such as the conventional RAKE receiver. Hence, the timing accuracy requirements are lessened for the precombining

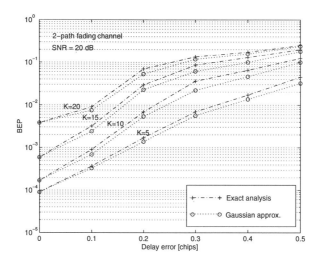

Figure 6.7 Bit error probabilities (BEP) as a function of the delay error for the precombining LMMSE receiver with exact analysis and the Gaussian approximation in two path fading channels at vehicle speeds 40 km/h and average SNR of 20 dB.

LMMSE receivers, and in particular, for the blind LS receivers. The timing accuracy requirements are of the order of 0.1 chips not to degrade performance significantly.

6.9.2 BEP Analysis for the Downlink of the FRAMES WCDMA Concept

The European third generation mobile communications proposal [102] based on wideband CDMA (WCDMA) techniques has mainly been developed within the European FRAMES project AC090. In the section, some of the results of the FRAMES WCDMA downlink performance analysis presented in [103, 104] are presented both for the conventional RAKE and the precombining LMMSE receivers.

The channel model used is a two path fading channel (equal energies) with maximum delay spread of 2 μs and velocity of 5 km/h for the spreading factors $G=2$ and 4; delay spread of 7 μs and velocity of 50 km/h for $G=8, 16, 32$. The actual delay values were randomly selected. The LMMSE detector length in symbol intervals was 17 symbols for $G=2$; 9 symbols for $G=4, 8$; 5 symbols for $G=16$; and 3 symbols for $G=32$. It should be noted that since the length of the optimum LMMSE detector is infinite, longer detectors would yield even better performance. Root raised cosine filtering with a roll-off factor of 0.22 was used and the number of samples per chip was 4. The analysis was perfomed by using 10000 different bit patterns of length 17 - 3 bits depending on the detector length, the carrier phase for the second multipath component was randomly selected for each bit pattern. The multipath delays were changed after every 100 bit patterns. The data channel (the Walsh code) was selected randomly for each bit pattern. The control channels [102] were not included in the analysis, which means that the data modulation scheme for the single data channel case was BPSK.

The bit error probability for QPSK modulation is the same as for BPSK, and hence, the results for BPSK can be applied to the QPSK case. The channel bit rates studied were 2.048 Mbit/s - 128 kbit/s ($G = 2 - 32$) for the single code channel. Neither forward error correcting (FEC) coding nor power control was included in the analysis.

The bit error probabilities as a function of the average SNR for different number of active users in single data channel case are presented in Figure 6.8 for the spreading factor $G = 4$. In the single-user case, the BEP of the conventional RAKE saturates with small spreading factors due to bad average autocorrelation properties of combined the Walsh code and the scrambling code. If the number of users is half of the spreading factor, the BEP of the conventional RAKE is the same regardless of the spreading factor. If the target raw BEP at 20 dB is 10^{-2}, the capacity with the conventional RAKE is half of the spreading factor, in the case of no near-far problem. The capacity with the precombining LMMSE detector is roughly 100 % of the spreading factor. Similarly the required SNR to support $G/2$ users is 20 dB with the conventional RAKE receiver and 10 dB with the LMMSE-RAKE receiver. RAKE receivers tolerate relatively small near-far ratios with small spreading factors (see Figure 6.9), which means that the power differences between data channels should be relatively small.

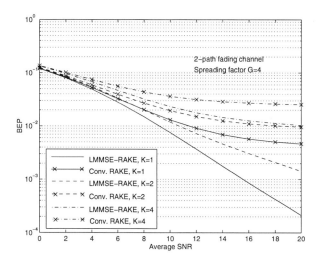

Figure 6.8 Bit error probabilities as a function of the average SNR for the conventional RAKE receiver and the precombining LMMSE (LMMSE-RAKE) receiver with different number of users ($K = 1, 2, 4$) in a two-path Rayleigh fading channel with maximum delay spreads of 2 μs. The data modulation is BPSK at a rate of 1.024 Mbit/s ($G=4$). The energy of all users is the same and no channel coding is assumed.

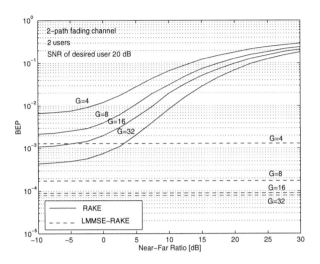

Figure 6.9 Bit error probabilities as a function of the near-far ratio for the conventional RAKE and the precombining LMMSE (LMMSE-RAKE) receiver with different spreading factors (G) in a two-path Rayleigh fading channel with maximum delay spreads of 2 μs for $G=4$, and 7 μs for other spreading factors. The average signal-to-noise ratio is 20 dB for the user of interest. The data modulation is BPSK and the number of users is 2. Data rates vary from 2.048 Mbit/s to 128 kbit/s. No FEC coding is assumed.

6.9.3 Comparisons of the Blind Adaptive Receivers

The adaptive algorithms are usually compared in terms of the rate of convergence and the steady-state MSE. The excess mean squared error[4] was simulated as a function of iterations for the blind adaptive receivers in a two-path fading channel with SNR of 20 dB with 10 and 20 active users. The results for the constant modulus, the Griffiths', the blind MOE, and the LMMSE-RAKE receivers are presented in Figures 6.10 and 6.11 with stepsizes of $\mu = 10^{-1}$, 100^{-1} (the NLMS algorithm was used with $\mu_{k,l}^{(n)} = \mu\big((2D+1)SG\bar{\mathbf{r}}_{k,l}^{H(n)}\bar{\mathbf{r}}_{k,l}^{(n)}\big)^{-1}$).

Based on the results, all schemes except the adaptive LMMSE-RAKE require a rather small stepsize for small enough excess MSE. With decreasing step-size, the excess MSE of all schemes studied approaches zero and their performance is equal.

---

[4] The excess MSE is $J_{[ex]k,l}^{(n)} = \mathrm{tr}[\Sigma_{\bar{\mathbf{r}}}\Gamma^{(n)}]$, where $\Gamma^{(n)} = \mathrm{E}\Big[(\mathbf{w}_{[MSE]k,l} - \mathbf{w}_{k,l}^{(n)})(\mathbf{w}_{[MSE]k,l} - \mathbf{w}_{k,l}^{(n)})^H\Big]$ is the tap weight error covariance matrix at iteration n. The filter coefficients have been normalized at every iteration when computing the excess MSE such that $\mathbf{w}_{k,l}^{H(n)}\mathbf{w}_{k,l}^{(n)} = 1$.

Using a filtered estimate (a recursive estimator with a forgetting factor of 0.9975 was used to average the absolute values of the channel estimates) of the received signal power instead of the instantaneous value did not improve the convergence of the constant modulus algorithm significantly. Neither did the addition of the constraint to the unconstrained algorithms. Therefore, these results are not shown in the figures. Based on the results, it may be concluded that the adaptive LMMSE-RAKE converges faster than the other known blind adaptive receivers based on stochastic gradient algorithms. The constant modulus algorithm requires the smallest step-size for the same final MSE and has the slowest convergence as a result. The Griffiths' algorithm and the blind MOE receivers have roughly the same convergence rate, although the latter is slightly better.

The bit error rates of different schemes were studied in a 30-user and 15-user case with different step-sizes at a SNR of 20 dB. The results are presented in Table 6.1. The performance of the diffrent algorithms is very similar with small step-sizes ($\mu = 100^{-1}$). When faster convergence is required and the step-size must be increased ($\mu = 10^{-1}$), only the adaptive LMMSE-RAKE receivers result in significant performance improvements with respect to the conventional RAKE receiver. With 15 users their performance is degraded only marginally when increasing the step-size from $\mu = 100^{-1}$ to $\mu = 10^{-1}$. The basic version of the adaptive LMMSE-RAKE receiver (6.22) has worse performance than the conventional RAKE receivers in the 30-user case. The orthogonalized (or constrained) LMMSE-RAKE receiver does not have such problems unless the step-size is too large.

The convergence curves (excess MSE) for the blind LS receivers are presented in Figure 6.12. Clearly, the blind LS receiver improves the convergence in comparison to the stochastic gradient algorithms.

Table 6.1 BERs of different blind adaptive receivers at the SNR of 20 dB in a two-path Rayleigh fading channel at vehicle speeds of 40 km/h.

Receiver	K = 30		K = 15		
	$\mu=100^{-1}$	$\mu=10^{-1}$	$\mu=100^{-1}$	$\mu=10^{-1}$	$\mu=2^{-1}$
LMMSE-RAKE	$4.5 \cdot 10^{-2}$	$3.9 \cdot 10^{-1}$	$6.3 \cdot 10^{-4}$	$7.2 \cdot 10^{-4}$	$3.0 \cdot 10^{-2}$
MOE	$2.8 \cdot 10^{-2}$	$4.2 \cdot 10^{-2}$	$6.0 \cdot 10^{-4}$	$2.1 \cdot 10^{-3}$	$9.1 \cdot 10^{-2}$
Griffiths'	$2.8 \cdot 10^{-2}$	$4.7 \cdot 10^{-2}$	$6.4 \cdot 10^{-4}$	$3.3 \cdot 10^{-3}$	$1.2 \cdot 10^{-1}$
CMA	$3.9 \cdot 10^{-2}$	$4.0 \cdot 10^{-1}$	$1.2 \cdot 10^{-3}$	$2.1 \cdot 10^{-2}$	$5.0 \cdot 10^{-1}$
CMA with av. power	$3.3 \cdot 10^{-2}$	$4.0 \cdot 10^{-1}$	$1.8 \cdot 10^{-3}$	$2.1 \cdot 10^{-2}$	$5.0 \cdot 10^{-1}$
Constr. LMMSE-RAKE	$3.2 \cdot 10^{-2}$	$4.2 \cdot 10^{-2}$	$6.3 \cdot 10^{-4}$	$6.4 \cdot 10^{-4}$	$1.9 \cdot 10^{-3}$
Constr. CMA	$3.3 \cdot 10^{-2}$	$5.0 \cdot 10^{-1}$	$6.1 \cdot 10^{-4}$	$3.8 \cdot 10^{-1}$	$5.0 \cdot 10^{-1}$
Constr. Griffiths'	$2.8 \cdot 10^{-2}$	$4.2 \cdot 10^{-2}$	$6.1 \cdot 10^{-4}$	$2.3 \cdot 10^{-3}$	$9.7 \cdot 10^{-2}$
Constr. CMA with av. power	$2.9 \cdot 10^{-2}$	$5.0 \cdot 10^{-1}$	$7.7 \cdot 10^{-4}$	$2.7 \cdot 10^{-1}$	$5.0 \cdot 10^{-1}$
Conventional RAKE	$3.1 \cdot 10^{-2}$	$3.1 \cdot 10^{-2}$	$7.1 \cdot 10^{-3}$	$7.1 \cdot 10^{-3}$	$7.1 \cdot 10^{-3}$

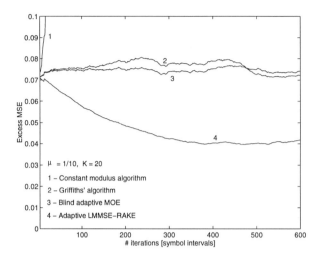

Figure 6.10 Excess mean squared error as a function of number of iterations for different blind adaptive receivers in a two-path fading channel with vehicle speeds of 40 km/h. The number of active users $K = 20$, SNR = 20 dB, $\mu = 10^{-1}$.

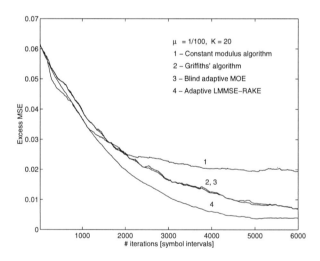

Figure 6.11 Excess mean squared error as a function of number of iterations for different blind adaptive receivers in a two-path fading channel with vehicle speed 40 km/h, the number of active users $K = 20$, SNR = 20 dB, $\mu = 100^{-1}$.

6.9.4 *Delay Acquisition Performance of the Precombining LMMSE Receivers*

The probabilities of detection for different acquisition schemes are presented in Figure 6.13 as a function of the number of users with the observation interval of 200 symbols.

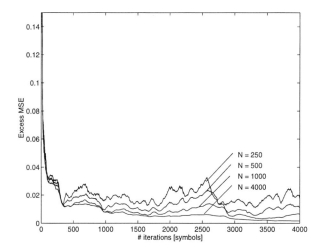

Figure 6.12 Excess mean squared error as a function of number of iterations for the one-shot blind adaptive LS receiver with different forgetting factors $(1 - 2/N$; forgetting factor $\gamma = 1 - 2/N)$ in a 10-user case at a SNR of 20 dB.

The results indicate that the minimum variance based method improves the acquisition performance in comparison to the conventional non-coherent acquisition. With larger system loads, both methods have relatively poor acquisition performance, whereas the conventional acquisition scheme with the extended despreading interval performs well in those cases. The subspace based delay estimators offer good performance when the number of users is relatively low. According to these results, the maximum number of users in the two-path channels is between 10 and 20 to obtain acceptable performance. After the number of signal components exceeds the dimensions of the sample-covariance matrix, the subspace based delay estimators do not exist at all. Similar results have been reported in [45].

6.9.5 Delay Tracking Performance

The tracking performance of the improved non-coherent DLL suitable to blind LS receivers is evaluated in this section. The parameters used are: root raised cosine waveform with roll-off of 0.5, 31 chip Gold sequences, two-path fading channels corresponding vehicle speeds of 40 km/h at 2.0 GHz, and two-sided normalized loop bandwidth of $2B_LT$=0.04.

The smaller the early-late difference, the better the interference can be suppressed in early and late correlator branches. Hence, it can be expected that the tracking performance can be improved by using the pre-processing described earlier. The tracking error analysis results are presented in Figure 6.14 for the SNR of 20 dB. According to the results, the pre-processing is useful with small early-late differences, as expected. If the inverse for the sample-covariance would be available seprately for the early and late phased correlators, the performance of the improved DLLs would be

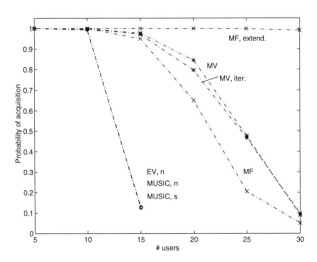

Figure 6.13 Probability of detection for different delay acquisition schemes as a function of the number of users in a two-path Rayleigh fading channel with vehicle speeds of 40 km/h, average SNR of 10 dB, and observation intervals of 200 symbols.

always better than with the conventional non-coherent DLLs. The improved tracking performance with smaller early-late differences results in decreased mean-time-to-lose-lock (MTLL), which is proportional to the s-curve area.

It is worth noting that with the sample-correlate-choose-largest (SCCL) [105] loops, the difference between the advanced and retarded correlators is usually one sample. Pre-processing with a sample-covariance matrix is well suited to such loops. Another issue to be mentioned is the small early-late differences required in multipath channels to diminish s-curve distortions caused by autocorrelation leaking. Therefore, the improved delay tracking scheme studied in this section could be applied in spread spectrum receivers operating in multipath channels.

6.9.6 Residual Blind Interference Suppression In PIC Receivers

Finally, the combination of the PIC and precombining LMMSE receivers is studied. The channel model used has been a two-path equal energy Rayleigh fading channel with the vehicle speeds 80 km/h and carrier frequency of 2.0 GHz. The data rate is assumed to be 16 kbit/s and the maximum delay spread is one symbol interval. Gold codes of length 31 chips are used, the receiver sampling rate used was 1 sample per chip. The simulations assume 4 - 32 known users at the SNR of 15 dB and one unknown user. The power of the unknown user has been -20 – 20 dB in comparison to the other users. The blind LS receiver described in Section 6.5.6 was used in the HD-PIC receiver to suppress residual interference due to unknown signal components. The sample-covariance was estimated recursively by using a forgetting factor value of $\gamma = 0.999$.

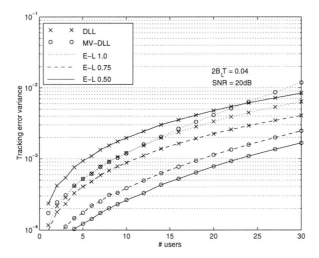

Figure 6.14 Tracking error variances as a function of the number of users for both the conventional (DLL) and the improved non-coherent DLLs with 2Δ = 1.0, 0.75 and 0.50 at the SNR of 20 dB and loop bandwidth $2B_L T$ = 0.04.

The BER results for the basic PIC and the hybrid LMMSE-PIC receiver are presented in Figure 6.15 for the average SNRs of 15 dB as a function of the unknown user power difference with respect to the synchronized users. If the blind LS receiver is used instead of the matched filter at the last receiver stage ("IC2" in the figure), which produces the channel estimates to all receiver stages, the BER degradation is significantly less. When the blind interference suppression filter is used both at the matched filter stage and the last cancellation stage ("MF & IC2" in the figure), the performance improvement is marginal compared to the "IC2" case. The blind LS receiver at the first stage alone did not improve the performance of the basic HD-PIC receiver due to relatively bad channel estimates. The best performance was obtained when the blind LS receiver was used at every receiver stage ("MF & IC1 & IC2" in the figure). In fact, the performance with a three-shot ($M = 3$) blind adaptive LS receiver is almost insensitive to the unknown user in the cases studied. The difference between the one-shot ($M = 1$) and three-shot receivers is quite small in the case when all receiver stages include blind interference suppression.

The BER of the basic HD-PIC receiver in the case of one unknown user of 10 dB higher power is quite large for all numbers of users studied (Figure 6.16). The LMMSE-PIC receiver results in good BER performance in all cases. In particular, the three-shot version is rather insensitive to the unknown signal components that cannot be cancelled.

174 CDMA TECHNIQUES FOR THIRD GENERATION MOBILE SYSTEMS

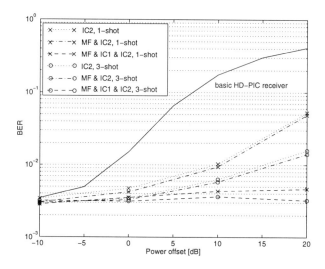

Figure 6.15 BER for the HD-PIC and LMMSE-PIC receivers in two-path fading channels at SNR of 15 dB as a function of the unknown user power offset with respect to the known users, $K = 16 + 1$.

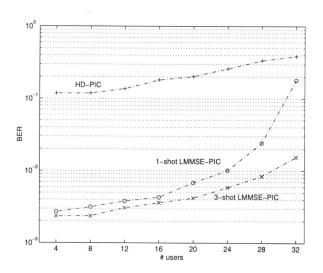

Figure 6.16 BER for the HD-PIC and LMMSE-PIC receivers in two-path fading channels at SNR of 15 dB as a function of the number of users with one unknown user of 10 dB higher power.

6.10 SUMMARY

LMMSE receivers for DS-CDMA communications in fading channels were studied in this chapter. The postcombining LMMSE receiver performs first multipath combining and subsequently interference suppression. The order of multipath combining and interference suppression is the opposite in the precombining LMMSE receivers. The receiver convergence problems caused by channel fading can be avoided by using the precombining LMMSE receivers. The precombining LMMSE receiver can be implemented without the use of training sequences. The fading channel analysis showed that the precombining LMMSE receivers offer significant performance and capacity improvements in comparison to the conventional RAKE receivers. The fading channel analysis was also performed for the FRAMES WCDMA concept. Based on the analysis, it may be concluded that the precombining LMMSE receivers significantly improve the receiver performance and system capacity of FRAMES WCDMA system.

The precombining LMMSE receiver results in separate interference suppression filters for each RAKE finger, hence it is called the LMMSE-RAKE receiver. Since the receiver adaptation is based on the decisions and the channel estimates obtained from a conventional RAKE receiver, no training sequences are needed. Based on the convergence studies, the adaptive LMMSE-RAKE receiver can provide significant improvements in receiver convergence in comparison to other known blind adaptive receivers based on LMS algorithms. If even faster convergence is required, one possibility is to use blind adaptive LS receivers.

The precombining LMMSE receivers require explicit delay estimation. The minimum variance method suitable for the blind adaptive LS receivers was shown to improve the performance of the conventional delay acquisition schemes in CDMA systems. The BER sensitivity of the precombining LMMSE receivers to the delay estimation errors showed that the timing requirements to the blind adaptive receiver can be met with practical delay estimators. The principle of precombining LMMSE receivers was also used to derive an improved delay tracking algorithm. The analysis showed improvements in comparison to the standard delay-locked loop techniques.

The hybrid LMMSE-PIC receiver was shown to give superior performance in comparison to the plain multistage PIC receiver in the presence of the unknown signal components. This property is important in a multi-cell environment when inter-cell interference can be relatively high. The hybrid LMMSE-PIC receiver is seen as one of the most promising alternatives for the base station receivers in future CDMA systems.

References

[1] S. Verdú, "Minimum probability of error for asynchronous Gaussian multiple-access channels," *IEEE Trans. Inform. Th.*, vol. 32, no. 1, pp. 85–96, Jan. 1986.

[2] M. Juntti, *Multiuser Demodulation for DS-CDMA Systems in Fading Channels*, vol. C106 of *Acta Universitatis Ouluensis, Doctoral thesis*, University of Oulu Press, Oulu, Finland, 1997.

[3] A. Duel-Hallen, J. Holtzman, and Z. Zvonar, "Multiuser detection for CDMA systems," *IEEE/ACM Pers. Commun.*, vol. 2, no. 2, pp. 46–58, Apr. 1995.

[4] S. Moshavi, "Multi-user detection for DS-CDMA communications," *IEEE Commun. Mag.*, vol. 34, no. 10, pp. 124–137, Oct. 1996.

[5] M. Juntti and S. Glisic, "Advanced CDMA for wireless communications," in *Wireless Communications: TDMA Versus CDMA*, S. G. Glisic and P. A. Leppänen, Eds., chapter 4, pp. 447–490. Kluwer, 1997.

[6] U. Madhow and M. L. Honig, "MMSE interference suppression for direct-sequence spread-spectrum CDMA," *IEEE Trans. Commun.*, vol. 42, no. 12, pp. 3178–3188, Dec. 1994.

[7] S. L. Miller, "An adaptive direct-sequence code-division multiple-access receiver for multiuser interference rejection," *IEEE Trans. Commun.*, vol. 43, no. 2/3/4, pp. 1746–1755, Feb./Mar./Apr. 1995.

[8] P. B. Rapajic and B. S. Vucetic, "Adaptive receiver structures for asynchronous CDMA systems," *IEEE J. Select. Areas Commun.*, vol. 12, no. 4, pp. 685–697, May 1994.

[9] I. Oppermann and M. Latva-aho, "Adaptive LMMSE receiver for wideband CDMA systems," in *Proc. Commun. Th. Mini-Conf. in conj. IEEE Glob. Telecommun. Conf.*, Phoenix, AZ, USA, Nov. 2–7 1997, pp. 133–138.

[10] M. Latva-aho and M. Juntti, "Modified LMMSE receiver for DS-CDMA – Part I: Performance analysis and adaptive implementations," in *Proc. IEEE Int. Symp. Spread Spectrum Techniques and Applications, to appear*, Sun City, South Africa, Sept. 2–4 1998.

[11] M. Latva-aho, *Advanced Receivers for Wideband CDMA Systems*, Doctoral thesis, University of Oulu. 1998.

[12] M. Honig, U. Madhow, and S. Verdú, "Blind adaptive multiuser detection," *IEEE Trans. Inform. Th.*, vol. 41, no. 3, pp. 944–960, July 1995.

[13] N. R. Mangalvedhe and J. H. Reed, "Blind CDMA interference rejection in multipath channels," in *Proc. IEEE Vehic. Tech. Conf.*, Phoenix, AZ, USA, May 4–7 1997, vol. 1, pp. 21–25.

[14] N. Zečević and J. Reed, "Blind adaptation algorithms for direct-sequence spread-spectrum CDMA single-user detection," in *Proc. IEEE Vehic. Tech. Conf.*, Phoenix, AZ, USA, May 4–7 1997, vol. 3, pp. 2133–2137.

[15] S. C. Park and J. F. Doherty, "Generalized projection algorithm for blind interference suppression in DS/CDMA communications," *IEEE Trans. Circ. Syst. — Part II*, vol. 44, no. 6, pp. 453–460, June 1997.

[16] J. B. Schodorf and D. B. Williams, "A constrained optimization approach to multiuser detection," *IEEE Trans. Sign. Proc.*, vol. 45, no. 1, pp. 258–262, Jan. 1997.

[17] M. K. Tsatsanis, "Time-varying system identification and channel equalization using wavelets and higher-order statistics," in *Control and Dynamic Systems: Advances in Theory and Applications*, C. T. Leondes, Ed., vol. 68, pp. 333–394. Academic Press, San Diego, CA, USA, 1995.

[18] M. Martone, "Blind multichannel deconvolution in multiple access spread spectrum communications using higher order statistics," in *Proc. IEEE Int. Conf. Commun.*, Seattle, WA, USA, June 18-22 1995, vol. 1, pp. 49–53.

[19] M. K. Tsatsanis, "Inverse filtering criteria for CDMA systems," *IEEE Trans. Sign. Proc.*, vol. 45, no. 1, pp. 102–112, Jan. 1997.

[20] R. de Gaudenzi, F. Giannetti, and M. Luise, "Design of a low-complexity adaptive interference-mitigating detector for DS/SS receivers in CDMA radio networks," *IEEE Trans. Commun.*, vol. 46, no. 1, pp. 125–134, Jan. 1998.

[21] M. Honig, M. Shensa, S. Miller, and L. Milstein, "Performance of adaptive linear interference suppression for DS-CDMA in the presence of flat Rayleigh fading," in *Proc. IEEE Vehic. Tech. Conf.*, Phoenix, AZ, USA, May 4–7 1997, vol. 3, pp. 2191–2195.

[22] X. Wang and H. V. Poor, "Blind equalization and multiuser detection in dispersive CDMA channels," *IEEE Trans. Commun.*, vol. 46, no. 1, pp. 91–103, Jan. 1998.

[23] M. Honig, "Performance of adaptive interference suppression for DS-CDMA with a time-varying user population," in *Proc. IEEE Int. Symp. Spread Spectrum Techniques and Applications*, Mainz, Germany, Sept. 22–25 1996, vol. 1, pp. 267–271.

[24] H. Ge, "Adaptive schemes of implementing the LMMSE multiuser detector for CDMA," in *Proc. IEEE Int. Conf. Commun.*, Montreal, Canada, June 8–12 1997.

[25] R. A. Iltis, "Performance of constrained and unconstrained adaptive multiuser detectors for quasi-synchronous CDMA," *IEEE Trans. Commun.*, vol. 46, no. 1, pp. 135–143, Jan. 1998.

[26] U. Madhow, "Blind adaptive interference suppression for the near-far resistant acquisition and demodulation of direct-sequence CDMA," *IEEE Trans. Sign. Proc.*, vol. 45, no. 1, pp. 124–136, Jan. 1997.

[27] H. C. Huang, *Combined Multipath Processing, Array Processing, and Multiuser Detection for DS-CDMA Channels*, Ph.D. thesis, Princeton University, Princeton, NJ, USA, Jan. 1996.

[28] A. Klein, *Multi-user Detection of CDMA Signals – Algorithms and Their Application to Cellular Mobile Radio*, Ph.D. thesis, University of Kaiserslautern, Kaiserslautern, Germany, 1996.

[29] Z. Zvonar, *Multiuser Detection for Rayleigh Fading Channels*, Ph.D. thesis, Northeastern University, Boston, MA, USA, Sept. 1993.

[30] S. Kay, *Fundamentals of Statistical Signal Processing: Estimation Theory*, Prentice-Hall, Englewood Cliffs, NJ, USA, 1993.

[31] M. J. Barrett, "Error probability for optimal and suboptimal quadratic receivers in rapid Rayleigh fading channels," *IEEE J. Select. Areas Commun.*, vol. 5, no. 2, pp. 302–304, Feb. 1987.

[32] M. J. Juntti, "Performance of decorrelating multiuser receiver with data-aided channel estimation," in *Proc. Commun. Th. Mini-Conf. in conj. IEEE Glob. Telecommun. Conf.*, Phoenix, AZ, USA, Nov. 5–7 1997, pp. 123–127.

[33] H. V. Poor and S. Verdú, "Probability of error in MMSE multiuser detection," *IEEE Trans. Inform. Th.*, vol. 43, no. 3, pp. 858–871, May 1997.

[34] S. Haykin, *Adaptive Filter Theory*, Prentice Hall, Upper Saddle River, NJ, USA, 3rd edition, 1996.

[35] J. R. Treichler, Jr. C. R. Johnson, and M. G. Larimore, *Theory and Design of Adaptive Filters*, John Wiley and Sons, New York, USA, 1987.

[36] L. L. Scharf, *Statistical Signal Processing: Detection, Estimation, and Time Series Analysis*, Addison-Wesley, Reading, MA, USA, 1991.

[37] D. H. Johnson and D. E. Dudgeon, *Array Signal Processing: Concepts and Techniques*, Prentice Hall, Englewood Cliffs, NJ, USA, 1993.

[38] R. A. Iltis and L. Mailaender, "An adaptive multiuser detector with joint amplitude and delay estimation," *IEEE J. Select. Areas Commun.*, vol. 12, no. 5, pp. 774–785, June 1994.

[39] J. Lilleberg, E. Nieminen, and M. Latva-aho, "Blind iterative multiuser delay estimator for CDMA," in *Proc. IEEE Int. Symp. Personal, Indoor and Mobile Radio Commun.*, Taipei, Taiwan, Oct. 15–18 1996, vol. 2, pp. 565–568.

[40] S. Y. Miller, *Detection and Estimation in Multiple-Access Channels*, Ph.D. thesis, Princeton University, Princeton, NJ, USA, 1989.

[41] S. E. Bensley and B. Aazhang, "Subspace-based channel estimation for code division multiple access communication systems," *IEEE Trans. Commun.*, vol. 44, no. 8, pp. 1009–1019, Aug. 1996.

[42] E. G. Ström, S. Parkvall, S. L. Miller, and B. E. Ottersen, "Propagation delay estimation in asynchronous direct-sequence code-division multiple access systems," *IEEE Trans. Commun.*, vol. 44, no. 1, pp. 84–93, Jan. 1996.

[43] S. M. Kay and S. L. Marple, "Spectrum analysis - a modern perspective," *Proc. IEEE*, vol. vol. 64, pp. 1380–1419, Nov. 1981.

[44] R. O. Schmidt, *A Signal Subspace Approach to Multiple Emitter Location and Spectral Estimation*, Ph.D. thesis, Stanford University, Stanford, CA, USA, 1981.

[45] D. Zheng, J. Li, S. L. Miller, and E. G. Ström, "An efficient code-timing estimator for DS-CDMA signals," *IEEE Trans. Sign. Proc.*, vol. 45, no. 1, pp. 82–89, Jan. 1997.

[46] D.H. Johnson and S.R. DeGraaf, "Improving the resolution of bearing in passive sonar arrays by eigenvalue analysis," *IEEE Trans. Acoust. Speech Sign. Proc.*, vol. 30, no. 4, pp. 638–647, Aug. 1982.

[47] M. K. Simon, J. K. Omura, R. A. Scholtz, and B. K. Levitt, *Spread Spectrum Communications Handbook*, McGraw-Hill, New York, USA, 1994.

[48] H. Meyr and G. Ascheid, *Synchronization in Digital Communications, Volume 1, Phase-, Frequency-Locked Loops and Amplitude Control*, John Wiley and Sons, New York, USA, 1990.

[49] H. Meyr, M. Moeneclaey, and S. A. Fechtel, *Digital Communication Receivers, Synchronization, Channel Estimation and Signal Processing*, John Wiley and Sons, New York, USA, 1998.

[50] F. M. Gardner, *Phaselock Techniques, 2nd edn.*, John Wiley and Sons, 1979.

[51] R. de Gaudenzi, M. Luise, and R. Viola, "A digital chip timing recovery loop for band-limited direct-sequence spread-spectrum signals," *IEEE Trans. Commun.*, vol. 41, no. 11, pp. 1760–1769, Nov. 1993.

[52] M. Moeneclaey and G. De Jonghe, "Tracking performance of digital chip synchronization algorithms for bandlimited direct-sequence spread-spectrum communication," *IEE Electronic Letters*, vol. 27, no. 13, pp. 1147–1149, June 1991.

[53] W.-H. Sheen and G.L. Stüber, "Effects of multipath fading on delay-locked loops for spread spectrum systems," *IEEE Trans. Commun.*, vol. 42, no. 2/3/4, pp. 1947–1956, Feb./Mar./Apr. 1994.

[54] B. S. Abrams, A. E. Zeger, and T. E. Jones, "Efficiently structured CDMA receiver with near-far immunity," *IEEE Trans. Vehic. Tech.*, vol. 44, no. 1, pp. 1–13, Feb. 1995.

[55] R. S. Mowbray, R. D. Pringle, and P. M. Grant, "Increased CDMA system capacity through adaptive cochannel interference regeneration and cancellation," *IEE Proc.-I*, vol. 139, no. 5, pp. 515–524, Oct. 1992.

[56] Y. Sanada and Q. Wang, "A co-channel interference cancellation technique using orthogonal convolutional codes," *IEEE Trans. Commun.*, vol. 44, no. 5, pp. 549–556, May 1996.

[57] S. Tachikawa, "Characteristics of M-ary/spread spectrum multiple access communication systems using co-channel interference cancellation techniques," *IEICE Trans. Commun.*, vol. E76-B, no. 8, pp. 941–946, June 1992.

[58] M. K. Varanasi and B. Aazhang, "Multistage detection in asynchronous code-division multiple-access communications," *IEEE Trans. Commun.*, vol. 38, no. 4, pp. 509–519, Apr. 1990.

[59] M. K. Varanasi and B. Aazhang, "Near-optimum detection in synchronous code-division multiple-access systems," *IEEE Trans. Commun.*, vol. 39, no. 5, pp. 725–736, May 1991.

[60] U. Fawer and B. Aazhang, "A multiuser receiver for code division multiple access communications over multipath channels," *IEEE Trans. Commun.*, vol. 43, no. 2/3/4, pp. 1556–1565, Feb./Mar./Apr. 1995.

[61] R. Kohno, H. Imai, M. Hatori, and S. Pasupathy, "Combination of an adaptive array antenna and a canceller of interference for direct-sequence spread-spectrum multiple-access system," *IEEE J. Select. Areas Commun.*, vol. 8, no. 4, pp. 675–682, May 1990.

[62] A. Saifuddin and R. Kohno, "Performance evaluation of near-far resistant receiver DS/CDMA cellular system over fading multipath channel," *IEICE Trans. Commun.*, vol. E78-B, no. 8, pp. 1136–1144, Aug. 1995.

[63] A. Saifuddin, R. Kohno, and H. Imai, "Integrated receiver structures of staged decoder and CCI canceller for CDMA with multilevel coded modulation," *European Trans. Telecommun.*, vol. 6, no. 1, pp. 9–19, Jan.–Feb. 1995.

[64] Y. Sanada and Q. Wang, "A co-channel interference cancellation technique using orthogonal convolutional codes on multipath Rayleigh fading channel," *IEEE Trans. Vehic. Tech.*, vol. 46, no. 1, pp. 114–128, Feb. 1997.

[65] Y. C. Yoon, R. Kohno, and H. Imai, "Cascaded co-channel interference cancellating and diversity combining for spread-spectrum multi-access system over multipath fading channels," *IEICE Trans. Commun.*, vol. E76-B, no. 2, pp. 163–168, Feb. 1993.

[66] Y. C. Yoon, R. Kohno, and H. Imai, "A spread-spectrum multiaccess system with cochannel interference cancellation for multipath fading channels," *IEEE J. Select. Areas Commun.*, vol. 11, no. 7, pp. 1067–1075, Sept. 1993.

[67] A. Hottinen, H. Holma, and A. Toskala, "Performance of multistage multiuser detection in a fading multipath channel," in *Proc. IEEE Int. Symp. Personal, Indoor and Mobile Radio Commun.*, Toronto, Canada, Sept. 27–29 1995, vol. 3, pp. 960–964.

[68] M. Latva-aho and J. Lilleberg, "Parallel interference cancellation in multiuser CDMA channel estimation," *Wireless Pers. Commun., Kluwer, to appear*, 1998.

[69] M. J. Juntti, M. Latva-aho, and M. Heikkilä, "Performance comparison of PIC and decorrelating multiuser receivers in fading channels," in *Proc. IEEE Glob. Telecommun. Conf.*, Phoenix, AZ, USA, Nov. 3–8 1997, vol. 2, pp. 609–613.

[70] A. Saifuddin and R. Kohno, "Performance evaluation of DS/CDMA scheme with diversity coding and MUI cancellation over fading multipath channel," *IEICE Trans. Fundamentals Elec., Commun. and Comp. Sc.*, vol. E79-A, no. 12, pp. 1994–2001, Dec. 1996.

[71] A. Hottinen, H. Holma, and A. Toskala, "Multiuser detection for multirate CDMA communications," in *Proc. IEEE Int. Conf. Commun.*, Dallas, TX, USA, June 24-28 1996.

[72] U. Fawer and B. Aazhang, "Multiuser receivers for code-division multiple-access systems with trellis-based modulation," *IEEE J. Select. Areas Commun.*, vol. 14, no. 8, pp. 1602–1609, Oct. 1996.

[73] S. D. Gray, M. Kocic, and D. Brady, "Multiuser detection in mismatched multiple-access channels," *IEEE Trans. Commun.*, vol. 43, no. 12, pp. 3080–3089, Dec. 1995.

[74] R. M. Buehrer, A. Kaul, S. Striglis, and B. D. Woerner, "Analysis of DS-CDMA parallel interference cancellation with phase and timing errors," *IEEE J. Select. Areas Commun.*, vol. 14, no. 8, pp. 1522–1535, Oct. 1996.

[75] P. Orten and T. Ottosson, "Robustness of DS-CDMA multiuser detectors," in *Proc. Commun. Th. Mini-Conf. in conj. IEEE Glob. Telecommun. Conf.*, Phoenix, AZ, USA, Nov. 2–7 1997.

[76] P. Agashe and B. Woerner, "Interference cancellation for a multicellular CDMA environment," *Wireless Pers. Commun., Kluwer*, vol. 3, no. 1-2, pp. 1–15, 1996.

[77] V. Vanghi and B. Vojcic, "Soft interference cancellation in multiuser communications," *Wireless Pers. Commun., Kluwer*, vol. 3, no. 1-2, pp. 111–128, 1996.

[78] D. Divsalar, M. K. Simon, and D. Raphaeli, "Improved parallel interference cancellation for CDMA," *IEEE Trans. Commun.*, vol. 46, no. 2, pp. 258–268, Feb. 1998.

[79] D. Dahlhaus, A. Jarosch, H. Fleury, and R. Heddergott, "Joint demodulation in DS/CDMA systems exploiting the space and time diversity of the mobile radio channel," in *Proc. IEEE Int. Symp. Personal, Indoor and Mobile Radio Commun.*, Helsinki, Finland, Sept. 1–4 1997, vol. 1, pp. 47–52.

[80] D. Dahlhaus, H. Fleury, and A. Radović, "A sequential algorithm for joint parameter estimation and multiuser detection in DS/CDMA systems with multipath propagation," *Wireless Pers. Commun., Kluwer, to appear*, 1998.

[81] L. B. Nelson and H. V. Poor, "Iterative multiuser receivers for CDMA channels: An EM-based approach," *IEEE Trans. Commun.*, vol. 44, no. 12, pp. 1700–1710, Dec. 1996.

[82] A. Radović and B. Aazhang, "Iterative algorithms for joint data detection and delay estimation for code division multiple access communication systems," in *Proc. Annual Allerton Conf. Communication Control and Computing*, Allerton House, Monticello, IL, USA, Sept. 29 – Oct. 1 1993.

[83] A. P. Dempster, N. M. Laird, and D. B. Rubin, "Maximum likelihood from incomplete data via the EM algorithm," *J. Royal Stat. Soc.*, vol. 39, no. 1, pp. 1–38, 1977.

[84] J. Fessler and A. Hero, "Space-alternating generalized expectation-maximization algorithm," *IEEE Trans. Sign. Proc.*, vol. 42, no. 10, pp. 2664–2677, Oct. 1994.

[85] P. Patel and J. Holtzman, "Analysis of a simple successive interference cancellation scheme in a DS/CDMA system," *IEEE J. Select. Areas Commun.*, vol. 12, no. 10, pp. 796–807, June 1994.

[86] A. J. Viterbi, "Very low rate convolutional codes for maximum theoretical performance of spread-spectrum multiple-access channels," *IEEE J. Select. Areas Commun.*, vol. 8, no. 4, pp. 641–649, May 1990.

[87] A. C. K. Soong and W. A. Krzymien, "Performance of a reference symbol assisted multistage successive interference cancelling receiver in a multi-cell CDMA wireless systems," in *Proc. IEEE Glob. Telecommun. Conf.*, Singapore, Nov. 13–17 1995, vol. 1, pp. 152–156.

[88] A. Hui and K. Letaief, "Multiuser asynchronous DS/CDMA detectors in multipath fading links," *IEEE Trans. Commun.*, vol. 46, no. 3, pp. 384–391, Mar. 1998.

[89] O. Nesper and P. Ho, "A pilot symbol assisted interference cancellation scheme for an asynchronous DS/CDMA system," in *Proc. IEEE Glob. Telecommun. Conf.*, London, U.K., Nov. 18–22 1996, vol. 3, pp. 1447–1451.

[90] A. C. K. Soong and W. A. Krzymien, "A novel CDMA multiuser interference cancellation receiver with reference symbol aided estimation of channel parameters," *IEEE J. Select. Areas Commun.*, vol. 14, no. 8, pp. 1536–1547, Oct. 1996.

[91] A.-L. Johansson and A. Svensson, "Successive interference cancellation in multiple data rate DS/CDMA systems," in *Proc. IEEE Vehic. Tech. Conf.*, Chicago, IL, USA, July 25–28 1995, pp. 704–708.

[92] A.-L. Johansson and A. Svensson, "Multistage interference cancellation in multirate DS/CDMA on a mobile radio channel," in *Proc. IEEE Vehic. Tech. Conf.*, Atlanta, GA, USA, Apr. 28 - May 1 1996, vol. 2, pp. 666–670.

[93] A. C. K. Soong and W. A. Krzymien, "Robustness of the reference symbol assisted multistage successive interference cancelling receiver with imperfect parameter estimates," in *Proc. IEEE Vehic. Tech. Conf.*, Atlanta, GA, USA, Apr. 28 - May 1 1996, vol. 2, pp. 676–680.

[94] T.-B. Oon, R. Steele, and Y. Li, "Performance of an adaptive successive serial-parallel CDMA cancellation scheme in flat Rayleigh fading channels," in *Proc. IEEE Vehic. Tech. Conf.*, Phoenix, AZ, USA, May 4–7 1997, vol. 1, pp. 193–197.

[95] R. M. Buehrer and B. D. Woerner, "The asymptotic multiuser efficiency of M-stage interference cancellation receivers," in *Proc. IEEE Int. Symp. Personal, Indoor and Mobile Radio Commun.*, Helsinki, Finland, Sept. 1–4 1997, vol. 2, pp. 570–574.

[96] F. van der Wijk, G. M. J. Janssen, and R. Prasad, "Groupwise successive interference cancellation in a DS/CDMA system," in *Proc. IEEE Int. Symp. Personal, Indoor and Mobile Radio Commun.*, Toronto, Canada, Sept. 27–29 1995, vol. 2, pp. 742–746.

[97] P. D. Alexander, L. K. Rasmussen, and C. Schlegel, "A linear receiver for coded multiuser CDMA," *IEEE Trans. Commun.*, vol. 45, no. 5, pp. 605–610, May 1997.

[98] W. Haifeng, J. Lilleberg, and K. Rikkinen, "A new sub-optimal multiuser detection approach for CDMA systems in Rayleigh fading channel," in *Proc. Conf. Inform. Sciences Systems*, The Johns Hopkins University, Baltimore, MD, USA, Mar. 19–21 1997, vol. 1, pp. 276–280.

[99] M. J. Juntti, "Multiuser detector performance comparisons in multirate CDMA systems," in *Proc. IEEE Vehic. Tech. Conf.*, Ottawa, Canada, May 18–21 1998, vol. 1, pp. 36–40.

[100] M. K. Varanasi, "Group detection in synchronous Gaussian code-division multiple-access channels," *IEEE Trans. Inform. Th.*, vol. 41, no. 3, pp. 1083–1096, July 1995.

[101] M. K. Varanasi, "Parallel group detection for synchronous CDMA communication over frequency-selective Rayleigh fading channels," *IEEE Trans. Inform. Th.*, vol. 42, no. 1, pp. 116–128, Jan. 1996.

[102] A. Toskala, J. Castro, E. Dahlman, M. Latva-aho, and T. Ojanperä, "FRAMES FMA2 wideband-CDMA for UMTS," *To appear in European Transaction on Telecommunications*.

[103] M. Latva-aho, "Bit error probability analysis for FRAMES W-CDMA downlink receivers," *IEEE Trans. Vehic. Tech., to appear*, 1998.

[104] M. Latva-aho, "Modified LMMSE receiver for DS-CDMA – Part II: Performance in FMA2 downlink," in *Proc. IEEE Int. Symp. Spread Spectrum Techniques and Applications, to appear*, Sun City, South Africa, Sept. 2–4 1998.

[105] K.-C. Chen and L. D. Davisson, "Analysis of SCCL as a PN-code tracking loop," *IEEE Trans. Commun.*, vol. 42, no. 11, pp. 2942–2946, Nov. 1994.

7 DETECTION STRATEGIES AND CANCELLATION SCHEMES IN A MC-CDMA SYSTEM

Frans Kleer, Shin Hara* and Ramjee Prasad

Centre for Wireless Personal Communications, IRCTR
Telecommunications and Traffic-Control Systems Group
Delft University of Technology
P.O Box 5031, 2600 GA Delft
The Netherlands
* Department of Communication Engineering, University of Osaka, Japan
R.Prasad@et.tudelft.nl

Abstract:
This paper presents a study of a MC-CDMA system. Different diversity combining techniques in an uplink and downlink channel; single and multi-user detection strategies; and an interference cancellation method have been considered. The performance of the system has been evaluated by simulations.

7.1 INTRODUCTION

This paper examines the performances of a MC-CDMA system in an indoor wireless radio environment, facing very important issues connected with spread spectrum techniques such as synchronisation and detection strategies. With MC-CDMA, a data symbol is transmitted over N narrowband subcarriers where each subcarrier is modulated by "1" or "-1" based on a spreading code; different users transmit over the same set of subcarriers but with a spreading code orthogonal to all the other codes. This method has shown to be robust in dispersive channels. The effect of channel dispersion is an important reason for using MC-CDMA rather than (Direct-Sequence) DS-CDMA. In DS-CDMA, channel dispersion leads to the reception of multiple resolvable paths. A typical RAKE receiver contains multiple correlators, each synchronised to a different resolvable path. However, no code exists to ensure that the partial correlation of a slightly delayed path is orthogonal to the dominant line-of-sight path of the wanted signal or interference [1].

The main issue of a MC-CDMA system is its capacity to spread the signal bandwidth without increasing the adverse effect of delay spread. In MC-CDMA, dispersion results in a different attenuation of different subcarriers, but each of the subcarriers experiences only narrowband fading. Narrowband communications have the desirable property of being relatively immune to intersymbol interference (ISI). An MC-CDMA receiver combines signals at different subcarriers to ensure strong attenuation of interference, while perceiving a reasonable signal to noise ratio [2].

If the symbol duration is relatively long, it is unlikely that the symbol energy completely vanishes during a signal fade. However, Orthogonal Frequency Division Multiplex (OFDM) subcarriers lose their mutual orthogonality if time variation of the channel occur. This typically leads to increased bit error rates. Also, phase jitter and/or receiver frequency offset leads to more interference. This sensitivity to frequency offset, as well as to non-linear amplification are believed to be the main disadvantages of MC-CDMA systems. A frequency error erodes the subcarrier orthogonality, and it also has a great impact on the synchronisation behaviour of the subcarriers. It is difficult to maintain and establish this synchronisation.

The important issue of timing is addressed in this report. For the downlink this is trivial because there is only one station that generates the codes; in the uplink different transmitters can have different clock offsets and propagation delays. We propose the use of a Time Division Duplex TDMA system to act as a time aligning framework for MC-CDMA transmission bursts in order to establish a quasi synchronous channel state, so reducing significantly multi-user interference.

Multiple access systems are interference limited so every technique which might reduce this interference (MAI) will cause the system capacity to increase. In a multi-user detection scheme, successive interference cancellation is a method which systematically reduces the multiple access interference and shows a substantial performance improvement. It is quite simple and it helps to solve near-far problems typical of a single-user detection scheme; increasing the complexity in the base station, it could eliminate the sophisticated high-precision power control schemes in the mobile units.

In this report, we will pay attention to four commonly used diversity techniques: Orthogonality Restoring Combining (ORC), Equal Gain Combining (EGC), Maximal Ratio Combining (MRC) and Minimum Mean Square Error Combining (MMSEC). The gain factors are applied in downlink and uplink. In the case of DS-CDMA, the gain factor is related to the amplification or attenuation of the received signals via different paths. For the case of MC-CDMA, this factor influences the relative strength of the subcarriers.

In order to see how a perfect synchronously MC-CDMA system behaves in terms of Bit Error Probability in the uplink and downlink of a Rayleigh fading channel, simulations with the previous diversity combining techniques are performed. Simulation results are given for the uplink as well as for the downlink.

7.2 MC/CDMA: SYSTEM DESCRIPTION

In serial systems like DS-CDMA, the symbols are transmitted sequentially. Each data symbol occupies the entire available spectrum. In such systems performance improvements e.g. higher data rates can be achieved at the cost of increasing bandwidth demands. Another drawback from serial systems is due to the bursty nature of the channel, several adjacent symbols might be destroyed during a deep fade. In contrast to Direct-Sequence Spread Spectrum (DS-SS), MC-CDMA does not multiply the user signal by a high speed (time variant) chip sequence, but the same user bit is transmitted on multiple subcarriers simultaneously. The pulseform of each sub-carrier signal is chosen to ensure orthogonality between subcarriers. Each subcarrier is narrowband in nature so no significant intersymbol interference (ISI) occurs between successive bits. For each user the subcarriers are phase shifted with a 0 or π phase offset. The set of subcarrier phase offsets follows a signature code sequence to distinguish different users. The resulting signal has a pn-coded structure in the frequency domain and multiple access will be possible using the orthogonality of the different codes.

If the number of spacing between subcarriers is appropriately chosen, it is unlikely as stated before that all of the subcarriers will be located in deep fade and consequently frequency diversity is achieved [3]. So Multi-Carrier Code Division Multiplexing or Multiple Access is an approach to intentionally disperse the signal over different subcarriers.

When an efficient use of bandwidth is not required, the most effective parallel systems use conventional frequency division multiplexing where the spectra of the different subchannels do not overlap. In such a system there is sufficient guard frequency band between adjacent subchannels to isolate them at the receiver using conventional filters. A much more efficient use of bandwidth can be obtained with a parallel system if the spectra of the individual subchannels are permitted to overlap [4].

We describe now the generation of a MC-CDMA signal with N parallel replicas of each data symbol, BPSK modulation, coherent detection and the use of subcarrier tones separated by the reciprocal of the signalling element duration ($\Delta f = 1/T_s$) so independent separation of the multiplexed tones is possible. The transmitted signal consists of the sum of all the outputs of the parallel branches.

If $\Delta f = 1/T_s$ is the subcarrier separation, $p_s(t)$ is the pulse waveform, which is unity in the interval $[0, T_s]$ and zero otherwise, $a^j(i)$ is the information bearing signal and in particular the i-th bit of the j-th user, c_m^j is the signature waveform of the j-th user, G_{MC} is the number of subcarriers or the length of the spreading code (processing gain), the composite transmitted signal for the j-th user can be expressed as follows:

$$S_{MC}^j(t) = \sum_{i=-\infty}^{+\infty} \sum_{m=1}^{G_{MC}} a^j(i) c_m^j p_s(t - iT_s) \cos\{2\pi(f_o + m\Delta f)t\} \quad (7.1)$$

In this case the carrier frequency is outside band. The spectrum of the MC-CDMA signal ranges from $f_o + \Delta f$ to $f_o + G_{MC}\Delta f$ and is composed of G_{MC} narrowband subcarrier signals each with a symbol duration, T_s, much larger than the delay spread, T_m, so that an MC-CDMA signal will not experience significant intersymbol interference (ISI). In the resulting MC-CDMA, each chip has a relatively long time duration (identical to the symbol time T_s) and a relatively narrow bandwidth of approximately $1/T_s$. This is in contrast to the short and wideband waveforms transmitted in DS-CDMA. As a bit is transmitted simultaneously on G_{MC} parallel subcarriers, each of which has one code element, we interpret this by saying that MC-CDMA spreads in the frequency domain, whereas DS-CDMA spreads in the time domain.

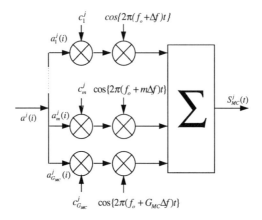

Figure 7.1 MC-CDMA transmitter for the j-th user

At the receiver a MC-CDMA signal through a frequency selective fading channel can be described as:

$$r(t) = \sum_{j=1}^{J} \sum_{i=-\infty}^{+\infty} \sum_{m=1}^{G_{MC}} \sum_{l=1}^{L} z_m^j \alpha^j(i) p(t - iT_s - \tau_l) c_m^j \cdot \cos\{2\pi(f_o + m\Delta f)t + \phi_m^j\} + n(t) \quad (7.2)$$

where z_m^j and ϕ_m^j are the received random amplitudes and phases of the j-th user of the m-th subcarrier, J is the number of users. We assume the amplitudes, z_m^j, iid Rayleigh random variables corresponding to the case in which the direct line-of-sight (LOS) path from the transmitter to the receiver is obstructed (Rayleigh fading). In an indoor environment, a Rician distribution could describe better the presence of a dominant component while a Rayleigh one can represent a conservative estimate of the performance; the absence of a line-of-sight path corresponds with a worst case propagation channel. The phases ϕ_m^j are assumed to be independent identically distributed (IID) uniform random variables on the interval $[0, 2\pi]$ for all users. The assumption that the random variables z_m^j and ϕ_m^j are independent of each other and between users is valid for channels where $(\Delta f)_c \ll 1/T_s$ ($(\Delta f)_c$ is the coherence bandwidth). L denotes the number of multipath signals. This term can be neglected, only if we can neglect the delay spread T_m. The Additive White Gaussian Noise (AWGN) is denoted by $n(t)$. The scheme for the receiver can be like this:

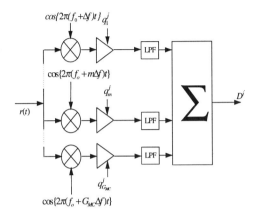

Figure 7.2   MC-CDMA receiver model for user j.

In the model of the channel there is a big difference between up and downlink: in a downlink channel we can assume that every envelope of the m-th subcarrier of every user is the same e.g. $z_m^1 = z_m^2 = ... = z_m^J = z_m$ because a terminal receives interfering signals designated for other users through the same channel as the wanted signal; on the other hand in an uplink channel depending on the location of the terminals, all envelopes of the m-th subcarrier are not the same because the base station receives each signal through different channels.

In the MC-CDMA receiver, the received signal is combined in the frequency domain. The receiver can always use all the energy scattered in the frequency domain [5], which is fed through G_{MC} parallel coherent detectors assuming that in demodulation a phase correction factor is applied. The resulting low-pass signal of each subcarrier is then multiplied by the gain factor $q_m^{j\prime}$ to combine the received signal scattered in the frequency domain. The decision variable $D^{j\prime}$ is given by (omitting the subscription i):

$$D^{j\prime} = \sum_{m=1}^{G_{MC}} q_m^{j\prime} y_m \; with \; y_m = \sum_{j=1}^{J} z_m^j a^j c_m^j + n_m \qquad (7.3)$$

where y_m and n_m are the baseband component of the received signal after down-conversion with subcarrier synchronisation and the Additive White Gaussian Noise at the m-th subcarrier, respectively, and z_m^j is the envelope of the m-th subcarrier for the j-th user. We have to notice that in MC-CDMA systems it is most important to have frequency non-selective fading over each subcarrier and this implies that each modulated subcarrier does not experience significant dispersion. In some other MC-schemes it is therefore sometimes necessary to convert the original high data rate signal from serial to parallel before spreading over the frequency domain, in order to prevent frequency selective fading. This method is called MC-DS-CDMA but it is not examined here. In our MC-CDMA system a proper choice of the number of subcarriers and guard interval is important in order to increase the robustness against frequency selective fading. There exists an optimal value in the number of subcarriers and the duration of the guard interval to minimise the bit error probability [6]. If we apply perfectly orthogonal Walsh-Hadamard code (WH-code) sequences over a linear, time invariant, frequency non-selective channel with perfect chip-synchronised transmission, the performance of DS-CDMA and MC-CDMA is equivalent, as the orthogonal multi-user interference completely vanishes [4]. In practice, multipath channels are less ideal and channel dispersion erodes the orthogonality of CDMA signals.

7.3 MULTIUSER INTERFERENCE IN CDMA SYSTEMS

In contrast to FDMA and TDMA techniques which are frequency bandwidth limited, CDMA systems are interference limited. In CDMA systems, each user's data is spread by a unique pseudorandom code. All users then transmit in the same frequency band and are distinguished at the receiver by the user specific spreading code. All other signals are not despread because they use different codes. These signals appear as interference to the desired user because of non-zero cross-correlation values between the spreading codes. As the number of users increases, the signal to interference ratio (SIR) decreases until the resulting performance is no longer acceptable. Thus, this multi-user interference must be reduced to achieve higher capacities and that is strictly connected to the cross-correlation factor between different users. We can reduce cross-correlation in spread spectrum systems by:

- Spreading the signal by orthogonal codes which have zero cross-correlation. This technique is very efficient in downlink transmission, because a base station can transmit to all users simultaneously and the signals are spread synchronously at chip level. Transmitting asynchronously in the uplink, to restore the orthogonality of the codes, the mobile users can be time-aligned by a synchronisation method.

- Cancellation schemes that usually work subtracting the interference caused by other users and require a significant processing power; they are very useful especially to solve near-far problems.

More in general other important techniques like power control schemes, voice activity detection or antenna sectorization are used to minimise the performance degradation caused by the total co-channel interference (intra and inter-cell) from other users.

7.4 SYNCHRONISATION IN A MC/CDMA SYSTEM

Synchronisation is one of the most determinant parameters to reduce multi-user interference. The main issue of synchronous MC-CDMA is timing. For the downlink, timing is not a problem because there is only one station that generates the codes [7]. The uplink is much more complicated, because different transmitters have different clock offsets and propagation delays, so uplink synchronisation seems to be difficult to establish when we face a fading multipath mobile radio channel. In an ideal radio channel the perfectly orthogonal spreading codes maintain orthogonality only if the codes remain strictly synchronised. Due to multipath propagation the orthogonality is partly lost; the delayed multipath signals are clearly not synchronised to the first arriving path and besides the dominant components are time-shifted for the different propagation delays. Because some synchronicity error exists, special orthogonal codes with small cross-correlation must be used. A great advantage of MC-CDMA is its relatively long symbol duration that could allow to establish a quasi-synchronous channel state.

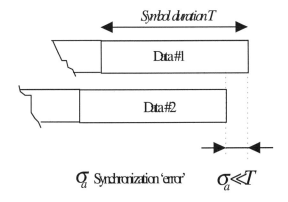

Figure 7.3 Symbol duration compared to propagation delay.

The synchronisation offsets should be a small fraction of the chip period if we want to minimise multi-user interference by orthogonal signals. Increasing the chip period would make it easier to synchronise the signals. However, for a given processing gain, an increase in the chip period results in an increase in the symbol period, hence a decrease in the data rate; the solution for such a problem is the adoption of a multi-carrier signalling scheme able to maintain the original data rate. Increasing the chip period by a factor M, the number of subcarriers, reduces the bandwidth of each of the subband signals by a factor M relative to the bandwidth of the original spread spectrum signal so that the overall bandwidth is approximately the same as that of the single-carrier signal.

7.4.1 Burst Format and Synchronisation Algorithm Establishing a Quasi-synchronous Uplink

In order to achieve a quasi-synchronous uplink we will suggest a particular signal burst placed in a time division duplex frame. Handling a signal burst like this makes possible to time align the active mobiles. The signal burst format fig.7.4 is comprised of two parts: one part contains the information bearing contents, and one part contains the pre-amble which consist of a code with a low cross-correlation value.

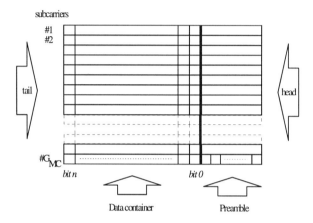

Figure 7.4   Signal burst format.

In the data container the information bearing bits are transported. These bits are first spread by means of an orthogonal spreading code (e.g. Walsh-Hadamard codes) and then are put in the container spread among the subcarriers. The pre-amble instead is used to obtain an estimation of the frequency responses (z_m^j) of the mobile radio channel. As already stated, the pre-amble should consist of a code which has a very low cross-correlation value, for instance a Goldcode generated by modulo-2 summation of maximum length series. An interesting issue is that such a pre-amble can be used to find the delay profile of a time dispersive, i.e., frequency selective channel. If such a sequence is transmitted in multi-carrier format, it can be used to find the Doppler components of the frequency dispersive channel. In a mobile multipath channel, signal waves coming from different paths often exhibit different Doppler shifts. A receiver can detect the individual components by searching for shifted versions of the sequence. The resulting correlation pattern can be used to control the Local Oscillator in order to track the signal. Detailed simulations should be performed to obtain the optimised size of the data block and which code can best be used in the pre-amble. The length of each transmission burst to and from the base station is constant, thereby allowing the mobile unit to estimate accurately the propagation delay and adjust its transmission time appropriately. In this way signals from other users will arrive at the base station almost synchronously. In TDD systems forward and reverse link communications are carried out in the same frequency bandwidth, with the base station and mobile units alternately transmitting and receiving information signals. The duration of all

DETECTION STRATEGIES AND CANCELLATION SCHEMES IN A MC-CDMA SYSTEM

the data bursts is controlled by the base station, which transmits to all the mobiles simultaneously. The mobile units transmit back after receiving the base station signal after a certain guard time. In fig.7.5 is depicted how a TDD frame is subdivided into a forward (down) and reverse (up) link data burst.

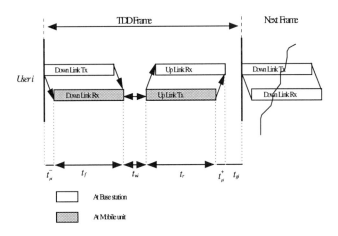

Figure 7.5 Time Division Duplex (TDD) frame timing structure.

These forward and reverse link data bursts have a duration of t_f and t_r respectively. The waiting time (e.g. guard time) at the mobile unit is t_{wi} and at the base station the guard time is t_{gi} Finally the propagation delays at the beginning and end of the TDD frame are denoted as t_{pi}^- and t_{pi}^+ respectively for the i-th user. In this system the guard time at the base station t_{gi} is constant, but the waiting time at the mobiles t_{wi} varies in response to changes in propagation delays for any user independently. All times are defined as integer multiples of the local mobile clock time T_m. This means that if t_{pi} is the actual propagation delay time, then $t_{pi} = [t_{pi}/T_m]$ indicates the integer part. From fig.7.5 it can be seen that the waiting time t_{wi} must be adjusted appropriately to compensate for the difference in propagation delay times of different users t_{pi}. This concept is depicted in fig.7.6. The waiting time is different for all the users to facilitate their quasi-synchronous arrival at the base station.

Let us look at the algorithm that permits the achievement of a quasi-synchronous state for all the mobile users (the guard time t_{gi} is equal for all active mobiles ($t_{gi} = t_g$)) in a general situation of no initial synchronisation with the base station, that is when an error has been accumulated till the present frame. The crux of this algorithm is how the waiting time correction factor, δ_{wi}^{n+2}, in the mobile is related to the control clock correction factor δ_{ci}^{n+1}. It is assumed that during the transition from frame $n+1$ to frame $n+2$ the propagation delay does not change significantly. We define the

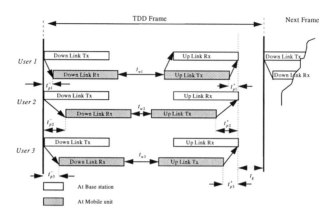

Figure 7.6 The waiting time t_{wi} changes with different propagation delays t_{pi}

summation of the total error due to the propagation delay times up to frame n,

$$\Delta_{gi}^n = \sum_{j=1}^n \delta_{gi}^j \qquad (7.4)$$

which is equally written as $\Delta_{gi}^{n+1} = \Delta_{gi}^n + \delta_{gi}^{n+1}$. In order to reduce the error to zero in (7.4), the waiting time t_{wi}^{n+2} must be properly adjusted by a correction factor $\delta_{wi}^{n+2} = t_{wi}^{n+2} - t_{wi}^{n+1}$. The mobile unit adjusts the guard time, δ_{gi}^{n+1}, using the control clock correction factor, $\delta_{ci}^{n+1} = t_{ci}^{n+1} - t_{ci}^n$ so that the propagation delay or advance times in the $(n+1)$-th frame should be neutralized in the next $(n+2)$-th frame forcing Δ_{gi}^{n+2} to zero. If we make $\Delta_{gi}^m = -\delta_{pi}^m$, by this the total error has been caused only by the error at step m and this will result in a non-cumulative error; in the next frame (TDD burst) this error will be removed by changing δ_{wi}^{m+1}. This deduction is taken partly from [8]. By using $\Delta_{gi}^m = -\delta_{pi}^m$ we can rewrite equation (7.4) as:

$$\delta_{gi}^m = \delta_{pi}^{m-1} - \delta_{pi}^m \qquad (7.5)$$

From fig.7.7 we can deduce $\delta_{ci}^m = 2\delta_{pi}^m + \delta_{gi}^m$ We can rewrite (7.5) as:

$$\delta_{ci}^m = \delta_{pi}^m - \delta_{pi}^{m-1} \qquad (7.6)$$

The waiting time in the $(n+2)$-th frame equals the waiting time in the $(n+1)$-th frame minus two times the propagation delay time, δ_{pi}^m, also from the $(n+1)$-th frame. We write $t_{wi}^{n+2} = t_{wi}^{n+1} - 2\delta_{pi}^{n+1}$ and this results in:

DETECTION STRATEGIES AND CANCELLATION SCHEMES IN A MC-CDMA SYSTEM

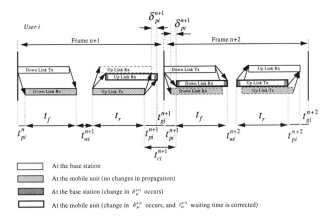

Figure 7.7 Quasi synchronous state in the uplink.

$$t_{wi}^{n+2} - t_{wi}^{n+1} = -2\delta_{pi}^{n+1} = \delta_{wi}^{n+2} \qquad (7.7)$$

If we substitute equation (7.7) in equation (7.6) having poned $m = n+1$, we obtain immediately the relation between the control clock and the waiting time correction factor:

$$\delta_{wi}^{n+2} = -2\delta_{ci}^{n+1} - \delta_{wi}^{n+1} \qquad (7.8)$$

We can see that the Δ_{gi}^m does not depends on past values of δ_{pi}^m, e.g. no accumulation of errors take place. The error is compensated immediately in the frame sequential to the one where it occurred. When we perform a discrete time domain analysis, by means of a Z transform, we conclude that all the poles in the feedback system are inside or at the edge of the unity circle. The system is stable which is confirmed in [8]. The quasi-synchronous system so established should mitigate negative effects of multi-user interference on the performance.

7.5 DETECTION STRATEGIES FOR MC-CDMA SYSTEMS

We will analyse and compare two main different detection strategies which can be used also in a TDMA/MC-CDMA system like this: single and multi-user detection strategies and about the latter in particular a kind of successive interference cancellation. The single user detector -also called the conventional detector- is the most straight forward detector. The total received signal is fed through a bank of correlators. Every signal is detected separately. In fig.8 the single user detector is depicted. In the conventional detection strategy the desired user is enhanced while the other users are suppressed like noise. The performance of the conventional single-user detector showed a large difference to the optimum detectors described in Gaussian noise channels. It was due to the near-far problem and the inability of the single user detectors to cope with the Multi Access Interference (MAI). A better performance can be achieved by not only suppressing the other users but jointly detecting them as well; the development of multi-user detection strategies, that can exploit the structure of the multi access interference, could eliminate the sophisticated high-precision power control schemes in the mobile units. Thus, an increase in complexity in the base station enables a reduction of complexity in mobile units. Also important is the near-far resistant property of multi user detection schemes that could allow a performance gain even with equal power reception (perfect power control). The performance gain should result in lower power consumption and processing gain requirements, which translate into increased longer battery life time.

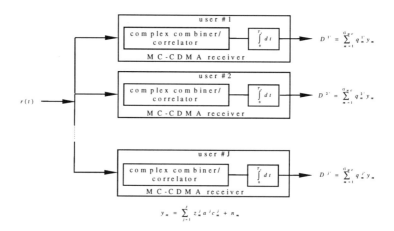

Figure 7.8 The single user detection method.

If we think about optimal detectors, multi-user detection schemes would request an exponential growth in complexity with the number of users; the challenge is to design detectors which do not have such complexity but they increase considerably the performances in comparison with the conventional ones, so we will talk about sub-optimal models. One of the simplest ideas in multi-user detectors is that of successive cancellation. The data of the strongest user is detected with a conventional detector and then its replica is subtracted from the total received waveform. This process can be repeated with the resulting waveform which contains no trace of the signal due to the strongest user assuming no error was made in its demodulation. In fig.7.9 a schematic diagram of a successive interference cancellation detector is depicted.

Now let us have a look in practice about how our successive interference cancellation algorithm can work, first of all recalling equation

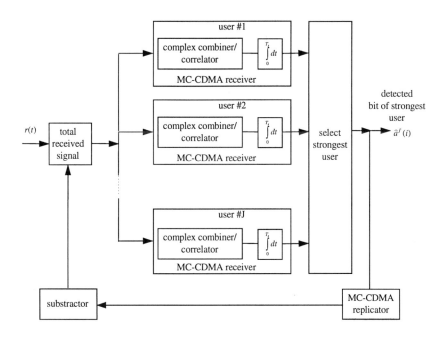

Figure 7.9 Model of a successive interference cancellation detector.

$$r(t) = \sum_{j=1}^{J} \sum_{i=-\infty}^{+\infty} \sum_{m=1}^{G_{MC}} z_m^j \alpha^j(i) c_m^j p(t - iT_s) \cos\{2\pi(f_o + m\Delta f)t + \phi_m^j\} + n(t) \tag{7.9}$$

Assuming the system operates quasi-synchronous, which is basically synchronous, we can isolate the symbol duration in the interval $[0, T_s]$ and the decision variable is:

$$D^{j\prime} = \sum_{j=1}^{J} \sum_{m=1}^{G_{MC}} z_m^j \alpha^j(i) c_m^j p(t - iT_s) \cos\{2\pi(f_o + m\Delta f)t + \phi_m^j\} \delta t \tag{7.10}$$

And putting in evidence the interference term:

$$D^{j\prime} = \sum_{m=1}^{G_{MC}} \sum_{i=-\infty}^{+\infty} z_{m\prime}^{j} \alpha^{j\prime}(i) c_m^{j\prime} q_m^{j\prime} + \sum_{m=1}^{G_{MC}} \sum_{i=-\infty}^{+\infty} \sum_{j=1\ j\neq j\prime}^{J} z_m^{j} a^{j}(i) c_m^{j} q_m^{j} + n(t)$$
(7.11)

So, the decision variable consists of three terms. The first term is the reconstruction of the desired signal, the second and third terms are the multi-user interference and the noise term respectively. The objective is to eliminate the second term without introducing errors. Of course, this is not possible, due to the uncertainty about the received envelop, z_m^j, so these channel characteristics should be estimated as well.

The cancellation scheme works as follows. First the initial stage is determined and then the remaining stages of processing follow. The initial stage is typically a conventional detector. Then followed by a scheme which successively subtracts the influence of the multiple access interference (MAI). The order of subtraction is determined by the relative strengths of the received signals. It is necessary to rank the received signals in order of their relative magnitude: one possibility is to look to the energy of the received signals but it is difficult to separate the energy contributions of the single users and besides accurate channel estimation is necessary. In [9],[10], another method of gaining information about the relative strength is suggested making use of a simple conventional detector or correlator Of course, an estimate of the channel characteristics is still necessary, but in [10] it is shown that the accuracy becomes less stringent to achieve the same performance compared to energy estimates. It is very important to cancel the strongest user before detection of the next signal, because this has the most negative effect. Another point is that the best estimate is most likely done on this signal, because the strongest signal has the minimum MAI. For example the negative influence due to other users is relatively speaking less when considering the strongest signal. So:

• Cancellation of the strongest user has the most benefit.
• Cancellation of the strongest user is most reliably performed.

In conclusion this cancellation method has significant potential of increasing capacity, and inherently good near/far resistance, being robust to received power variance. The aim is to design a nearly optimal detector which is not too complex to implement but it should be cost effective to build in practical systems and improve the performances over conventional detector schemes. Taking in consideration security aspects, that is the mobiles only know their own chip sequences while the base station has the knowledge of all the sequences, a multi-user detection scheme is of course more suitable to apply to the uplink rather than to the downlink; besides a mobile can usually tolerate minor complexity than the base station. From all these general considerations we have to note that, since the uplink is usually more limiting than the forward link [9], increasing the up link capacity will improve the overall system performance, but it cannot increase capacity beyond the capacity of the downlink. Because of this the multi-user detection must have a good performance/cost trade-off advantage compared with the single user detection scheme. Another aspect important to stress out is that the interference from outside the cell will cause limitation to the improvements to be expected by a multi-user detection scheme. In fact the algorithm provides only cancellation of intracell interference and this yields that intercell interference will

DETECTION STRATEGIES AND CANCELLATION SCHEMES IN A MC-CDMA SYSTEM

bound the capacity increase; so in areas where two or more operators are present the inter-network interference will contribute to this limiting effect. An approximation given for the possible enhancement factor, that is the capacity increase bound, will be $(1+f)/f$, with all the intracell interference cancelled, where f is the fraction between the intercell interference and the total intracell interference.

7.6 DIVERSITY COMBINING TECHNIQUES

By setting the gain factor q_m of the detectors equal to a certain value, the chosen diversity strategy is determined. Here we will consider four common diversity techniques:

7.6.1 Orthogonality Restoring Combining (ORC)

In an uplink channel, we can set the gain as:

$$q_m^{j'} = c_m^{j'} z_m^{j'*}/|z_m^j|^2 \qquad (7.12)$$

and the decision variable for the j'-th user is:

$$D^{j'} = \alpha^{j'} + \sum_{m=1}^{G_{MC}} \frac{c_m^{j'} z_m^{j'*}}{|z_m^j|^2} n_m \qquad (7.13)$$

where $\alpha^{j'}$ is the estimated bit and $c_m^{j'}$ is the spreading code for the j'-th user. In the first equation, weak subcarrier tends to be multiplied by high gain, but also the noise components are multiplied by this same gain factor. Even if the receiver can eliminate in theory the multi-user interference completely, this noise amplification degrades the BEP performance. Estimation of the received envelope of the subcarrier z_m^j is required. Due to the excessive noise amplification the perfect orthogonality restored by ORC will not be enough. In the downlink there are slight differences since all the received envelopes are the same being $z_m^j = z_m$.

7.6.2 Equal Gain Combining (EGC)

The gain for the uplink for Equal Gain Combining (EGC) is given by:

$$q_m^{j'} = c_m^{j'} \frac{z_m^{j'*}}{|z_m^j|} \qquad (7.14)$$

and the decision variable will be:

$$D^{j'} = \alpha^{j'} \sum_{m=1}^{G_{MC}} |z_m^{j'}| + \sum_{m=1}^{G_{MC}} c_m^{j'} z_m^{*j} n_m / |z_m^j| \qquad (7.15)$$

With Equal Gain Combining, the gain correction removes the code structure but does not try to equalise the effects of the amplitude scaling introduced by different channel attenuation at different subcarrier frequencies; EGC smoothes the effect of unequal received signal envelopes by using the square root of the received envelope preventing small noisy envelopes from becoming excessively amplified.

7.6.3 Maximal Ratio Combining (MRC)

The gain in the uplink for Maximal Ratio Combining (MRC) is given by:

$$q_m^{j'} = c_m^{j'} z_m^{j'*} \tag{7.16}$$

and the decision variable:

$$D^{j'} = \alpha^{j'} \sum_{m=1}^{G_{MC}} |z_m^{j'}|^2 + \sum_{m=1}^{G_{MC}} c_m^{j'} z_m^{*j'} n_m \tag{7.17}$$

In the case for one user, the Maximal Ratio Combining method can minimise the BEP [5]. The motivation behind Maximal Ratio Combining is that the components of the received signal with large amplitudes are likely to contain relatively less noise, so their effect on the decision process is increased by squaring their amplitudes; even if the performances are similar with EGC, the former has a greater implementation complexity because of the requirement of more precise correct weighting factors.

7.6.4 Minimum Mean Square Error Combining (MMSEC)

In a down-link channel, this method estimates the transmitted symbol $\alpha^{j'}$ by the linear sum [11],[12]:

$$\tilde{a}^{j'} = \sum_{m=1}^{G_{MC}} q_m^{j'} y_m \tag{7.18}$$

Based on the minimum mean square estimation (MMSEC) criterion, the error must be orthogonal to all the baseband components of the received subcarriers [13],[5]. Thus,

$$(\alpha^{j'} - \tilde{a}^{j'}).y_m^* = 0 (m = 1, 2, .., G_{MC}) \tag{7.19}$$

For the uplink we get the following result for the gain factor:

$$q_m^{j'} = \frac{z_m^{*j'} c_m^{j'}}{\sum_{j=1}^{J} |z_m^j|^2 + N_o} \tag{7.20}$$

where N_o is the noise power. Estimation of z_m^j, the number of active users J and the noise power is required. We can see that for small $|z_m^j|$ the gain becomes small to avoid large noise amplification, while for large $|z_m^j|$ it increases in proportion to the inverse of the subcarrier envelope $z_m^{*j}/|z_m^j|^2$ in order to restore orthogonality among the active users. As already issued by [12] mean square error combining (Wiener filtering) is optimal when we consider both noise and interference since this technique directly addresses the effects of the interference on the BEP. Finally we have to note that the expressions for the downlink are very similar for all these techniques and a main point common to all of them that is the great importance of a correct channel estimation to obtain acceptable performances.

7.7 SIMULATIONS

The simulations are performed based on the Monte Carlo method. The simulation program is written in C and can be divided in three parts: the transmitter, the mobile radio channel and the receiver or detection part. In the following sections, these parts are described.

7.7.1 MC-CDMA Transmitter Model.

In this part of the simulator the MC-CDMA subcarriers are generated. First the input or information bearing signal -1,+1 is generated by a random sequence generator. This source signal is BPSK modulated, copied, and afterwards spread by means of a Walsh Hadamard code in the MC-CDMA modulator. In fig.7.10, the transmitter is depicted.

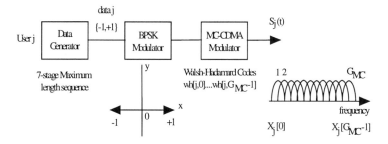

Figure 7.10 Model of the MC-CDMA transmitter.

The model is suitable to use in uplink as well as downlink simulations. The difference in the downlink is that all the modulators output of individual users are multiplexed, and afterwards fed through one mobile radio channel. However, for the uplink all the users are fed through separate channels and then combined at the base station. In fig.7.11 the uplink model is depicted. In the base station the received signals will be detected. In the next section we will describe the receiver.

The model for the downlink is simply one multiplexer to combine the separate modulator outputs to one sum signal. This signal is fed through one channel.

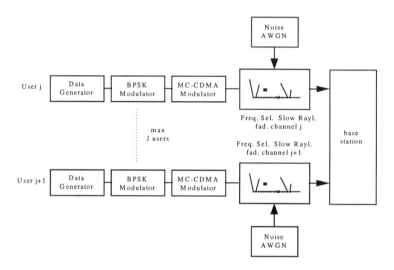

Figure 7.11 Simulation model for the uplink.

7.7.2 Mobile Radio Channel Model

In the simulation we use 32 subcarriers with carrier spacing $1/T_s$. We use a uniform Delay Profile in all simulations. The RMS delay spread is τ_{RMS} (RMS = 20 ns, the symbol rate is about 100 kbps. By increasing the number of subcarriers the symbol rate can be increased. We perform the calculations in the frequency domain. We need only one operation, the FFT of the instantaneous channel response. This frequency channel response is then multiplied with all independent inphase and quadrature components of the subcarriers.

It is essential for multi carrier transmission to guarantee the frequency nonselective fading over each subcarrier. In this report, we assume 32 subcarriers for the processing gain of 32. When the delay spread is too large for the guard time, we cannot guarantee the frequency nonselective fading over each subcarrier. However, if this is the case, we can apply more subcarriers, for instance, 1024 subcarriers with the same processing gain. In [6], it was shown that with 1024 subcarriers frequency nonselective fading over each subcarrier can be guaranteed. The performance is completely the same.

An instantaneous channel frequency response is obtained by taking the Fourier Transform of an instantaneous channel response. The inphase and quadrature components of the instantaneous impulse response are generated randomly according to a Gaussian distribution. The FFT is used to obtain 32 correlated Gaussian random variables each of which corresponds to an inphase and quadrature component of the path. In fig.7.11 the channel is represented by a number of vectors. For the downlink

DETECTION STRATEGIES AND CANCELLATION SCHEMES IN A MC-CDMA SYSTEM

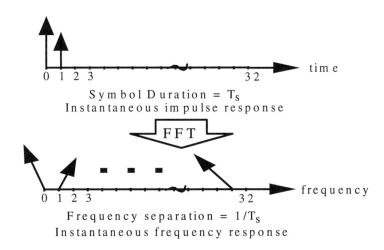

Figure 7.12 Symbolic representation of the mobile radio channel.

this model is applied once e.g. it is the same for every user. However, for the uplink the model is applied separately to every user. The channel characteristic is given as $c(t)$ whereas in the simulator it is treated as $cx + jcy$. Both variables are independent Gaussian random variables. The vectors are also contaminated with noise. The received signal through the channel $r(t)$ can be written as:

$$r(t) = c(t) * s(t) + complexnoise = rx + jry \qquad (7.21)$$

Where $rx = (x[j] * cx[j] - y[j] * cy[j]) + noise$ and $ry = (x[j] * cy[j] + y[j] * cx[j]) + noise$. Feeding a MC-CDMA signal through this channel will distort the orthogonality among the carriers. This orthogonality must be retained as much as possible.

7.7.3 MC-CDMA Receiver Model

Several receiver structures are used in the simulation. The BEP of the system depends on the type of diversity combining techniques which are applied. The gain q_m^i determines the diversity combining techniques and therefore the bit error probability. Further, perfect synchronisation is assumed. The general model for the receiver is depicted in fig. 7.13.

7.7.4 Simulation Flow

Here the simplified flowcharts are outlined. Basically there are two kinds of simulators built one for the single user detection and one for the multi-user detection. The

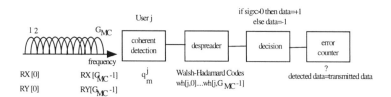

Figure 7.13 Simulation model of the single user MC-CDMA detector.

simulator is based on the Monte Carlo method e.g. the maximum allowable number of errors for an individual users is set. The algorithm is iterated until this number of errors is reached, after which the Bit Error Probability is calculated. The simulation runs are performed for different signal to noise values as well as for different loads (e.g. the number of simultaneous users).

In the single-user detection algorithm every user is detected separately. The diversity combining strategy is set in the detector or receiver. After all active users are detected, the simulator runs again with a newly created received baseband signal. For the multi-user detection algorithm this is slightly different. After all active users are detected and cancelled successively (in sequence of their relative strengths) the simulation starts again with a newly created received baseband signal. In figs.7.14 and 7.15 the flow diagrams of the named detection schemes are depicted respectively.

The simulations were performed on a Pentium PC and/or VAX4100. The VAX4100 is fast when we compare this with the PC run times. Generally, for more accurate simulation results (e.g. smooth BEP graphs) the number of allowable errors in the Monte Carlo simulation must be set higher. All the simulation are completed with 100 errors as a maximum value. With 100 errors, most of the uplink simulations took several hours on the VAX and several days on the PC. This suggests that, if we increase the accuracy in the model by taking more parameters (e.g. phase jitter synchronisation errors, subcarrier tracking etc.) into account, we should conduct our computer analysis on really fast computers.

We simulated several diversity combining techniques. The techniques we investigated were Orthogonality Restoring Combining (ORC), Equal Gain Combining (EGC), Maximal Ratio Combining (MRC) and Minimum Mean Square Error Combining (MMSEC). Most of these techniques are applied to single and multi-user detection schemes. The performance is compared. For all the systems perfect synchronisation is assumed. In addition, no error correction codes are used as well. All simulations are performed with a BPSK modulation scheme with rectangular waveforms. For all mobile units perfect power control is assumed e.g. no near-far effects is investigated. For the uplink we give the results for every diversity combining technique, started with a single user approach, followed by a multi-user detection scheme. In all multi-user

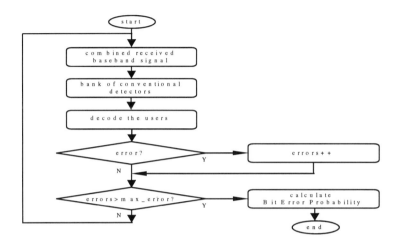

Figure 7.14 Flow diagram of the conventional single user detector

detection schemes the method of successive interference cancellation is applied. For the downlink only the single user detection techniques are applied, because applying multi-user reception technique is questionable when we take security issues into consideration. Besides in order to inquire on the sensitivity and robustness of the system when there are channel response estimation errors, the received amplitudes were intentionally contaminated with noise in some simulations. An error in the estimation of an envelope is modelled as Gaussian noise, therefore at each envelope a random figure with a Gaussian distribution is added.

7.8 RESULTS

7.8.1 Downlink

ORC does not perform very well due to the excessive noise amplification which degrades the BEP; even if we choose a threshold of 10^{-2} as a maximal bit error probability value, only slow data rate can be achieved (fig.7.16). In a noise limited channel MRC should perform better than EGC but from the simulations we see that for a large number of users EGC outperforms MRC; this reflects the observation that MRC distorts the orthogonality between users while we could achieve enhanced performances only by restoring the orthogonality of the interfering signals (fig.7.17). Both methods EGC and MRC showed that for a downlink situation approximately 50MMSEC (Wiener filtering) outperforms substantially the other conventional detection methods because of its great properties in combating interference. Also under conditions of a full load Minimum Mean Square Error Combining performs very well. For instance, when we

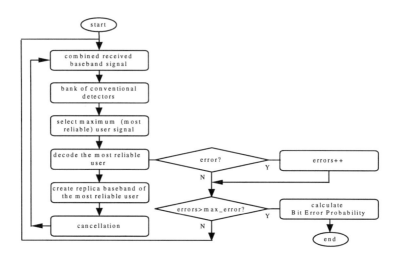

Figure 7.15 Flow diagram of the (successive interference cancellation) multi-user detector.

transmit coded voice at a rate of 13 kbit/s (depending on the used speech codec) [14] the minimum acceptable BEP must be around 10^{-3} in order to have acceptable quality. This is feasible for a fully loaded system with a E_b/N_o larger than 18 dB (fig.7.18).

7.8.2 Uplink

In the uplink we have considered two different techniques, single and multi-user detection schemes. With a conventional detector (single-user), when we take a BEP of 10^{-2} as a minimum value required for acceptable performance for low data rates, we can conclude that ORC is not suitable to be used in an uplink situation. Using a EGC strategy and taking again a BEP of 10^{-2} as maximum bound in consideration, it needs at least an average SNR of 18 dB to accommodate 8 users (fig.7.19). This is 1/4 of the maximum capacity of our system. The results of MRC in MC-CDMA uplink differ not much from EGC and are optimal when there is only one active user (in this case the minimised BEP approaches the lower theoretical bound). The results for EGC are 1 dB worse than the results for MRC. Like in the downlink MMSEC performances are considerably better then ORC, EGC and MRC, especially when we do not operate at the maximum capacity. When we consider a successive interference cancellation scheme the general performances are much improved above all in a full loaded system where the interference plays a determinant role. This method is, according to [15], less sensitive to near-far effects; the capacity of the system is enhanced, because the requested SNR bound with a given Bit Error Probability and the same number of users is lower. This technique does not give satisfying improvements with ORC (fig.7.20); yet we obtain acceptable performances with EGC and MRC because we need an average SNR of 10 dB to accommodate 8 users with the same BEP of 10^{-2} (fig.7.21), even if we do not reach the best performances compared with MMSEC. Successive interference cancellation using a Minimum Mean Square Error Combining (MMSEC) method as a diversity combining technique in a MC-CDMA system performs well and much better in comparison with a single-user detection scheme with a high number of active users. When we consider the number of active users in a fully loaded system the Bit Error Probability remains below 10^{-3} at a signal to noise ratio of around 15 dB (fig.7.22). This is in contrast to a fully loaded MC-CDMA system using a conventional detector, which experiences severe performance degradation (fig.7.23)

Finally we have to stress how essential it is to have a perfect channel estimation even with a successive interference cancellation scheme; the sensitivity to errors is higher of course in a conventional detector but since this conventional receiver is also used in the successive interference cancellation scheme in order to rank the received signals, the performance of this Multi-user Detection scheme will also degrade (fig.7.24). The simulation results must be interpreted as an indication of the performance degradation, because the relative power of the added noise might be out of proportion in carefully designed systems.

One last note is that the simulation results can always be compared with a reference value, namely the lower bound. The lower bound expresses the lowest possible bit error probability (BEP) when there is only one active user in the system. In [16] this lower bound value is derived for a binary system with L diversity channels. For each channel is assumed frequency non-selective and slowly fading with Rayleigh-distributed envelope statistics. This value holds also for MC-CDMA systems. In [6] it is proven that a frequency domain MC-CDMA Rake receiver based on maximal ratio combining can achieve the best BEP performance. This is because the lower bound is determined by the power delay profile and is therefore valid for the uplink as well as for the downlink. We can write (with E_b/N_0 the average signal to noise ratio):

$$BEP_{lowerbound} = (\frac{1-\mu}{2})^L \sum_{l=0}^{L-1} \binom{L-1+l}{l}(\frac{1+\mu}{2})^l \quad (7.22)$$

where $\mu = \sqrt{\frac{E_b/N_o}{L+E_b/N_o}}$

7.9 CONCLUSIONS AND RECOMMENDATIONS

Code division multiple access systems are interference limited (MAI). Any method of reducing this interference will increase the systems capacity. With capacity we mean the number of simultaneous active users which can be accommodated by the system before severe performance degradation in terms of higher Bit Error Probability (BEP) occurs. In this report we investigated different methods to reduce MAI. Applying diversity combining techniques which utilise the signal received over multi paths is one aspect of the whole problem. Simulation showed that with a minimum mean square error combining strategy the best performance can be achieved. Other strategies (maximal ratio combining and equal gain combining) perform less well because these techniques do not address the effect of interference to the Bit Error Probability directly.

Synchronisation on the downlink is performed by the base station which transmits simultaneously towards all the active users. Because there is only one downlink transmitter, carrier synchronisation is common to all the active users. Also, on the downlink we make use of orthogonal code sequences (Walsh-Hadamard codes). This orthogonality is easily maintained, because there is only one transmitter. On the uplink, synchronisation can be achieved by time alignment of the active mobile transmitters. A method for establishing and maintaining synchronisation is to force all active users to transmit in predefined time slots. A Time Division Duplex frame is subdivided into a transmit and a receive burst. A relatively simple algorithm -which manages a feedback control system in every mobile unit -will force the mobiles in a quasi-synchronous state. It is shown that with this system a very high close quasi-synchronisation state can be achieved. With this system, only a feedback unit is required at the mobiles to compensate for any propagation delay changes.

In order to gain further enhancement of the system a successive interference cancellation method is applied. This techniques cancels users from the total received signal in sequence to their relative receiving strength. Simulation showed a performance enhancement. This cancellation algorithm in MC-CDMA environment in combination with minimum mean square error combining technique realise a system which achieves an acceptable low Bit Error Probability even under full load. Simulations have also shown that in comparable circumstances the above method outperforms a Direct-Sequence CDMA system substantially.

The ease of synchronisation achievable in a Time Division Duplex system in combination with the robustness of MC-CDMA against multipath fading and multi-user interference, should deliver a system which can be truly applied in third generation mobile communication networks. Further, a thorough analysis of the proposed TDD/MC-CDMA system is necessary to come to the right choice for implementation of a particular service in a particular environment.

The sensitivity of the successive interference cancellation scheme to imperfect channel estimation can be considered as a serious issue. Further work has to be done to determine a method for optimum channel estimation. It is certainly worth while pursuing to optimise and improve this multi user detection strategy, because in [10] the capability of handling near-far problems was also shown. It relaxed the demands of perfect power control in the mobile units, which of course will reduce the cost of the units. Successive interference cancellation is technically speaking suitable to apply in the downlink as well as the uplink. However using it in the downlink is questionable when taking security into consideration. Using it in the uplink, will in practice result in a continuous real-time signal processing in the base station. Effects of performance in terms of throughput due to this processing delay has to be investigated further. Propagation of errors in a successive interference cancellation scheme merits further research. If for instance a decision error will propagate (e.g. sequential errors will be generated out of previous errors) the algorithm will probably perform worse than a conventional single user detection scheme. It is recommended to perform simulations with different number of subcarriers, different number of delay spread values and investigate the effects on the bit error probability. Various bit rate values are interesting to study and see the effects on the overall performance. Another issue to investigate further is the performance degradation due to synchronisation errors that still are present since we work in a quasi-synchronous state and due to subcarrier tracking errors since we made only a burst or symbol synchronisation.

References

[1] Y. Bar-Ness, J.P. Linnartz and X. Liu: "Synchronous Multi-user multi-carrier CDMA communication system with decorrelating interference canceler," Proc. of IEEE PIMRC'94, pp. 184-188, The Hague, The Netherlands, September 1994.

[2] F. Kleer, S. Hara and R. Prasad: "Performance evaluation of a successive interference cancellation scheme in a quasi-synchronous MC-CDMA system", Proc. of ICC' 98, June 1998, Atlanta, USA.

[3] N. Yee, J.P.M.G. Linnartz and G. Fettweis: "Multi-Carrier CDMA in Indoor Wireless Radio Networks," IEICE Trans. commun., vol. E77-B, No. 7 July 1994.

[4] L.J. Cimini: "Analysis and simulation of a digital mobile channel using orthogonal frequency division," IEEE transactions on communications vol. COM-33, no. 7, July 1985.

[5] R. Prasad and S. Hara: "An overview of multi-carrier CDMA,", Proc. of IEEE ISSSTA'96, Germany, September 1996.

[6] S. Hara, T-H. Lee and R. Prasad: "BER comparison of DS-CDMA and MC-CDMA for frequency selective fading channels," 7th Tyrrhenian International workshop on digital communications, 10-14 September 1995.

[7] R.D.J. van Nee: "Timing aspects of synchronous CDMA," Proc. of IEEE PIMRC'94, pp. 439-443, The Hague, The Netherlands, September 1994.

[8] R. Esmailzadeh and M. Nakagawa: "Quasi-Synchronous Time Division Duplex CDMA,", IEICE Transactions on Communications, Vol. E78-A no. 9, pp. 1201-1204, September 1995

[9] S.G. Glisic and P. A. Leppanen (Eds.): "Code Division Multiple Access Communications,", Kluwer Academic Publishers, 1995.

[10] P.R. Patel and J.M. Holtzman: "Analysis of a DS-CDMA Successive Interference Cancellation Scheme using correlations," Proc. of IEEE GLOBECOM '93, pp. 76-80, Houston, USA, November 1993.

[11] Chouly, A. Brajal and S. Jourdan: "Orthogonal Multicarrier Techniques Applied to Direct-Sequence Spread Spectrum CDMA Systems," Proc. of IEEE GLOBECOM '93, pp. 1723-1728, Houston, USA, November 1993.

[12] N. Yee and J-P. Linnartz: "Wiener filtering of Multicarrier CDMA in a Rayleigh fading channel," Proc. of IEEE PIMRC'94, pp. 1344-1347, The Hague, The Netherlands, September 1994.

[13] Leon-Garcia: "Probability and Random Processes for the electrical engineering," Addison Wesley,. 2nd Ed. May 1994.

[14] Mehrotra: "Cellular Radio Performance Engineering," Artech House, Norwood MA, 1994.

[15] P. Patel and J. Holtzman: "Analysis of a Simple Successive Interference Cancellation Scheme in a DS-CDMA System," IEEE journal on selected areas in communications, Vol. 12, No. 5, pp. 796-807, June 1994.

[16] J.G. Proakis: "Digital Communications," Mc-Graw Hill, 3th Ed. 1995.

DETECTION STRATEGIES AND CANCELLATION SCHEMES IN A MC-CDMA SYSTEM 211

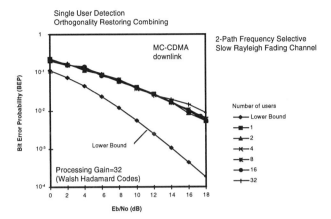

Figure 7.16   Single user detection with ORC in a MC-CDMA downlink.

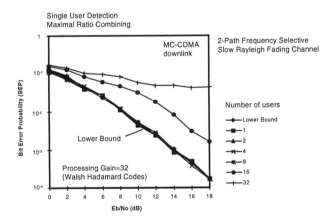

Figure 7.17   Single User Detection MRC in a MC-CDMA downlink.

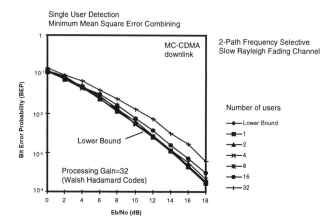

Figure 7.18 Single user detection with MMSEC in a MC-CDMA downlink.

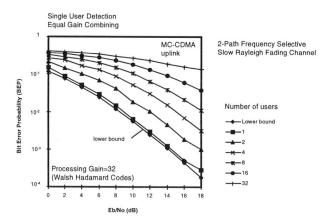

Figure 7.19 Single User Detection EGC in a MC-CDMA uplink.

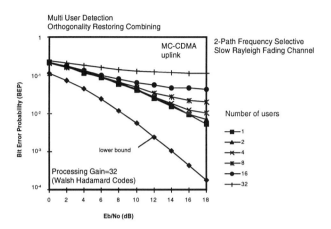

Figure 7.20 Multi User Detection ORC in a MC-CDMA uplink.

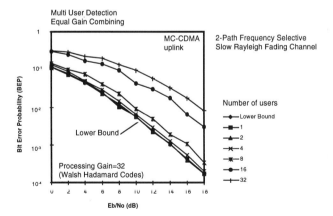

Figure 7.21 Multi-user Detection with EGC in a MC-CDMA uplink.

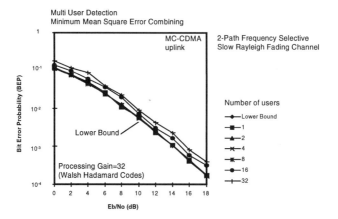

Figure 7.22 Multi-user Detection with MMSEC in a MC-CDMA uplink.

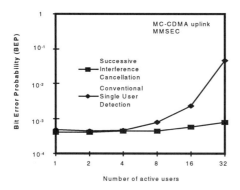

Figure 7.23 Performance enhancement of Multi User Detection compared to Single User Detection.

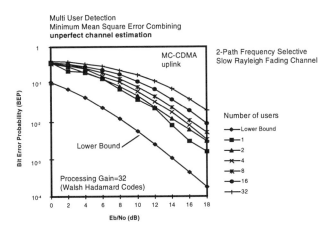

Figure 7.24 Imperfect channel estimation breaks down the performance of a Multi User Detection scheme using MMSEC.

8 CODING VS. SPREADING OVER BLOCK FADING CHANNELS

Ezio Biglieri, Giuseppe Caire, and Giorgio Taricco

Dipartimento di Elettronica
Politecnico di Torino
Corso Duca degli Abruzzi 24
I-10129 Torino, Italy*
<name>@polito.it

Abstract: In this chapter we study the optimum tradeoff of coding vs. spreading in a single-cell CDMA mobile communication system with block-fading, block-synchronous (but symbol-asynchronous) transmission, and slow power control. The optimization criterion we choose is based on system capacity, as measured in users/cell×bit/s/Hz,

We adopt an information-theoretic definition of *outage*: this is the event that the mutual information experienced by a user code word falls below the actual user code rate. The system capacity is then defined as above under decoding-delay and outage-probability constraints. We examine the conventional single-user receiver and a linear MMSE multiuser receiver. Our results show that, with ideal power control and optimum coding/spreading tradeoff, capacities close to 1 user/cell×bit/s/Hz are achievable by the conventional receiver, while the capacity gain offered by MMSE multiuser detection is moderate. With non-ideal power control MMSE multiuser detection is more attractive: in fact, it proves to be very robust to residual power-control errors, while conventional detection suffers from a large capacity degradation.

8.1 INTRODUCTION

In this contribution we consider the following problem: Given a system whose bandwidth W and bit rate R_b are assigned, what fraction of the overall bandwidth-expansion factor W/R_b should be allocated to coding, and what to spreading in order to optimize performance? As we shall see, the solution to this problem depends heavily on system assumptions. Hence, we describe our approach to the problem through a case study

---

*This research was supported by the Italian Space Agency (ASI).

which, although involving a special system model, outlines the basic solution philosophy. Specifically, we consider here a single-cell CDMA mobile communication system with block-fading, block-synchronous (but symbol-asynchronous) transmission, and slow power control. The optimization criterion is based on system capacity, measured in users/cell×bit/s/Hz and defined under decoding-delay and outage-probability constraints.

8.1.1 Channel model and its constraints

The Gaussian multiple-access channel (MAC), as well as its generalization to a channel in which users may have different signaling waveforms while transmission may be symbol-asynchronous, have been extensively studied [16]. Typical mobile communication channels differ considerably from the Gaussian MAC, since they suffer from time-varying propagation vagaries. These can be either frequency-flat, like shadowing and path attenuation due to changing distance, or frequency-selective, like multipath fading.

In mobile telephony, a strict constraint on the decoding delay is imposed and channel variations are normally rather slow with respect to the maximum allowed delay ΔT (usually a few tens of milliseconds). This situation makes it appropriate to use a channel model introduced in [7, 10] (see also [8, 9]) and known as *block fading* model. It is motivated by the fact that, in many mobile radio situations, the channel coherence time is much longer than one symbol interval, and hence several transmitted symbols are affected by the same fading value. The transmission of a user code word is characterized by a small number of "channel states," intended as the realization of the random variables which determine the channel. For example, the fading time-variation is characterized by its Doppler bandwidth B_d [12, Ch. 14]. The number of "almost independent" realizations of the fading channel during a time span ΔT is, roughly speaking, $\lfloor B_d \Delta T \rfloor + 1$. For a portable handset held by a pedestrian, which is by far the most common case in today's personal communications, $B_d \simeq 0$ so that the channel is constant (but random) during the transmission of a code word [1].

Use of this channel model allows one to introduce a delay constraint for transmission, which is realistic whenever infinite-depth interleaving is not a reasonable assumption. In fact, the block-fading model assumes that a code word of length $n = MN_s$ spans M blocks of N_s symbols each (a group of M blocks will be referred to as a *frame*.) The value of the fading in each block is constant, and each block is sent through an independent channel. An interleaver spreads the code symbols over the M blocks. M is now a measure of the interleaving *delay* of the system: in fact, $M = 1$ (or $N_s = n$) corresponds to no interleaving, while $M = n$ (or $N_s = 1$) corresponds to perfect interleaving. Thus, results obtained for different values of M illustrate the effects of finite decoding delay.

As discussed in [10], with the block fading model if $M < \infty$ the notions of *average mutual information* and *average capacity* have no practical operational meaning. On

---

[1] Normal Doppler bandwidths with moving vehicles and carrier frequencies around 1 GHz are about 100 Hz, so that with $\Delta T = 100$ ms we get not more than 10 fading "blocks".

the contrary, the "instantaneous" mutual information of the channel is a random variable and the Shannon capacity is zero, since the probability that the mutual information is below any specified code rate $\mathcal{R} > 0$ is strictly positive. In this framework, it makes sense to study the capacity for an assigned outage probability, the latter being defined as the probability that the mutual information of the random channel experienced by the transmission of a user code word falls below the actual code rate [10].

In real mobile communication channels synchronization among users has a major impact on the receiver design. For example, consider a transmission scheme where users transmit signal blocks possibly separated in time and assume that a block of user 0 overlaps partially with a block of user 1 and partially with a block of user 2. These are characterized by different average signal levels, carrier phases, transmission delays and linear distortion due to the multipath fading channel. The channel "seen by user 0" exhibits fast (almost instantaneous) variations of the interference parameters which are difficult, if not impossible, to track. Hence, the use of a multiuser detection scheme (e.g., an adaptive interference canceler [6]) may not be possible with block-asynchronous transmission. On the other hand, synchronization at the block level is not difficult to achieve and implies neither frame nor symbol synchronization: it is just a coarse quantization of the time axis, obtained by distributing a common clock to all users in the same cell through the downlink channel (from the base station to the mobile).

8.1.2 Outline of this chapter

In this chapter we consider a simplified model for the link from mobile to base station (the "uplink") in a single-cell CDMA mobile communication system with block fading, block-synchronous (possibly frame and symbol-asynchronous) transmission, and slow power control. We study the capacity of this system, expressed as the maximum of the product of the number of users per cell times the user spectral efficiency, under a constraint on delay and on outage probability. We examine the conventional single-user receiver as well as the linear minimum mean-square error (MMSE) multiuser receiver [14]. With ideal power control, by choosing a target outage probability 10^{-2}, we find that capacities close to 1 user/cell×bit/s/Hz can be achieved by the single-user receiver even with moderate interleaving depth. In this case, the MMSE multiuser receiver offers only a moderate capacity increase with respect to the conventional single-user receiver. On the other hand, with non-ideal power control the latter receiver suffers from a very large capacity degradation, while the MMSE multiuser receiver proves to be considerably robust.

8.2 SYSTEM MODEL

Let the uplink channel of a single-cell mobile communication system be characterized by a total (two-sided) system bandwidth W, user information bit rate R_b (bit/s) and maximum allowable decoding delay ΔT. The main assumptions underlying our model are:

- Access is time-slotted, with slots of duration T_s. Each user signal block occupies one time slot and spans the whole system bandwidth.

- A control channel from the receiver to the transmitters implements slow power control. No form of channelized FDMA/TDMA with dynamic centralized allocation is allowed, nor do the transmitters have knowledge of the channel state.

- User codes are selected independently, and decoding is strictly single-user and feedforward. No joint decoding and/or feedback decoding (viz., any form of "onion peeling") is considered.

- The propagation channel is modeled as a frequency-selective Rayleigh block-fading channel. Frequency-flat attenuation due to propagation distance is assumed to be perfectly compensated by power control. However, because of non-ideal power control, a residual log-normal shadowing may be present. This is modeled as the power gain $10^{\sigma_{\text{sh}} s/10}$, where s is normally distributed with mean zero and unit variance and σ_{sh} (expressed in dB) is the residual *shadowing factor* [4].

Coding and interleaving. In order to introduce time diversity, code words are interleaved and transmitted over M separate signal blocks. We assume that guard times longer than the maximum channel memory are inserted, so that the channel can be considered blockwise memoryless. Encoders produce sequences of length $N_s M$ with elements in the complex signal set $\mathcal{X}$ with unit energy per symbol (as defined before, N denotes the number of symbols per block.) The overall code rate is $\mathcal{R}$ bit/symbol. Following [10], we assume that M (the interleaving depth) is a small integer while the number of symbols per block N_s is large, i.e., we are interested in analyzing the system in the limit for $N_s \to \infty$ and finite (small) M.

Fading channel. Because of the block-fading assumption, during the transmission of a code word the channel is described by the sequence of M impulse responses $\{c_m(\tau) : m = 0, \ldots, M-1\}$ or equivalently by the sequence of M frequency responses $\{C_m(f) : m = 0, \ldots, M-1\}$, where

$$C_m(f) \triangleq \int_{-\infty}^{\infty} c_m(\tau) e^{-j2\pi f \tau} \, d\tau$$

is the (continuous-time) Fourier transform of $c_m(\tau)$. The support of $c_m(\tau)$ is $[0, T_d]$, where T_d is referred to as the *channel delay spread* [12, Ch. 14]-[2, Ch. 13]. For each positive integer n and for all epochs $\{\tau_i : i = 0, \ldots, n-1\}$, the complex random variables $\{c_m(\tau_i) : i = 0, \ldots, n-1\}$ are zero-mean jointly Gaussian with i.i.d. real and imaginary parts. Finally, we assume that $c_m(\tau)$ and $c_m(\tau')$ are independent for $\tau \neq \tau'$ (independent scattering assumption). The channel is characterized by the *multipath intensity profile* [12, Ch. 7] $\sigma^2(\tau) = E[|c_m(\tau)|^2]$ with normalized total power $\int_0^{T_d} \sigma^2(\tau) d\tau = 1$. The fading time-frequency second-order statistics is characterized by the time-frequency autocorrelation function

$$\Phi_C(m, \Delta\theta) = E[C_n(\theta) C_{n-m}(\theta - \Delta\theta)^*] \tag{8.1}$$

where we assume implicitly that the fading channel is wide-sense stationary both in time and in frequency.

Equivalent discrete-time channel. Let N denote the number of users transmitting at the same time. We assume that the signal transmitted by user i over block m is linearly modulated and characterized by the unit-energy *signature waveform* $s_m^i(t)$, for $i = 0, \ldots, N-1$ and $m = 0, \ldots, M-1$. Since $s_m^i(t)$ is allowed to change from block to block, we are considering the possibility of time-varying signature waveforms. The number of symbols (i.e., independent complex Shannon dimensions) for a block of duration T_s is given by $N_s = WT_s/L$, where $1/T$ is the symbol rate and $L = WT$ is the *spreading factor* of the waveform $s_m^i(t)$ (assumed to be the same for all i and m).

The signal received at the base station during the m-th block can be written as

$$y_m(t) = \sum_{i=0}^{N-1} A_m^i c_m^i(t) \star x_m^i(t) + n(t) \qquad (8.2)$$

where A_m^i and $c_m^i(t)$ are the amplitude gain and the fading channel from transmitter i to the base station, $\star$ denotes convolution, $n(t)$ is a complex Gaussian white noise with i.i.d. real and imaginary parts and autocorrelation function $E[n(t)n(t-\tau)^*] = N_0 \delta(\tau)$, and $x_m^i(t)$ is the signal of user i transmitted during the m-th block, given by

$$x_m^i(t) = \sum_{k=0}^{N_s-1} x_m^i[k] s_m^i(t - kT - \tau_m^i - mT_f) e^{j\phi_m^i} \qquad (8.3)$$

Here, $x_m^i[k] \in \mathcal{X}$, τ_m^i and ϕ_m^i are the delay and carrier phase of user i during block m, and T_f is the block time separation.

Let user 0 be the reference user. Receiver 0 is assumed to be a linear time-varying piecewise-constant filter characterized by the sequence of impulse responses $\{w_m(-\tau)^* : m = 0, \ldots, M-1\}$, followed by a sampler at the symbol rate $1/T$ and by a decoder matched to the reference-user code (decoder 0). The sequence of output samples in block m is given by

$$y_m[k] = \sum_{i=0}^{N-1} \sum_j A_m^i p_m^i[j] x_m^i[k-j] + \nu_m[k] \qquad (8.4)$$

where we define

$$\begin{aligned} h_m^i(t) &\triangleq c_m^i(t) \star s_m^i(t) e^{j\phi_m^i} \\ p_m^i[k] &\triangleq w_m^*(-t) \star h_m^i(t - \tau_m^i)|_{t=kT+\tau_m^0+mT_f} \\ \nu_m[k] &\triangleq w_m^*(-t) \star n(t)|_{t=kT+\tau_m^0+mT_f} \end{aligned} \qquad (8.5)$$

The noise sequence $\nu_m[k]$ has autocorrelation function

$$E[\nu_m[j] \nu_m^*[j-k]] = N_0 r_{w,m}[k] \qquad (8.6)$$

where

$$r_{w,m}[k] \triangleq \int_{-\infty}^{\infty} w_m(\tau) w_m^*(\tau - kT) \, d\tau$$

Decoder 0 makes a decision on the user 0 message based on the observation

$$\{y_m[k] : k = 0, \ldots, N_s - 1\} \quad \text{for} \quad m = 0, \ldots, M - 1$$

8.3 SYSTEM CAPACITY VERSUS OUTAGE PROBABILITY

Outage probability. Let the *channel state* S_m denote the set of random variables $A_m^i, \tau_m^i, \phi_m^i$, of random impulse responses $c_m^i(\tau)$ and, in the case of random signature waveform selection, of the waveforms $s_m^i(t)$ which determine the discrete-time channel (8.4). For the sake of notational simplicity, let $\mathbf{S} = \{S_m : m = 0, \ldots, M-1\}$ denote the sequence of channel states over the M blocks spanned by a user code word and denote by $I_M(\mathbf{S})$ the instantaneous conditional mutual information (in bit/symbol) as $N_s \to \infty$:

$$I_M(\mathbf{S}) \triangleq \lim_{N_s \to \infty} \frac{1}{MN_s} I\left(\bigcup_{m=0}^{M-1} \{x_m^0[k]\}_{k=0}^{N_s-1}; \bigcup_{m=0}^{M-1} \{y_m[k]\}_{k=0}^{N_s-1} \middle| \mathbf{S} = \mathbf{S}\right) \quad (8.7)$$

where (with a slight abuse of notation [10]) we indicate by $I(\mathbf{X}; \mathbf{Y} | \mathbf{S} = \mathbf{S})$ the functional

$$I(\mathbf{X}; \mathbf{Y} | \mathbf{S} = \mathbf{S}) \triangleq \sum_{\mathbf{x} \in \mathcal{X}} \sum_{\mathbf{y} \in \mathcal{Y}} p(\mathbf{x}, \mathbf{y} | \mathbf{S}) \log_2 \left(\frac{p(\mathbf{x}, \mathbf{y} | \mathbf{S})}{p(\mathbf{x} | \mathbf{S}) p(\mathbf{y} | \mathbf{S})}\right)$$

where $(\mathbf{X}, \mathbf{Y}) \in \mathcal{X} \times \mathcal{Y}$ are random vectors jointly distributed according to $p(\mathbf{x}, \mathbf{y} | \mathbf{S})$ with marginals $p(\mathbf{x} | \mathbf{S})$ and $p(\mathbf{y} | \mathbf{S})$ conditionally on $\mathbf{S}$ (note that if $\mathbf{S}$ is a random vector, $I(\mathbf{X}; \mathbf{Y} | \mathbf{S} = \mathbf{S})$ is a random variable. The standard conditional average mutual information is obtained by averaging $I_M(\mathbf{S})$ with respect to $\mathbf{S}$.

Outage probability is defined as

$$P_{\text{out}}(\mathcal{R}) \triangleq P(I_M(\mathbf{S}) < \mathcal{R}) \quad (8.8)$$

An operational motivation of the above definition of outage probability is the following. Assume a code rate $\mathcal{R}$ bit/symbol, and let $\overline{P}_{e|\mathbf{S}}(\mathcal{R})$ denote the message error probability averaged over the code ensemble of all codes with rate $\mathcal{R}$ and length MN_s, randomly generated according to a certain input probability distribution and conditioned with respect to the sequence of channel realizations $\mathbf{S}$. From the channel coding theorem and its strong converse (see [3] and references therein) we can write

$$\lim_{N_s \to \infty} \overline{P}_{e|\mathbf{S}}(\mathcal{R}) = \mathbb{J}_{\{I_M(\mathbf{S}) < \mathcal{R}\}} = \begin{cases} 0 & \text{if } I_M(\mathbf{S}) \geq \mathcal{R} \\ 1 & \text{if } I_M(\mathbf{S}) < \mathcal{R} \end{cases} \quad (8.9)$$

($\mathbb{J}_\mathcal{A}$ denotes the indicator function of the event $\mathcal{A}$). By averaging $\overline{P}_{e|\mathbf{S}}(\mathcal{R})$ with respect to $\mathbf{S}$ and exchanging limit with expectation, we can write

$$\lim_{N_s \to \infty} \overline{P}_e(\mathcal{R}) = \lim_{N_s \to \infty} E[\overline{P}_{e|\mathbf{S}}(\mathcal{R})] = E[\mathbb{J}_{\{I_M(\mathbf{S}) < \mathcal{R}\}}] = P(I_M(\mathbf{S}) < \mathcal{R}) = P_{\text{out}}(\mathcal{R}) \quad (8.10)$$

Hence, the information-theoretic outage probability defined by (8.8) is equal to the message error probability averaged over the random code ensemble and over all the possible channel realizations $\mathbf{S}$, in the limit for large N_s.

System capacity. Let R_b denote the user information bit-rate, i.e., the number of bit/s transmitted by a user during a block of duration T_s [2]. The user spectral efficiency (measured in bit/s/Hz) is given by $R_b/W = \mathcal{R}N_s/(WT_s) = \mathcal{R}/L$. Hence, for fixed R_b and W and for a desired outage probability P_{out}, the system capacity is defined by

$$C_{\text{sys}} \triangleq \frac{R_b}{W} \max\{N : P(I_M(\mathbf{S})/L < R_b/W) < P_{\text{out}}\} \quad \text{users/cell} \times \text{bit/s/Hz} \quad (8.11)$$

In the following we assume that user codes are randomly generated with i.i.d. components according to a complex Gaussian distribution with i.i.d. real and imaginary parts (i.e., with flat power spectral density). An argument supporting this choice is that transmitters have no knowledge of the fading channels and fading amplitudes are identically distributed for all frequencies. However, we do not claim that this choice leads to any kind of optimality in the present setting.

8.4 RECEIVER DESIGN AND MUTUAL INFORMATION

The receiver filters $w_m^*(-t)$ can be designed according to several criteria. Here we examine the conventional single-user receiver and the linear MMSE multiuser receiver. In both cases we assume that the receiver has perfect channel-state information, i.e., that it knows A_m^i, τ_m^i, ϕ_m^i, $c_m^i(\tau)$ and $s_m^i(t)$ for all i and m.

Conventional single-user receiver. In this case the receiving filter is the single-user matched filter $w_m^*(-t) = [h_m^0(-t)]^*$.

Linear MMSE multiuser receiver. Linear MMSE multiuser receiver has been independently rediscovered by many after [14]. The goal here is to design $w_m^*(-t)$ such that the mean-square error $\varepsilon^2(w_m) = E[|y_m[k] - x_m^0[k]|^2]$ is minimized. It can be shown that the linear MMSE multiuser receiver is formed by a bank of N filters with the i-th filter matched to the i-th user signal, followed by a (possibly infinite) vector tapped delay line with coefficients $\mathbf{w}_m^\dagger[j]$. By assuming very long blocks ($N_s \to \infty$) we obtain the D-transform $\mathbf{w}_m(D) = \sum_j \mathbf{w}_m[j]D^j$ of the sequence of coefficients $\mathbf{w}_m[j]$ as [3]

$$\mathbf{w}_m(D) = \frac{1}{A_m^0}\text{col}_0\left\{\left[\mathbf{R}_m(D^{-1}) + \frac{1}{\gamma_m}\mathbf{E}_m^{-1}\right]^{-1}\right\} \quad (8.12)$$

where we have defined

- The signal-to-noise ratio of user 0 over block m, $\gamma_m \triangleq (A_m^0)^2/N_0$.

---

[2] Note that in a slotted transmission the average user bit-rate depends also on the number of transmitted blocks per second. Here, we assume that users transmit continuously, so that the burst bit-rate coincides with the information bit-rate R_b. In the case of a variable rate system, blocks can be transmitted with a duty cycle $<$ 100% in order to accommodate lower rates.
[3] We number row and columns of N-vectors and $N \times N$ matrices from 0 to $N-1$ and row$_i\{\mathbf{M}\}$, col$_i\{\mathbf{M}\}$ and $\{\mathbf{M}\}_{ij}$ denote the i-th row, the i-th column and the (i,j) element of $\mathbf{M}$, respectively.

- The diagonal matrix of the received energies per symbol, normalized to the energy of signal 0
$$\mathbf{E}_m \triangleq \operatorname{diag}\left(1, \mathcal{E}_m^1, \ldots, \mathcal{E}_m^{N-1}\right)$$
where $\mathcal{E}_m^i = (A_m^i/A_m^0)^2$.

- The cross-correlation matrix sequence $\mathbf{R}_m[k]$ whose entry (i,j) is
$$\{\mathbf{R}_m[k]\}_{ij} \triangleq e^{-j\phi_m^{ij}} \int_{-\infty}^{\infty} \int_{-\infty}^{\infty} c_m^i(\tau)^* c_m^j(\tau') r_{s,m}^{ij}(kT + \tau - \tau' + \tau_m^{ij}) d\tau d\tau' \tag{8.13}$$

where $\phi_m^{ij} \triangleq \phi_m^i - \phi_m^j$ and $\tau_m^{ij} \triangleq \tau_m^i - \tau_m^j$, and where
$$r_{s,m}^{ij}(\tau) \triangleq \int_{-\infty}^{\infty} s_m^i(t-\tau)^* s_m^j(t) dt \tag{8.14}$$

- The D-transform $\mathbf{R}_m(D) \triangleq \sum_k \mathbf{R}_m[k] D^k$.

By using (discrete-time) inverse Fourier transform, we obtain the resulting minimum MSE as
$$\varepsilon_{\min,m}^2 = \int_{-1/2}^{1/2} \left\{\left[\gamma_m \mathbf{R}_m(\lambda) + \mathbf{E}_m^{-1}\right]^{-1}\right\}_{00} d\lambda \tag{8.15}$$

where, with a slight abuse of notation, we let $\mathbf{R}_m(\lambda) \triangleq \mathbf{R}_m(D)|_{D=e^{j2\pi\lambda}}$.

8.4.1 Mutual information

Because of the channel blockwise memoryless assumption we can write the mutual information $I_M(\mathbf{S})$ as the sum of M contributions, one for each block. By assuming i.i.d. complex Gaussian symbols for all users, the channel (8.4) with input $x_m^0[k]$ and output $y_m[k]$ conditioned on $\mathbf{S}$ is a single-user discrete-time channel with ISI and correlated Gaussian noise. Moreover, we assume that the impulse responses $p_m^i[k]$ are square-summable with probability 1 (this is always met in practice since the channel impulse responses have finite energy). Hence, the mutual information can be obtained as an application of the Toeplitz distribution theorem [5]. Let $Y_m(f)$ denote the power spectral density of $y_m[k]$ conditioned on S_m and let $Z_m(f)$ denote the power spectral density of $y_m[k]$ conditioned on both the input sequence $x_m^0[k]$ and S_m. Hence we can write
$$I_M(\mathbf{S}) = \frac{1}{M} \sum_{m=0}^{M-1} \int_{-1/2}^{1/2} \log_2 \frac{Y_m(f)}{Z_m(f)} df \tag{8.16}$$

With the single-user receiver, assuming $\{\mathbf{R}_m(\lambda)\}_{00} > 0$ for all $\lambda \in [-1/2, 1/2]$ we obtain
$$I_M(\mathbf{S}) = \frac{1}{M} \sum_{m=0}^{M-1} \int_{-1/2}^{1/2} \log_2 \left(1 + \frac{\{\mathbf{R}_m(\lambda)\}_{00}}{\sum_{i=1}^{N-1} \mathcal{E}_m^i \frac{|\{\mathbf{R}_m(\lambda)\}_{0i}|^2}{\{\mathbf{R}_m(\lambda)\}_{00}} + 1/\gamma_m}\right) d\lambda \tag{8.17}$$

With the linear MMSE multiuser receiver, after some algebra we obtain

$$I_M(\mathbf{S}) = \frac{1}{M} \sum_{m=0}^{M-1} \int_{-1/2}^{1/2} -\log_2\left(\left\{[\gamma_m \mathbf{R}_m(\lambda) + \mathbf{E}_m^{-1}]^{-1}\right\}_{00}\right) d\lambda \quad (8.18)$$

By comparing the MMSE given in (8.15) with the above expression, application of Jensen's inequality to the convex function $-\log_2(x)$ yields the inequality

$$I_M(\mathbf{S}) \geq -\log_2 \left[\prod_{m=0}^{M-1} \varepsilon_{\min,m}^2\right]^{1/M} \quad (8.19)$$

where the argument of the logarithm in the RHS is the geometric mean of the MMSE of the M blocks. Equality is obtained when the argument of $\log_2(\cdot)$ in (8.18) does not depend on frequency, i.e., for a memoryless channel. This is the case, for example, of symbol-synchronous transmission and frequency-flat fading.

Strictly Nyquist band-limited waveforms. In the following we limit ourselves to strictly Nyquist band-limited waveforms of the type

$$s_m^i(t) = \frac{\operatorname{sinc}(t/T)}{\sqrt{TL}} \sum_{\ell=0}^{L-1} e^{j(2\pi \ell t/T + \psi_m^i[\ell])} \quad (8.20)$$

where $\operatorname{sinc}(t) \triangleq \sin(\pi t)/(\pi t)$ and where $\{\psi_m^i[\ell] : \ell = 0, \ldots, L-1\}$ is the spreading sequence of user i over block m. Using waveforms (8.20) is tantamount to orthogonal frequency division multiplexing (OFDM) [1] where the same symbol $x_m^i[k]$ is transmitted over L adjacent subbands spaced by $1/T$ and where a different phase $\psi_m^i[\ell]$ is assigned to each ℓ-th subcarrier. This choice of the user waveforms allows us to write $\mathbf{R}_m(\lambda)$ in a very simple way in terms of the fading channel frequency responses $C_m^i(f) = \int_0^{T_d} c_m^i(\tau) e^{-j2\pi f\tau} d\tau$. Although choice (8.20) is admittedly restrictive, we believe it is not overly so, inasmuch as it allows us to reach conclusions that are expected to hold for other waveforms as well.

In this case, after long but straightforward calculations we can write

$$\mathbf{R}_m(\lambda) = \frac{1}{L} \mathbf{C}_m(\lambda)^\dagger \mathbf{C}_m(\lambda)$$

where $\mathbf{C}(\lambda)$ is the $L \times N$ matrix whose entry (ℓ, i) is

$$\{\mathbf{C}_m(\lambda)\}_{\ell i} \triangleq e^{j(\phi_m^i - 2\pi \lambda \tau_m^i/T)} C_m^i\left(\frac{\lambda + \ell}{T}\right) e^{j(\psi_m^i[\ell] - 2\pi \ell \tau_m^i/T)} \quad (8.21)$$

Note that $\operatorname{rank}(\mathbf{R}_m(\lambda)) \leq \min\{L, N\}$ and, for $L < N$, $\mathbf{R}_m(\lambda)$ is not invertible.

Tradeoff between coding and spreading. If $L = 1$, then $\mathbf{R}_m(\lambda)$ has rank 1. Thus, we write $\mathbf{R}_m(\lambda) = \mathbf{a}(\lambda)\mathbf{a}^\dagger(\lambda)$ and use the matrix identity

$$[\mathbf{I} + \mathbf{a}\mathbf{a}^\dagger]^{-1} = \mathbf{I} - \frac{1}{1 + |\mathbf{a}|^2} \mathbf{a}\mathbf{a}^\dagger$$

(where **a** is a column N-vector) to see that for $L = 1$ the two mutual informations (8.17) and (8.18) coincide. Hence, for signaling waveforms of the type (8.20) with spreading $L = 1$ the conventional single-user receiver and the linear MMSE multiuser receiver give exactly the same result in terms of outage probability and system capacity. Moreover, in this case the mutual information does not depend on the user relative delays (so that symbol-synchronous and symbol-asynchronous transmission are equivalent in terms of outage and system capacity). In order to exploit the ability of the MMSE receiver to cancel interference we need some spreading ($L > 1$). Hence, for a given spectral efficiency $R_b/W = \mathcal{R}/L$ coding and spreading must be traded off: if we want to increase the waveform spreading factor L we must increase the code rate $\mathcal{R}$ accordingly. The best solution is the one that maximizes system capacity for given channel statistics.

Asymptotic analysis. We can gain more intuition on the coding-spreading trade-off from the asymptotic characterization of the system capacity of a very idealized CDMA single-cell system presented by D. Tse and S. Hanly in [15]. Consider the uplink of a symbol-synchronous Direct-Sequence (DS) CDMA system with a single cell, N users and spreading sequences of length L chips. No fading and no shadowing or power control errors are taken into account, so that the channel is purely AWGN. The spreading sequences are real or complex and randomly generated i.i.d. sequences with mean zero and variance $1/L$ (classical examples are BPSK and QPSK spreading sequences). Once the sequences are generated, they are assigned to the users permanently. Although for every finite L the system capacity as defined by (8.11) is a random variable depending on the random spreading sequence assignment, a very strong characterization of system capacity is possible for very large systems, i.e., those in which both L and N grow to infinity while keeping constant the ratio $\alpha = N/L$.

Specifically, assume independent random coding [3] for each user, where users' symbols $x_m^i[k]$ are generated i.i.d. according to a complex circularly symmetric Gaussian distribution with mean zero and per-component variance $1/2$. Moreover, assume that all users are received with the same amplitude $A_m^i = A$ (perfect power control and no fading). From the results of [15] we obtain the asymptotic system capacity as $L \to \infty$ and $N/L \to \alpha$ with a conventional single-user receiver as

$$C_{\text{sys}} = \log_2(1 + \beta) \max\left\{\left(\frac{1}{\beta} - \frac{N_0}{A^2}\right), 0\right\} \qquad (8.22)$$

where β is the desired signal-to-interference plus noise ratio (SINR) at the receiver output. Under the same assumptions, the asymptotic system capacity with a linear MMSE multiuser receiver is given by

$$C_{\text{sys}} = \log_2(1 + \beta)(1 + \beta) \max\left\{\left(\frac{1}{\beta} - \frac{N_0}{A^2}\right), 0\right\} \qquad (8.23)$$

Two comments are in order to illustrate the above formulae:

- Since the channel is AWGN and the received power from each user is kept constant, the mutual information as derived previously is constant if conditioned on the choice of the random spreading sequences. For $L \to \infty$, it is possible to prove that the mutual information converges in probability to the constant $\log_2(1 + \beta)$ (where β depends on α and on the SNR A^2/N_0). Then, by letting the code rate $\mathcal{R}$ equal to $\log_2(1 + \beta)$, the outage probability is zero.

- It is convenient to parameterize C_{sys} in terms of β, which plays the role of a design parameter. We can write $C_{\text{sys}} = NR_b/W = N\mathcal{R}/L = \alpha\mathcal{R}$. Since in order to have zero outage probability, we let $\mathcal{R} = \log_2(1 + \beta)$, then the number α of users×cell per dimension is given by

$$\alpha = \max\left\{\left(\frac{1}{\beta} - \frac{N_0}{A^2}\right), 0\right\} \quad (8.24)$$

for the conventional single-user receiver, and by

$$\alpha = (1 + \beta)\max\left\{\left(\frac{1}{\beta} - \frac{N_0}{A^2}\right), 0\right\} \quad (8.25)$$

for the linear MMSE receiver.

Fig. 8.1 shows C_{sys} vs. β, for SNR= 10 dB. We notice that the maximum system

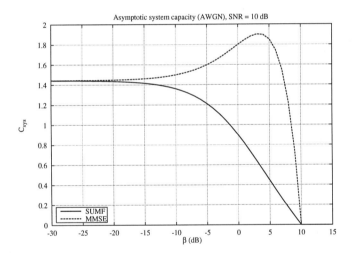

Figure 8.1 System capacity vs. β, the signal-plus-interference to noise ratio at the receiver output for SNR= 10 dB, a single-user matched-filter receiver (continuous line) and a linear MMSE receiver (dotted line). The channel is ideal AWGN.

capacity with the conventional receiver is obtained as $\beta \to 0$, i.e., for a very large number of users×cell per dimension ($\alpha \to \infty$) and very low-rate coding ($\mathcal{R} \to 0$). This is in agreement with Viterbi's findings and with the IS-95 return link philosophy [17, 13]. The system capacity attained by the linear MMSE receiver is larger than that of the conventional receiver for all values of β. We notice that in this case there exists an optimal β maximizing the system capacity. This represents (at least asymptotically and in the case of AWGN and perfect power control) the best trade-off between coding and spreading. Let β_{opt} be the capacity maximizing β, and assume a system with $N \gg 1$ users. Then, the optimal coding rate is approximately given by $\mathcal{R}_{\text{opt}} = \log_2(1 + \beta_{\text{opt}})$ and the optimal spreading length is approximately given by $L_{\text{opt}} = N/\alpha$, where α is given by (8.25) and evaluated for $\beta = \beta_{\text{opt}}$.

8.5 NUMERICAL RESULTS

To substantiate the above results with numerical examples, we borrow some system parameters from the cellular CDMA standard IS-95 [13]. We consider system bandwidth $W = 1.25$ MHz, user bit-rate $R_b = 9.6$ kb/s and interleaving depths $M = 1, 2$, and 4. The resulting user spectral efficiency is $\mathcal{R}/L = R_b/W \simeq 7.7 \cdot 10^{-3}$. We make the simplifying assumption that all fading channels $c_m^i(\tau)$ are i.i.d. for all i and m. We consider the Rayleigh fading channel model given in [18] for a typical urban environment, with a multipath intensity profile

$$\sigma^2(\tau) = \begin{cases} \dfrac{e^{-\tau/t_0}}{t_0(1 - e^{-T_d/t_0})} & 0 \leq \tau \leq T_d \\ 0 & \text{elsewhere} \end{cases}$$

with $t_0 = 1$ μs and $T_d = 7$ μs.

Since outage probability does not seem to be amenable to a closed-form expression, we resorted to Monte Carlo simulation. In order to compute mutual information, we discretize the integration domain $[-1/2, 1/2]$ into frequency intervals, in each of which the matrix $\mathbf{R}_m(\lambda)$ does not vary appreciably with λ. The number of discretization intervals D is chosen as $D = \lfloor W/(LB_c) \rfloor + 1$ where B_c is the channel coherence bandwidth [12, Ch. 14]-[2, Ch. 13]. We found that accurate results for the fading model considered here can be obtained by using $B_c = 66.6$ kHz.

Spreading sequences $\{\psi_m^i[\ell] : \ell = 0, \ldots, L-1\}$ are assumed to be i.i.d. randomly generated for all users according to a uniform distribution over $[-\pi, \pi]$. Independently generated sequences are used in different blocks.

The received energy per symbol $(A_m^i)^2$ is modeled as a log-normal random variable with residual shadowing factor σ_{sh}. We considered the values $\sigma_{\text{sh}} = 0, 2$ and 8 dB. For $\sigma_{\text{sh}} = 0$ we have ideal power control. In our simulations we consider a signal-to-noise ratio in the absence of residual shadowing (or, equivalently, with ideal power control) equal to $1/N_0 = 10$ dB (recall that the symbols in $\mathcal{X}$ have unit average energy). We assume shadowing to be a process so slow that it can be considered as constant over all the M blocks spanned by a user code word. Hence, A_m^0 does not depend on m. However, we assume that a slot hopping scheme is applied so that two users can interfere in at most one block out of M. Then, the A_m^i, for $i > 0$, are i.i.d. for different i's and m's, since user 0 "sees" different interfering users in each block. A method based on orthogonal latin squares for designing hopping schemes with the above property is advocated in [11]. Use of such hopping schemes is highly desirable since in this way the *interferer diversity* of the system is equal to the interleaving depth

M, and the probability of worst case situations where a user experiences persistently strong interference over all the blocks spanned by a code word is reduced.

Finally, we note that in order to reduce the amount of computations in the Monte Carlo simulation of the MMSE multiuser outage probability we can compute the argument of the logarithm in (8.18) as

$$\left\{ \left[\gamma_m \mathbf{R}_m(\lambda) + \mathbf{E}_m^{-1} \right]^{-1} \right\}_{00} = (\{\mathbf{U}(\lambda)\}_{00})^{-2}$$

where $\mathbf{U}(\lambda)\mathbf{U}^\dagger(\lambda)$ is the Cholesky factorization of $[\gamma_m \mathbf{R}(\lambda) + \mathbf{E}_m^{-1}]$ and $\mathbf{U}(\lambda)$ is upper triangular.

8.5.1 Outage probability results

The outage probability $P_{\text{out}}(\mathcal{R}) = P(I_M(\mathbf{S}) < \mathcal{R})$ was computed for $M = 1, 2, 4$, $L = 1, 2, 4, 8, 16$, $N = 10, 20, 30, 40, 50, 60, 70, 80, 90, 100$ and $\sigma_{\text{sh}} = 0, 2$ and 8 dB, for both the single-user receiver and the linear MMSE multiuser receiver. Because of space limitations we include here only a subset of our results.

Fig. 8.2 shows $P_{\text{out}}(\mathcal{R})$ vs. the user spectral efficiency $\mathcal{R}/L$ for $M = 1, L = 1$ (above) and for $M = 1, L = 16$ (below), in the case of ideal power control ($\sigma_{\text{sh}} = 0$ dB). As expected, the performance of the two receivers for $L = 1$ is the same. On the contrary, for $L > 1$ the MMSE receiver performs uniformly better than the single-user receiver.

Fig. 8.3 shows $P_{\text{out}}(\mathcal{R})$ vs. $\mathcal{R}/L$ for $\sigma_{\text{sh}} = 2$ dB (above) and $\sigma_{\text{sh}} = 8$ dB (below), for $M = 1, L = 16$. Fig. 8.4 shows analogous results for $M = 4, L = 16$. We note that with non-ideal power control the single-user receiver suffers from a large performance degradation in terms of outage probability. For example, in the case $\sigma_{\text{sh}} = 8$ dB, the single-user yields $P_{\text{out}}(\mathcal{R}) > 10^{-1}$ already with $N = 10$ users. On the contrary, the linear MMSE multiuser receiver is much more robust to power control errors and achieves low outage probabilities even for $\sigma_{\text{sh}} = 8$ dB. This fact may be viewed as a redefinition in terms of outage probability of the *near-far resistance* of MMSE multiuser detectors (see [6] and references therein): the MMSE multiuser receiver is able to cope with unbalanced signal power situations.

As expected, interleaving depth $M > 1$ provides a benefit. This effect is particularly worthy of notice when hopping schemes with maximum interferer diversity are employed, as in these simulations. In this way, a diversity order equal to the interleaving depth M is achieved against the residual log-normal interference.

8.5.2 System capacity results

Fig. 8.5 shows P_{out} as a function of N, for $\mathcal{R}/L = 7.7 \cdot 10^{-3}$, $M = 1, 2, 4$, $L = 1, 2, 4, 8, 16$ and $\sigma_{\text{sh}} = 0$ dB, for the single-user (above) and for the linear MMSE multiuser receivers (below), respectively. A usual value for the desired outage probability is $P_{\text{out}} = 10^{-2}$. In this case we see that with $M = 4$ and $L = 16$ the single-user and the MMSE multiuser receiver can accommodate $N \simeq 105$ and $N \simeq 125$ user/cell, respectively. This corresponds to $C_{\text{sys}} \simeq 0.8$ (single-user) and $C_{\text{sys}} \simeq 0.96$ (multiuser) user/cell×bit/s/Hz.

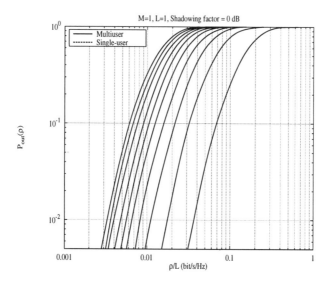

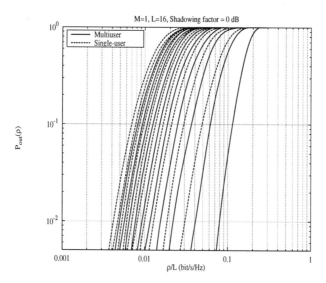

Figure 8.2 $P_{\text{out}}(\rho)$ vs. ρ/L for $M = 1$, $L = 1$ (above) and $L = 16$ (below). Curves for $N = 10, 20, \ldots, 100$ users are shown. For each family of curves, the rightmost corresponds to $N = 10$ and the leftmost to $N = 100$.

CODING VS. SPREADING OVER BLOCK FADING CHANNELS   231

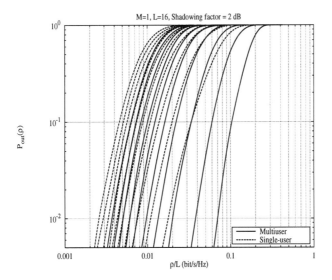

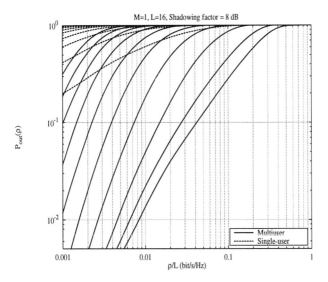

Figure 8.3 $P_{\text{out}}(\rho)$ vs. ρ/L for $M = 1$, $L = 16$, with residual shadowing factor $\sigma_{\text{sh}} = 2$ dB (above) and $\sigma_{\text{sh}} = 8$ dB (below). Curves for $N = 10, 20, \ldots, 100$ users are shown. For each family of curves, the rightmost corresponds to $N = 10$ and the leftmost to $N = 100$.

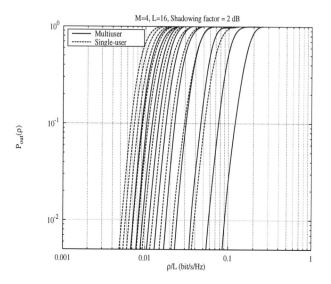

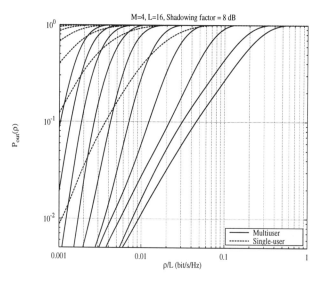

Figure 8.4 $P_{\text{out}}(\rho)$ vs. ρ/L for $M = 4$, $L = 16$, with residual shadowing factor $\sigma_{\text{sh}} = 2$ dB (above) and $\sigma_{\text{sh}} = 8$ dB (below). Curves for $N = 10, 20, \ldots, 100$ users are shown. For each family of curves, the rightmost corresponds to $N = 10$ and the leftmost to $N = 100$.

Fig. 8.6 shows P_{out} as a function of N, for $\mathcal{R}/L = 7.7 \cdot 10^{-3}$, $M = 1, 2, 4$, $L = 2, 16$ and $\sigma_{\text{sh}} = 2$ dB, for the single-user (above) and for the linear MMSE multiuser receivers (below), respectively. We observe that an increase of waveform spreading from 2 to 16 yields a large improvement only with the MMSE multiuser receiver. For a desired outage probability $P_{\text{out}} = 10^{-2}$, with $M = 4$ and $L = 16$ the single-user and the MMSE multiuser receiver can accommodate $N \simeq 66$ and $N \simeq 95$ user/cell, respectively, corresponding to $C_{\text{sys}} \simeq 0.5$ (single-user) and $C_{\text{sys}} \simeq 0.73$ (multiuser) user/cell×bit/s/Hz.

Finally, Fig. 8.7 shows P_{out} as a function of N, for $\mathcal{R}/L = 7.7 \cdot 10^{-3}$, $M = 1, 2, 4$, $L = 2, 16$ and $\sigma_{\text{sh}} = 8$ dB, for the single-user (above) and for the linear MMSE multiuser receivers (below), respectively. Here we see that the single-user receiver cannot achieve outage probabilities smaller than 10^{-1}. On the contrary, with the MMSE multiuser receiver it is possible to accommodate up to 20 users (with $M = 4$ and $L = 16$), even in this very adverse residual shadowing conditions. The resulting system capacity is $C_{\text{sys}} \simeq 0.15$ user/cell×bit/s/Hz.

8.6 CONCLUSIONS

In this paper we examined the tradeoff between coding and spreading in a single-cell CDMA mobile communication system with block-fading, block-synchronous transmission and slow power control. The cost function selected for optimization was *system capacity*, expressed as the maximum of the product of the number of users per cell times the user spectral efficiency, under an outage probability constraint. Despite some necessary simplifications, the model chosen here takes into account many features of real-world systems, such as time-varying random spreading waveforms, non-ideal power control, symbol-asynchronous transmission, and multipath fading.

We derived expressions for the mutual information characterizing the M-block random channel spanned by the transmission of a user code word in the cases of conventional single-user receiver of the linear MMSE multiuser receiver. >From these expressions it was possible to compute outage probability and system capacity by Monte Carlo simulation. The tradeoff between coding and spreading and interleaving depth was examined for different receiver structures. Notice finally that our analysis is independent of the particular coding and modulation scheme adopted.

Our results show that with ideal power control and moderate interleaving depth ($M = 4$) and spreading ($L = 16$), capacities close to 1 user/cell×bit/s/Hz can be obtained by both the single-user and the MMSE multiuser receivers, for outage probability $P_{\text{out}} = 10^{-2}$. On the contrary, the MMSE multiuser receiver proves to be very robust to power control inaccuracies while the single-user receiver breaks down. Hence, a precise power control algorithm is a key issue in conventional CDMA systems, while MMSE multiuser detection allows for (moderately) unbalanced signal powers. This fact may justify the implementation of linear MMSE multiuser detection. We note in passing that, with block-synchronous transmission, adaptive MMSE multiuser detection may be implemented also with random time-varying spreading waveforms.

Two major aspects that have not been taken into account here are voice activity and inter-cell interference [4]. On one hand, voice activity detection in the transmitter increase the system capacity, since the number of users simultaneously transmitting

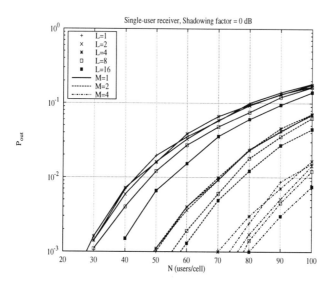

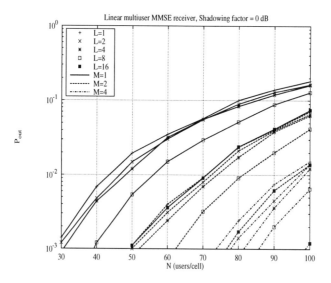

Figure 8.5 P_{out} vs. N for fixed $\rho/L = 7.7 \cdot 10^{-3}$, for the single-user receiver (above) and for the linear MMSE multiuser receiver (below) with interleaving depths $M = 1, 2, 4$, spreading factors $L = 1, 2, 4, 8, 16$ and ideal power control ($\sigma_{\text{sh}} = 0$ dB).

CODING VS. SPREADING OVER BLOCK FADING CHANNELS

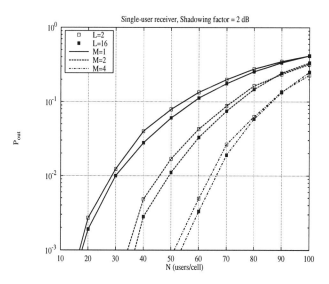

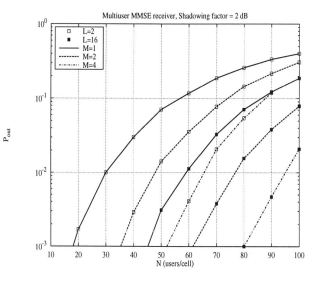

Figure 8.6 P_{out} vs. N for fixed $\rho/L = 7.7 \cdot 10^{-3}$, for the single-user receiver (above) and for the linear MMSE multiuser receiver (below) with interleaving depths $M = 1, 2, 4$, spreading factors $L = 2, 16$ and residual shadowing factor $\sigma_{\text{sh}} = 2$ dB.

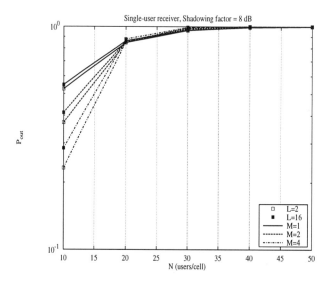

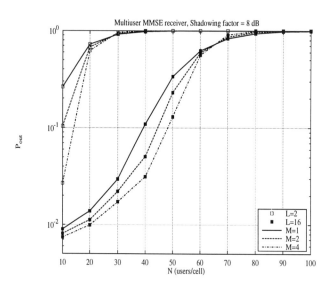

Figure 8.7 P_{out} vs. N for fixed $\rho/L = 7.7 \cdot 10^{-3}$, for the single-user receiver (above) and for the linear MMSE multiuser receiver (below) with interleaving depths $M = 1, 2, 4$, spreading factors $L = 2, 16$ and residual shadowing factor $\sigma_{\text{sh}} = 8$ dB.

is just a fraction of the total number of users per cell N. On the other hand, intercell interference decreases system capacity as it increase the interference total power. Moreover, also cell sectorization should be taken into account in order to asses the system capacity of an actual system.

References

[1] M. Alard and R. Lassalle, "Principles of modulation and channel coding for digital broadcasting for mobile receivers," *European Broadcasting Union Review,* No. 224, Aug. 1987.

[2] S. Benedetto and E. Biglieri, *Principles of Digital Transmission with Wireless Applications.* New York: Plenum, 1998.

[3] T. Cover and J. Thomas, *Elements of Information Theory.* New York: J. Wiley & Sons, 1991.

[4] K. Gilhousen, I. Jacobs, R. Padovani, A. Viterbi, L. Weaver and C. Wheatly III, "On the capacity of a cellular CDMA system," *IEEE Trans. on Vehic. Tech.,* Vol. 40, No. 2, pp. 303 – 312, May 1991.

[5] U. Grenander and G. Szegö, *Toeplitz Forms and Their Applications.* New York: Chelsea, 1983.

[6] M. Honig, U. Madhow and S. Verdu, "Blind adaptive multiuser detection," *IEEE Trans. on Inform. Theory,* Vol. 41, No. 4, pp. 944 – 960, July 1995.

[7] G. Kaplan and S. Shamai (Shitz), "Error probabilities for the block-fading Gaussian channel," *A.E.Ü.,* Vol. 49, No. 4, pp. 192 – 205, 1995.

[8] R. Knopp, *Coding and Multiple Access over Fading Channels,* Ph.D. Thesis, Ecole Polytechnique Fédérale de Lausanne, Lausanne, Switzerland, 1997.

[9] R. Knopp, P. A. Humblet, "Maximizing Diversity on Block Fading Channels," *Proceedings of ICC '97,* Montréal, Canada, 8 – 12 June, 1997.

[10] L. Ozarow, S. Shamai, and A. D. Wyner, "Information theoretic considerations for cellular mobile radio," *IEEE Trans. on Vehic. Tech.,* Vol. 43, No. 2, May 1994.

[11] G. Pottie and R. Calderbank, "Channel coding strategies for cellular mobile radio," *IEEE Trans. on Vehic. Tech.,* Vol. 44, No. 4, pp. 763 –769, Nov. 1995.

[12] J. Proakis, *Digital Communications.* New York: McGraw-Hill, 1983.

[13] T. Rappaport, *Wireless Communications.* Englewood Cliffs, NJ: Prentice-Hall, 1996.

[14] J. Salz, "Digital transmission over cross-coupled linear channels," AT&T *Tech. J.,* Vol. 64, no. 6, July-Aug. 1985.

[15] D. Tse and S. Hanly, "Linear Multiuser Receivers: Effective Interference, Effective Bandwidth and Capacity," Memorandum No. UCB/ERL M98/1, UC Berkeley, Jan. 1998. Also: submitted for publication to *IEEE Trans. on Inform. Theory.*

[16] S. Verdú, "The capacity region of the symbol-asynchronous Gaussian multiple-access channel," *IEEE Trans. on Inform. Theory,* Vol. 35, No. 4, pp. 733 – 751, July 1989.

[17] A. J. Viterbi, *CDMA – Principles of Spread-Spectrum Communications,* Reading, MA: Addison-Wesley, 1995.

[18] COST 207 Management Committee, *COST 207: Digital land mobile radio communications (Final Report).* Commission of the European Communities, 1989.

9 TURBO-CODES FOR FUTURE MOBILE RADIO APPLICATIONS

Peter Jung, Jörg Plechinger and Markus Doetsch

Siemens AG Semiconductors Group
Signal Processing
Cellular Innovation (HL SP CIN)
Sankt-Martin-Strasse 76
D-81541 Munich
Germany

{ Peter.Jung, Jörg.Plechinger, Markus.Doetsch } @siemens-scg.com

Abstract: Turbo-Codes have recently been introduced to the communications community. Being parallel concatenated recursive systematic convolutional (RSC) codes which can be decoded in an iterative manner by exploiting the decoding result of previous iterations, their name "Turbo-Codes" has been coined in analogy to the turbo engine principle. In this communication, an illustrative introduction to the area of Turbo-Codes shall be given with particular focus on mobile radio applications.

9.1 INTRODUCTION

Future mobile radio applications will be services driven, thus extending the focus of present-day systems to more flexible user-friendly communications encompassing a wide range of application areas, bearer services and different deployment scenarios. Recently, the European standardization activities toward UMTS (Universal Mobile Telecommunications System) [1] resulted in the agreement to employ CDMA (code division multiple access) with a bandwidth of 5 MHz in both outdoor and indoor scenarios because it was found that an evolutionary radio interface setting out from second generation TDMA (time division multiple access) cannot provide the desired services portfolio.

In accordance with the general purpose of UMTS as a third generation mobile radio system, UMTS terrestrial radio access (UTRA) must offer a multi-level and open platform, i.e. a concept to cater for current and future developments of UMTS radio interface standards. Furthermore, certain services might not be made available in all environments. It is foreseen that high bearer data rate services will be restricted to a limited coverage, possibly provided in micro and pico cells. Speech and narrow band ISDN services will be available in a wide area coverage, i.e. in macro, micro and pico cells. Therefore, UTRA comprises two modes, namely

- WCDMA (wideband code division multiple access)for frequenzy domain duplex (FDD) and wide area coverage and

- TD/CDMA (time division code division multiple access) for time domain duplexing (TDD), asymmetrical services and small area coverage with high bearer data rate services provision.

The anticipated chip rate is agreed to be 4.096 Mchips/s in both UTRA FDD and TDD modes. Furthermore, UTRA requires the support of coexisting bearer services with packet and circuit switching, having bearer data rates of up to 2 Mbit/s and striving for the best possible spectral efficiency [1].

The mapping of services is done as follows. Finite bit blocks, comprising between a few hundred and several thousand bits, are fed into the channel encoder which introduces a given amount of redundancy into the data stream. Then, rate matching by a combination of repetition and puncturing is done. Afterwards, the encoded and rate matched data stream is multiplexed onto the physical channels together with signaling information.

Services are characterized by a particular quality of service (QoS) criterion. An important component of the QoS criterion is the bearer data rate R. The QoS also comprises a maximum admissible bit error ratio P_b^G or a packet loss ratio P_l^G in combination with a maximum outage probability P_{out}^G. In the case of circuit switched services, the probability $Pr\{P_b > P_b^G\}$ that the instantaneous bit error ratio P_b exceeds P_b^G must not be greater than P_{out}^G:

$$Pr\{P_b > P_b^G\} < P_{out}^G.$$

For speech transmission, usually $P_b^G = 10^{-3}$ and $P_{out}^G = 0.05$ are assumed. For packet services a similar expression holds for the instantaneous packet loss ratio P_l:

$$Pr\{P_l > P_l^G\} < P_{out}^G.$$

To meet the above-mentioned performance requirement, powerful error correction coding (ECC) must be used. ECC must be adaptable to the needs of different bearer services. In particular, varying code rates R_c must be handled efficiently. Such an adaptivity can be provided by deploying rate compatible punctured convolutional (RCPC) codes. No wonder, that efficient and flexible coding techniques such as the recently invented Turbo-Codes, providing rate compatible puncturing capabilities, are under discussion for the inclusion in UTRA and also Japanese and American mobile radio systems of the future. Meanwhile, at least two important special issues of journals [2, 3] and several conferences, e.g. [4], have been devoted to Turbo-Codes. In this communication, we will report several results published in these publications together with our own findings.

In the case of circuit switched services and delay sensitive packet services, QoS criteria also comprise delay constraints. For instance, in speech transmission, a delay larger than 40 ms is usually not acceptable.

In circuit switching, the bearer data rate R is usually fixed. However, regarding flexible exploitation of transmission resources, variable bearer data rate services are anticipated. The bearer data rate R is either set according to a user's needs or radio network requirements, e.g. depending on the instantaneous transmission quality.

To facilitate delay constrained services, the data stream is divided into bit blocks of finite duration which are handled independently of each other. The number of bits per block varies with service. Block size, i.e. number of bits per block, block duration and block rate are determined by both the delay criterion and the bearer data rate.

The QoS parameters P_b^G, P_l^G and P_{out}^G determine the choice of the ECC. Furthermore, the multiple access scheme and the modulation are chosen according to these parameters. ECC, multiple access scheme, modulation and block parameters essentially determine the code rate R_c. Moreover, the chosen multiple access scheme, the modulation, the block parameters and the code rate R_c govern the shape of the signals on the physical channel. In the case of e.g. multilevel modulation, the ECC also influences these signals. Fig. 9.1 summarizes the above discussion graphically.

The paper is organized into five further sections. In Sect. 9.2, the basic principles of code concatenation will be treated and Turbo-Codes will be discussed as a representative of parallel code concatenation. Sect. 9.3 will discuss the iterative decoding of Turbo-Codes. A possible design paradigm of rate compatible punctured Turbo-Codes (RCPTC's), which generalize the classical Turbo-Codes, will be illustrated in Sect. 9.4. The decoding complexity of Turbo-Codes will be discussed in Sect. 9.5. Finally, Sect. 9.6 will present some performance results.

9.2 CODE CONCATENATION

9.2.1 General principles

A most important goal of information theory is the striving for channel capacity. All codes that we presently know have been designed to fulfil this purpose. However, none of them has been capable of performing sufficiently well. Therefore, the concatenation of codes has been considered in the recent past. The design idea is to combine known codes, e.g. block codes and convolutional codes to facilitate the implementation of hardware inexpensive decoders and still allow for improved performance.

Generally speaking, two basic classes of code concatenation exist, namely serial and parallel code concatenation. Since the decoding of the constituent codes is based on the assumption of uncorrelated input samples, interleaving is employed between the codes to be concatenated. In UTRA, code concatenation will be employed for high quality services, which provide e.g. a very low maximum bit error ratio of e.g. 10^{-6} or below at the lowest possibly signal-to-noise ratio.

In serial concatenation, the constituent codes are usually not identical. First, the information stream is encoded by an outer code which is often a high rate systematic block code like a Reed-Solomon code. Then, the data stream encoded by this outer code is fed into a second so-called inner code. Usually, this inner code is a low rate

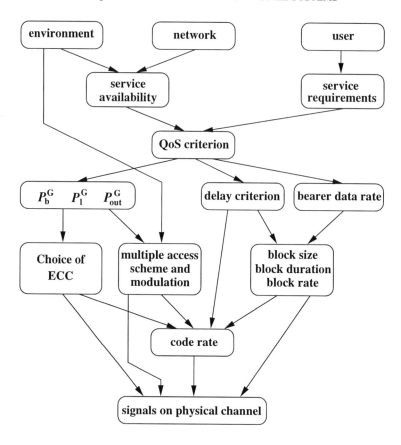

Figure 9.1 Dependence of physical signals on enviroment, network and user requirements

convolutional code. The finally encoded data stream is considered for transmission. The reason for this manner of combining outer block codes and inner convolutional codes has the following reason. The low rate inner convolutional code can be decoded very efficiently and powerfully by soft input decoders such as Viterbi decoders, which can be implemented rather easily. The strong inner code corrects as many transmission errors as possible to provide a bit error ratio of below 10^{-3} at the lowest possible average signal-to-noise ratio. However, to achieve the desired performance target, e.g. a maximum bit error ratio of 10^{-6} at the lowest possible average signal-to-noise ratio, the inner code is not sufficient. The outer high rate block code is thus exploited to achieve the desired performance.

At the beginning of the 1990's, a paramount break through in parallel code concatenation has occurred when Turbo-Codes were introduced. In parallel code concatenation, the information stream is fed into a first encoder uninterleaved and the interleaved version of the information stream is fed into the second encoder. The encoded data stream is generated by multiplexing and puncturing of the encoded sequences stemming from both encoding processes.

9.2.2 Turbo-Code structure

Fig. 9.2 shows the basic structure of a Turbo-Code encoder [5]. It consists of $N_e = 2$ binary RSC encoders with small constraint lengths usually set between 3 and 5, which are concatenated by a Turbo-Code interleaver, and a puncturing and multiplexing device. The minimum code rate $R_{c,min}$ is therefore $1/(N_e + 1)$ equal to $1/3$. Clearly, $R_{c,min}$ can be further reduced by introducing additional RSC encoders which will, however, not be considered in this commmunication.

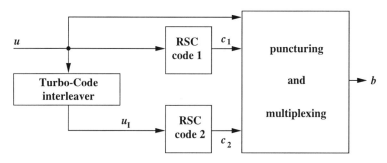

Figure 9.2 Turbo-Code encoder structure [5]

The binary input sequence u which has finite duration is fed into the first RSC encoder yielding the redundancy sequence c_1 with the same finite duration as u. The sequence u_I which is an interleaved replica of u is put into RSC encoder 2. The encoding process in the second RSC encoder yields the redundancy sequence . The redundancy sequences c_1 and c_2 as well as u are punctured and multiplexed to form the output sequence b. Obviously, the Turbo-Code encoder is systematic with u being the basis of the systematic information contained in b.

Since Turbo-Codes are based on the parallel concatenation of convolutional codes, Turbo-Codes are often called PCCC's (parallel concatenated convolutional codes). Extensions of the classical Turbo-Code to the concatenation of non-identical RSC codes or even to block codes are also conceivable and have already been studied in some reports. In this communication, we will renounce on the discussion of these issues.

9.2.3 RSC codes

In order to understand the performance of Turbo-Codes, some features of RSC codes will be discussed, first. It is well-known that RSC codes have the same free distance as the corresponding classical non-systematic convolutional codes with a minimum value of 2. Owing to their recursive structure of RSC codes, an RSC encoder, initially being in the zero state, will not return to the zero state in the case of an input sequence with weight one. This feature is in contrast to non-recursive codes. Instead, an input sequence with at least weight two is required. Moreover, there has to be a certain separation between the two ones in this input sequence, otherwise the encoded sequence will have an infinitely large weight.

Furthermore, systematic codes such as RSC codes transmit both redundancy and uncoded information and therefore allow a better performance at low signal-to-noise ratios than the corresponding non-systematic codes. Since a proper performance at low signal-to-noise ratios is desirable, RSC codes lend themselves as viable constituent codes for parallel code concatenation.

9.2.4 Turbo-Code interleaver

The Turbo-Code interleaver has a strong impact on the minimum free distance of the Turbo-Code. It was explained previously that RSC codes require input sequences with at least weight two and a certain separation between the two ones to generate output sequences with finite and low weights. A low weight of the output sequence usually results in a weak error protection and hence in a poor bit error ratio at a given signal-to-noise ratio. Hence, we seek to avoid such low weights of the output sequence which can be accomplished by properly choosing the Turbo-Code interleaver. Assume that the uniterleaved information sequence would yield a low weight output sequence at the first RSC encoder. Now, it is the task to guarantee that the interleaved version of the information sequence will not lead to a low weight of the output sequence generated by RSC encoder 2.

Several Turbo-Code interleaver design techniques have been proposed since 1993. It has been found in general that non-uniform interleavers, having a random look in contrast to classical block or convolutional interleavers, usually provide a good overall code performance. A promising technique has been put into a public domain piece of software available at http://www.sworld.com.au/. This systematic approach is particularly viable for mobile radio applications because the interleaver generation should be done in software executing in mobile terminals and in base stations.

However, in order to achieve the capacity limit, it is not sufficient to strive for maximization of the minimum free distance of the code. Rather, the whole distance spectrum of the code must be considered. In his famous communication, Claude Shannon proved his capacity theorem by introducing the random coding argument. It is thus reasonable to device codes which exhibit a distance spectrum similar to that of random codes. In fact, the distance spectra of Turbo-Codes show similar features like random codes. The authors believe that it is this essential feature which is the reason for the remarkable performance of Turbo-Codes.

9.3 ITERATIVE TURBO-CODE DECODING

Fig. 9.3 shows the basic structure of the iterative Turbo-Code decoder. The Turbo-Code decoder consists of two constituent soft input/soft output decoders, one for each RSC encoder, a Turbo-Code interleaver and a Turbo-Code deinterleaver. The demodulator delivers estimates x_n of the systematic information u_n contained in u as well as estimates $y_{1,n}$ and $y_{2,n}$ of the transmitted redundancy bits generated by the RSC encoders 1 and 2, respectively. The constituent soft input/soft output decoders require channel state information (CSI), comprised of instantaneous signal amplitudes and noise variance. Each constituent decoder processes systematic information, redundancy and a-priori information $L_{e1,n}$ and $L_{e2,n}$ by exploiting CSI, generating the extrinsic information $L_{e2,n}$ and $L_{e1,n}$ which is then used as a-priori knowledge at the successive constituent decoder. The decoding is iterative and improves with each iteration. However, the improvement gradually diminishes with each further iteration. After a certain number of iterations, the output of the Turbo-Code decoder is fed into a detector.

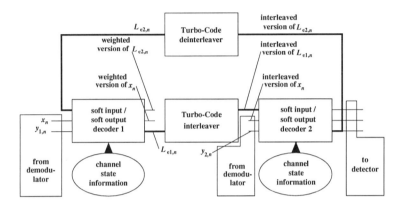

Figure 9.3 Turbo-Code decoder [5]

The adaptation of Turbo-Codes to service requirements can be done in four different ways. Firstly, the RSC encoders can be varied. This would call for an appropriate variation of the constituent decoders at the receivers, however, leading to an unfavorable impact on the hardware expense. Therefore, varying the RSC encoders is not regarded as a viable measure.

Secondly, the Turbo-Code interleaver size and structure can be adjusted. In the case of delay constrained services, however, the Turbo-Code interleaver size must be limited. Limiting the Turbo-Code interleaver size is also favorable with respect to packet services, thus, allowing a quick reaction of the radio network in the case of detected packet errors. It is obvious that various services require different block sizes of uncoded and consequently of encoded bits. The number of uncoded bits per block determines the size of the Turbo-Code interleaver [5]. Varying the interleaver size is considered mandatory for future mobile radio systems. Clearly, Turbo-Codes provide this features.

When one of the redundancy sequences c_1 or c_2 is fully suppressed by the puncturing, letting the second one pass the puncturing together with u, the Turbo-Code encoder becomes a conventional RSC encoder. The Turbo-Code decoder reduces to a conventional RSC decoder which is realized by performing half an iteration. However, a-priori knowledge based on extrinsic information is not available in this case. The code rate R_c can be varied between $1/2$ and 1, depending on the QoS criterion. Since the $N_e = 2$ RSC encoders can be based on different codes, the QoS criterion and code complexity can also be varied by suppressing a particular redundancy sequence c_1 or c_2 without changing the code rate R_c.

The aforementioned options, however, prohibit the Turbo-Code operation which is only available if bits from both redundancy sequences c_1 and c_2 are transmitted and

$$u_n \neq u_{I,n}$$

holds where u_n and $u_{I,n}$ are contained in u and u_I, respectively. In this case, we have

$$R_{c,min} \leq R_c < 1.$$

The minimum code rate $R_{c,min} = 1/(N_e + 1)$ is realized when puncturing is not used at all. In this case, either conventional RSC decoding or Turbo-Code decoding can be realized depending on the QoS criterion and the transmission channel conditions which are time varying in mobile radio applications. When exploiting the Turbo-Code properties, the performance only depends on the chosen Turbo-Code interleaver for a given number of decoding iterations.

Finally, the number of decoding iterations can be set according to the QoS criterion, adjusting the total code complexity. Two ways of exploiting this Turbo-Code property are readily available at the receiver. For a given QoS, the number of iterations can be reduced with increasing E_b/N_0. This feature is particularly beneficial for fading channels such as the mobile radio channel. Furthermore, the number of iterations can be varied with time varying QoS criterion. Table 9.1 summarizes the above discussion by presenting six different readily available options.

The most appealing feature of RCPTC's is the possibility to adaptively change the code rate R_c without having to transmit the whole encoded block. Rather, the transmission of an additional piece of information which makes up for the difference in code rate is necessary [10]. This feature can be exploited when RCPTC's are combined with ARQ (Automatic Repeat Request) protocols in packet switching.

Clearly, RCPTC's facilitate the migration from fixed ECC to flexible ECC strategies which are adaptable to time varying requirements, thereby facilitating flexible link layer and medium access controls. Furthermore, the easy introduction of various terminal types ranging from low-cost basic terminals for voice service to multi-service terminals, for e.g. multimedia applications, is supported.

9.4 DESIGNING RATE COMPATIBLE PUNCTURED TURBO-CODES

With respect to an optimum hardware reuse, the ECC circuitry could be fixed whereas the ECC configuration would be software controlled allowing a high flexibility. It will be argued that RCPTC's [7] possess the required flexibility.

When designing an RCPTC, constituent RSC encoders, Turbo-Code interleaver and puncturing should be jointly optimized in order to allow a best possible performance. However, requirements imposed by technology might not allow this approach. Therefore, the design procedure illustrated here is divided into three design steps [7].

In the first design step, the two constituent RSC encoders are chosen. A major requirement for this choice could be the hardware expense of the constituent RSC decoders, which should be kept as small as possible. For this reason, the two constituent

Table 9.1 RCPTC options for $N_e = 2$ constituent RSC encoders

Option	Code Structure	Operation as RSC code	Operation as Turbo code	Number of decoding iterations	Code Rate R_c
1	do not puncture u, fully puncture c_1 and c_2	no	no	not adjustable	1
2	do not puncture u, fully puncture c_1, partly puncture c_2	yes	no	not adjustable	$\frac{1}{2} \ldots 1$
3	do not puncture u, fully puncture c_2, partly puncture c_1	yes	no	not adjustable	$\frac{1}{2} \ldots 1$
4	do not puncture u, partly puncture c_1 and c_2, see [11] (Berrou's puncturing)	yes	yes	adjustable	$\frac{1}{3} \ldots 1$
5	partly puncture u, c_1 and c_2 (UKL puncturing)	no	yes	adjustable	$\frac{1}{3} \ldots 1$
6	no puncturing	yes	yes	adjustable	$\frac{1}{3}$

RSC encoders are identical, having constraint length three and octal generators 7 and 5. Although these RSC encoders have only four states, they still facilitate good performance at low values of E_b/N_0. Therefore the RCPTC performance will also be favorable at low signal-to-noise ratios. In Fig. 9.4, the structure of the chosen constituent RSC encoders is depicted.

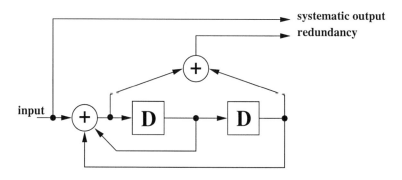

Figure 9.4 Structure of constituent RSC encoders

After having chosen the RSC encoders, the design of the Turbo-Code interleaver is most critical for the overall performance [11, 12, 13]. Therefore, the second design step is devoted to the Turbo-Code interleaver search. To facilitate the best possible performance for varying code rates, the Turbo-Code interleaver should be designed independent of the particular Turbo-Code puncturing. Thus, the Turbo-Code interleavers for the RCPTC's, which shall be designed, are searched for code rate 1/3, i.e. without any puncturing. The Turbo-Code interleaver search used by the authors is based on Berrou's design procedure [11]. An almost square block interleaver with I_r rows and I_c columns is considered where $I_r * I_c$ is equal to the number of bits per uncoded block. The bits contained in u are input row by row. The output bit index is determined by an elaborate combination of row and column numbers [11], where row and column numbers are linked by an appropriately chosen column number offset. The column number offset is relative prime to the row number. For this reason, the Turbo-Code interleaver is non-uniform and appears to be random in nature.

When designing the RCPTC's, the column number offset has been determined for different block sizes, i.e. for various RCPTC's, and optimized by computer search similar to [12, 13]. The design criterion has been the weight distribution of the overall Turbo-Code. In accordance with performance bound calculations [14] it was found that the Turbo-Code performance is governed by the frequency of code words with low and medium low weights. Several Turbo-Code interleavers with promising weight distributions were considered for each block size and then simulated in AWGN (additive white Gaussian noise) channels. Those particular Turbo-Code interleavers allowing the most favorable performance at bit error ratios between 10^{-6} and 10^{-3} for each block size were then kept.

After having chosen RSC encoders and Turbo-Code interleavers, the puncturing has to be optimized in the third design step for each block size. Firstly, the block size of u is divided into subblocks of size P. P is the puncturing period. Within this puncturing period P puncturing patterns are chosen to meet the code rate requirement. Again, the criterion for the choice is the weight distribution of the overall Turbo-Code. For given valid block size and code rate R_c, several puncturing patterns with promising weight distributions have been considered in simulations of the transmission over an AWGN channel. That particular puncturing pattern allowing the most favorable performance at bit error ratios between 10^{-6} and $10-3$ has been kept.

9.5 DECODING COMPLEXITY

This section extends the discussion of [8] which assessed the decoding complexity in a simple and enlightening fashion, however, underestimating the actual decoder complexity in terms of operations per uncoded bit. In what follows, we assume that the RCPTC uses two identical constituent RSC encoders each having $S = 4$ states. The number of operations for the branch metric computation is upper bounded by

$$N_{br} \leq 4 * \left(\frac{1}{R_c} - 1\right) = \frac{4}{R_c} - 4$$

per state.

Table 9.2 Decoding complexity in terms of operations per uncoded bit

	Code Rate R_c				
	$\frac{1}{3}$	$\frac{1}{2}$	$\frac{2}{3}$	$\frac{3}{4}$	$\frac{4}{5}$
Complexity	$86I$	$78I$	$74I$	$\frac{218I}{3}$	$72I$
$I = 1$	86	78	74	73	72
$I = 2$	172	156	148	146	144
$I = 3$	258	234	222	218	216
$I = 4$	344	312	296	291	288
$I = 5$	430	390	370	364	360
$I = 10$	860	780	740	627	720

In the case of MAP based constituent decoders, we need 6 operations per state to determine the quantities α and β. Since we have $S = 4$ states, the number of computations yields

$$N_{\alpha\beta} = 6S = 24.$$

The computation of γ requires

$$N_\gamma = S = 4$$

operations. For the determining the log likelihood ratios

$$N_{\text{LLR}} = 2(S - 1) + 1 = 2S - 1 = 7$$

operations are necessary. The total number of operations is therefore upper bounded by

$$N_{\text{op}} \leq 2I \left\{ N_{\text{br}} + N_{\alpha\beta} + N_\gamma + N_{\text{LLR}} \right\},$$

where I is the number of decoding iterations. Thus,

$$N_{\text{op}} \leq I \left\{ 62 + \frac{8}{R_c} \right\}$$

holds. Table 9.2 summarizes the above discussion for various code rates. For five decoding iterations, the designed RCPTC's are less complex than a conventional nonsystematic convolutional (NSC) code with constraint length nine [8].

9.6 PERFORMANCE RESULTS

In this section, performance results obtained for the transmission over two different types of discrete memoryless channels (DMC's), namely the AWGN and the fully interleaved Rayleigh fading channels, shall be presented. These binary input DMC's are assumed to be unlimited in bandwidth. Different RCPTC's which have been devised for the application in FMA bearer services will be considered. The constituent decoders were MAP symbol by symbol estimators also used in [5, 8]. The second constituent code was terminated by using two tail bits.

Fig. 9.5 shows the performance of an RCPTC, which was designed for speech transmission at a bearer data rate of 8 kbit/s, in terms of the bit error ratio versus E_b/N_0 at a code rate of approximately 1/3. Furthermore, the uncoded performance is given as a reference. The parameter of the simulations is the number of decoding iterations which varies between 1 and 5. After the first decoding iteration, the minimum signal-to-noise ratio required to obtain a bit error ratio of $< 10^{-3}$ is about 3.5 dB. After the second decoding iteration, about 1.3 dB less is necessary. The next decoding iteration allows a further gain of 0.2 dB. The next iterations facilitate additional gains of less than 0.1 dB. After five iterations the minimum signal-to-noise ratio required for a bit error ratio of $< 10^{-3}$ is approximately 1.8 dB which is also in accordance with [8]. It follows that the performance improvement diminishes. Also, the well known flattening of the bit error ratio curves due to the free distance lower bound [14] can be seen in Fig. 9.5.

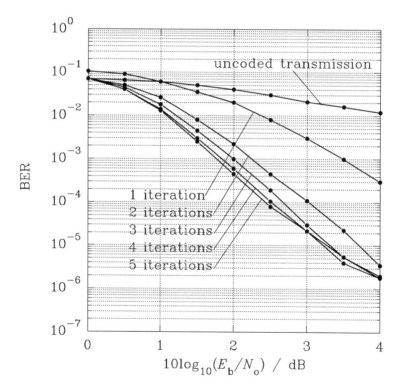

Figure 9.5 Performance of an RCPTC in terms of bit error ratio versus E_b/N_0 for speech transmission; block size 150 bit, code rate $\approx 1/3$, AWGN channel

According to [8], an NSC code with code rate 1/3 and comparable decoding complexity has either constraint length eight or nine. With respect to [8] the best NSC with constraint length eight requires 2 dB and the best NSC with constraint length nine would need about 1.9 dB to achieve the same bit error ratio of $< 10^{-3}$. Still, the RCPTC is slightly more powerful than conventional codes even at block sizes as small as 150 bit which is in accordance with the results of [8].

Fig. 9.6 shows obtained performance results using an RCPTC designed for narrow band ISDN transmission in fully interleaved Rayleigh fading channels. This RCPTC was obtained by setting out from a basic Turbo-Code with code rate 1/3. The number of decoding iterations is a simulation parameter. After four decoding iterations a bit error ratio of $< 10^{-3}$ requires a minimum signal-to-noise ratio of 3.8 dB. After ten iterations, only about 3.4 dB is necessary. The conventional NSC code with similar decoding complexity as four decoding iterations has constraint length eight and requires a 1.1 dB higher signal-to-noise ratio. The designed RCPTC obtains a bit error ratio of $< 10^{-6}$ at a signal-to noise ratio larger than about 5.6 dB. Further six decoding iterations facilitate only a gain of about 0.1 dB. The aforementioned NSC code would however require more than 8 dB of signal-to-noise ratio.

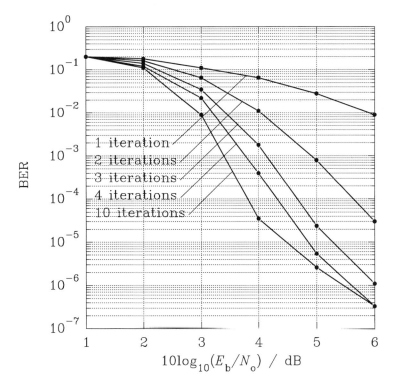

Figure 9.6 Performance of an RCPTC in terms of bit error ratio versus E_b/N_0 for narrow band ISDN at a bearer data rate of 144 kbit/s; block size 672 bit, code rate $\approx 1/2$, fully interleaved Rayleigh fading channel

Fig. 9.7 and Fig. 9.8 show performance results in terms of bit and frame error ratio versus E_b/N_0 for varying code rates of the RCPTC with block size 672 bit for transmission over both AWGN and fully interleaved Rayleigh fading channels, respectively. Ten decoding iterations were considered.

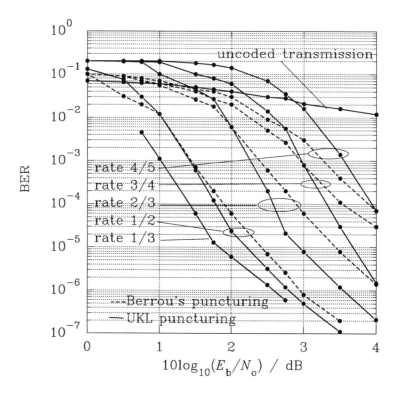

Figure 9.7 Performance of an RCPTC in terms of bit error ratio versus E_b/N_0; block size 672 bit, ten decoding iterations, AWGN channel

Furthermore, two different puncturing approaches called Berrou's puncturing and UKL puncturing are considered. The approach called Berrou's puncturing prohibits the puncturing of u [11]. The second approach which was developed by the authors does not have such a constraint. The performance results obtained for these two puncturing approaches are compared in Fig. 9.7 and Fig. 9.8. It is obvious, that Berrou's puncturing facilitates a better performance at lower values of E_b/N_0 whereas UKL puncturing is more beneficial for higher E_b/N_0 and therefore for bit error ratios of $< 10^{-4}$. The cross-over points move towards lower bit error ratios for increasing code rates. The reason for this behavior is the fact that Berrou's puncturing does not allow puncturing of the systematic information whereas UKL puncturing is more in favor of transmitting redundancy. The design strategy described in Sect. 9.4, which is based on an optimization for transmission over the AWGN channel, leads to RCPTC's with favorable performance also in fully interleaved Rayleigh fading channels.

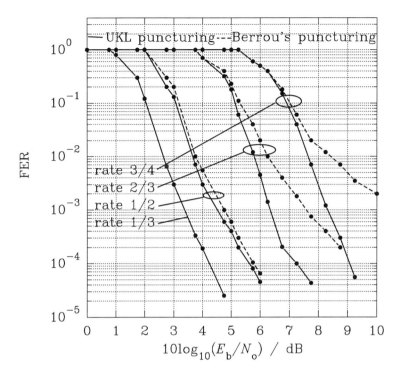

Figure 9.8 Performance of an RCPTC in terms of frame eror ratio versus E_b/N_0; block size 672 bit, ten decoding iterations, fully interleaved Rayleigh fading channel

Fig. 9.9 shows the bit error ratio versus the variance σ_{LLR}^2 of the log likelihood ratios at the output of the second constituent decoder for the RCPTC considered in Fig. 9.7. The code rate does not have any impact on the relation between the bit error ratio and σ_{LLR}^2 because these quantities are similarly dependent on E_b/N_0. Obviously, when σ_{LLR}^2 is known, an estimate of the bit error ratio can be easily generated supporting the possible interaction between radio interface and link level control.

A similar argument is valid for the dependence of σ_{LLR}^2 on the occurrence of frame errors, cf. Fig. 9.10. Clearly, σ_{LLR}^2 for erroneously decoded frames is always greater than σ_{LLR}^2 in the case of correctly decoded frames. Hence, when the E_b/N_0 and σ_{LLR}^2 for a frame under test are known, a soft decision variable related to the probability of a frame error can be easily generated and delivered to the link layer control.

Acknowledgments

This work has been performed partly in the framework of the project ACTS AC090 FRAMES, which is partly funded by the European Community. Although the presented results do not necessarily represent the opinions of their fellow scientists in FRAMES, the authors would like to acknowledge the contributions of their colleagues from Siemens AG, Roke Manor Research Limited, Ericsson Radio Systems AB, Nokia Corporation, Technical University of Delft, Uni-

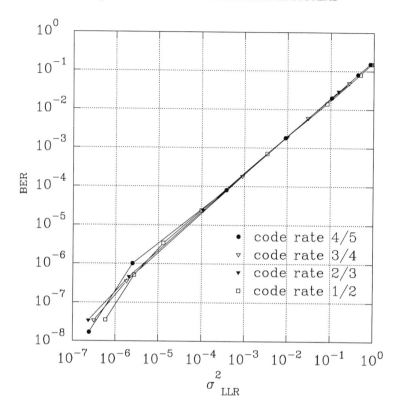

Figure 9.9 Bit eror ratio versus variance of the log likelihood ratios at the output of the second constituent decoder; RCPTC with block size 672 bit; ten decoding iterations; AWGN channel

versity of Oulu, France Telecom CNET, Centre Suisse d'Electronique et de Microtechnique SA, ETHZ, University of Kaiserslautern, Chalmers University of Technology AB, The Royal Institute of Technology, Instituto Superior Tecnico and Integracion y Sistema. Furthermore, the authors wish to acknowledge fruitful discussions with H. Koorapaty, Y.-P. Eric Wang and K. Balachandran of Ericsson, Inc., Research Triangle Park, NC, USA. Finally, the authors wish to acknowledge valuable comments by F. Berens, University of Kaiserslautern, R. Pirhonen, Nokia Research Center, and P.-O. Anderson, Ericsson Radio Systems.

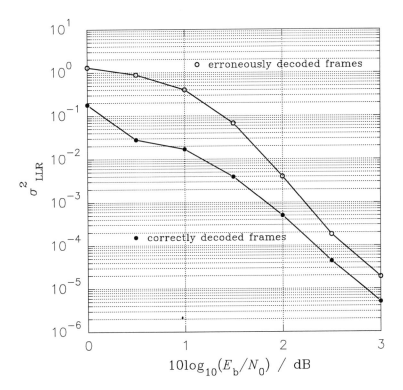

Figure 9.10 Variance of the log likelihood ratios at the output of the second constituent decoder versus E_b/N_0; RCPTC with block size 600 bit; code rate approximately $5/9$, ten decoding iterations; AWGN channel

References

[1] T. Ojanperä, *GSM: Evolution towards 3rd generation systems.* UMTS Data Services, Kluwer Academic Publishers, 1998.

[2] E. Biglieri and J. Hagenauer, "Focus on iterative and "turbo" decoding," *European Transactions on Telecommunications*, vol. 6, no. 5, 1995.

[3] S. Benedetto, D. Divsalar, and J. Hagenauer, "Sailing toward channel capacity," *IEEE Journal on Selected Areas in Communication, Special Issue on Concatenated Coding Techniques and Iterative Decoding*, vol. 16, no. 2, 1998.

[4] "Proceedings of the International Symposium on Turbo-Codes & related topics," 3-5 September 1997, Brest/France.

[5] P. Jung, "Comparison of turbo-code decoders applied to short frame transmission systems," *IEEE Journal on Selected Areas in Communications*, vol. 14, pp. 530–537, 1996.

[6] P. Jung and M. Naßhan, "Results on Turbo-Codes for speech transmission in a joint detection CDMA mobile radio system with coherent receiver antenna diversity," *IEEE Transactions on Vehicular Technology*, November 1997.

[7] P. Jung, J. Plechinger, M. Doetsch, and F. Berens, "A pragmatic approach to rate compatible punctured turbo-codes for mobile radio applications," in *Proceedings of the 6th International Conference on Advances in Communications and Control: Telecommunications/Signal Processing, Corfu*, 1997.

[8] H. Koorapaty, Y.-P. E. Wang, and K. Balachandran, "Performance of turbo-codes with short frame size," in *Proceedings of the 47th IEEE Vehicular Technology Conference VTC'97, Phoenix*, pp. 329–333.

[9] B. Friedrichs, "Kanalcodierung," *Berlin, Springer,* 1996.

[10] J. Hagenauer, "Rate compatible punctured convolutional codes (rcpc-codes) and their applications," *IEEE Transactions on Communications*, vol. 36, pp. 389–400, 1988.

[11] C. Berrou and A. Glavieux, "Near optimum error correcting coding and decoding: Turbo-codes," *IEEE Transactions on Communications*, vol. 44, pp. 1261–1271, 1996.

[12] P. Jung and M. Naßhan, "Performance evaluation of turbo-codes for short frame transmission systems," *Electronics Letters*, vol. 30, pp. 111–113, 1994.

[13] P. Jung and M. Naßhan, "Dependence of the error performance of turbo-codes on the interleaver structure in short frame transmission systems," *Electronic Letters*, vol. 30, pp. 287–288, 1994.

[14] C. Schlegel, "Trellis coding," *Piscataway, IEEE Press*, 1997.

10 SOFTWARE RADIO RECEIVERS

Tim Hentschel and Gerhard Fettweis

Dresden University of Technology
Mobile Communications Systems Chair
D-01062 Dresden
Germany

hentsch@ifn.et.tu-dresden.de

10.1 INTRODUCTION

From the experiences made one can easily extrapolate that the foreseeable mobile communications market as well as the communications devices will allow for heterogeneous plurality, i.e. there will be no common standard. At least today's standards will continue as new standards are introduced. Different operators will deploy different standards in different areas of the world, always in order to try to exploit the forces of market to their own profit. We cannot expect a unification of the mobile communications market organized by the network operators. On the other hand there is a demand for unification of the mobile communications market from the equipment manufacturer's and user's point-of-view. In the future few users of mobile communications services will accept to carry dedicated terminals for different services in different networks. Moreover, the equipment manufacturers can reduce the cost of their products by unifying the hardware platform.

Looking at existing and newly emerging standards in the mobile telecommunications market we recognize a limited number of different principles of sharing the limited resource bandwidth. These basic principles are CDMA, TDMA, and FDMA, as well as combinations thereof. However, the different mobile communications standards are not only to be separated by the channel access scheme but by many other characteristics, such as modulation scheme, antenna beam-forming, or error correction. Having in mind this diversity as well as the demand for common terminal equipment the need for software parameterizable and programmable terminals - i.e. software radio - arises. The aim is to have a general hardware that can principally cope with the strongest constraints of all the mobile communications standards to be supported by this hardware, which is programmed by means of software and thus temporarily made standard and service specific.

The unification of the mobile communications market will certainly not take place at the network side but on the users side. And this is the place where we meet the network operators again. Since they would like to react to changes in user's behaviour or to technical progress very quickly (if there is profit to be made) the network operators surely will like to have software programmable base-stations in their networks. However, as in the PC area: Software runs on Hardware. The term "Software Radio" may be confusing, since the main effort to be undertaken today is to get hardware to support a Software Radio. It must cope with the strong diversity of existing and future mobile communications systems. If, moreover, the mobile terminals are concerned there is one constraint more that resides above all other constraints: the power consumption. This is the most restrictive constraint that forces the designer to make compromises. Hence, there will be no software programmable terminal overnight but the design of such hardware will be a process of continuous improvement slowly evolving upwards, starting at a relatively low level, e.g. with dual- and multi-mode terminals.

In this paper we shall develop a concept of software radio enabling the design of software radio terminals. This will be done by analyzing carefully the candidates of mobile communications standards to be implemented in a software radio terminal and thus deriving a system architecture. With respect to the availability of technologies we shall name the most critical functionalities of a software radio receiver. These critical functionalities are investigated. Finally a system is suggested that is realizable with existing technologies today. This suggestion could be a starting point of the development towards an advanced software radio which will be guided and limited by the available technology at that time.

10.2 SOFTWARE RADIO CONCEPT

10.2.1 Signal Characteristics

When designing a receiver, regardless if standard-specific or software-programmable, the signals to be received and processed have to be characterized first. By signal characteristics we mean such basic features as dynamic range (determined by the interferer characteristics) and some stochastic properties. The receiver has to be designed in a way that it is able to process the signals without losing necessary information carried by the signal. Tackling the problem most generally the mobile communications signal comprises several MHz of bandwidth (split or continuous in frequency), each service (standard-specific frequency-band of a single operator) covers a part of the whole mobile communications signal band. This is principally sketched in Figure 10.1. Each service's band is again split into several channels in case of FDMA or is wholly occupied by every channel of the service (spread spectrum systems).

Interference Properties. The interference characteristics of spread spectrum and FDMA systems are different. Since in spread spectrum systems every channel occupies the total or at least a major part of the system bandwidth, there are no standard-inherent

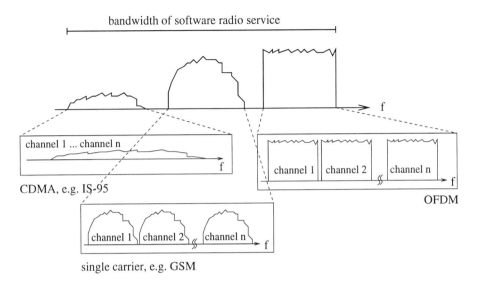

Figure 10.1 Software Radio supported services sketched over frequency

adjacent channel interferers. However, signals of other services adjacent to the current service-of-interest might appear as adjacent channel interferers. On the other hand all channels of a spread-spectrum service act as cochannel interferers. Still, this is a fundamental spread spectrum systems characteristic and is not a specific issue in software programmable terminals.

In FDMA systems several channels coexist adjacent in frequency. In standard specific terminals for FDMA the receiver has to 'get' the signal-of-interest by step-by-step attenuating the adjacent channels and amplifying the signal-of-interest. Since a software radio is basically a wide-band receiver, several channels are received and processed simultaneously. The dynamic range of such a "multi-channel FDMA signal" is determined by the minimum required power level (often referred to as reference sensitivity level) and the maximum allowed power levels of the adjacent interferers (adjacent interference characteristics, blocking characteristics).

As an example these interference characteristics of the GSM system are given in Figure 10.2. It can be seen that the dynamic range of a GSM signal received by a wide-band receiver e.g. of 4 MHz bandwidth is more than 70 dB. Hence as long as the interferers are present the terminal components have to cope with this high dynamic range. Such strong dynamic range constraints apply to every mobile communications standard where an FDMA scheme is used.

Statistical Properties. For estimating effects such as quantization in analog-to-digital converters it is useful to have knowledge about the statistical properties of the signals. Mobile communications signals are mainly influenced by the mobile communications channel with its properties time variance and multi-path propagation. A huge effort has been made in order to model the mobile communications channel and to estimate the statistical distribution of the signal depending on several parameters.

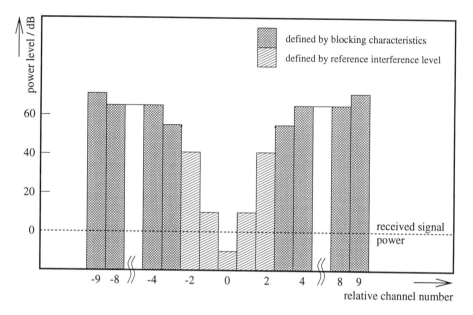

Figure 10.2 GSM Interferer Mask

Here we just assume every propagation path of every channel to be a signal (random function X_i) having any distribution D_i with finite mean and finite non-zero power. All these N signals are to be statistically independent. Applying the central limit theorem to the sum X of these random functions

$$X = \sum_{i=1}^{N} X_i \qquad (10.1)$$

we can approximate the statistical properties of a mobile communications signal comprising several channels for large values of N to have a Normal distribution.

10.2.2 Receiver Architecture

Communications Receivers. In the beginning of wireless communications direct conversion was the first and only receiver architecture. Due to several draw-backs of this architecture the super-heterodyne principle took over as soon as it was realizable. However, because of the necessary stages of mixers and analog filters in a super-heterodyne receiver, that cause increasing cost and effort in mass-production, designers of mobile communications receivers have remembered the direct conversion principle for its simplicity [1]. The main problems such as DC offset and Tx-Rx-coupling must then be cancelled with digital signal processing techniques. There are modulation-scheme dependent solutions to these problems, especially for constant-envelope signals. However, general solutions have not been found yet, although the direct-conversion technique bears a high potential for the design of future software-radio terminals.

The image-rejection receiver [6, 13] is a trade-off between the two mentioned receiver types. Down-conversion to a low IF instead of a zero IF circumvents the main problems of the direct-conversion technique. However, for the necessary image rejection of the two overlapping frequency bands Hilbert transformers have to be used.

These three types of receiver architecture are shown in Figure 10.3. The advantages

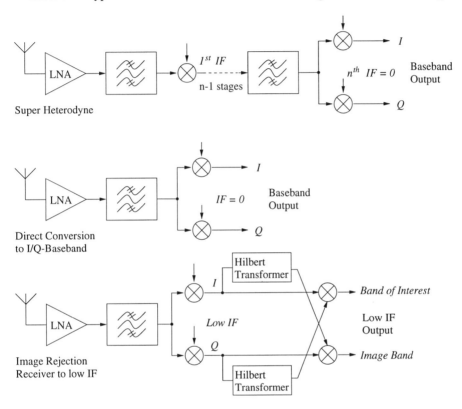

Figure 10.3 Architectures of Communications Receivers

of the individual architectures are

- one-step down-conversion leading to small amount of analog components,
- down-conversion to IF avoiding problems like DC offset or Rx-Tx coupling, and
- down-conversion to 'higher' IF avoiding I-Q down-conversion and image rejection effort.

Combining these advantages leads to a receiver with down-conversion to a fixed IF followed by IF sampling. This suggestion, which is sketched in Figure 10.4, is based upon considering the pros and cons of different communications receiver types and has been derived by tackling the problem of designing a Software Radio receiver from an architectural point-of-view. In section 10.3.1 a selection will be made from a functional point-of-view that fortunately will not be contradictory to the current selection.

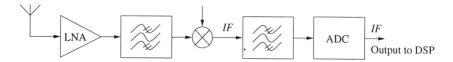

Figure 10.4   Suggested Receiver Architecture for Mobile Communications

Generic Receiver. The 'Generic Receiver' is a model of a communications receiver that consists of all necessary functionalities. It is 'generic', since different functionalities can be implemented intermeshed and in different order. A generic receiver is sketched in Figure 10.5. The different boxes and symbols stand for different functionalities and tasks to be performed in a mobile communications receiver. Obviously there are inter-dependencies between these boxes and symbols which will be looked at later. First the different functionalities are investigated in more detail with respect to a possibly necessary parameterizability or programmability of their characteristics. This is summarized in Table 10.1, where the parameter 'algorithm' stands for different algorithms performing different versions of the same functionality. The main parameter of

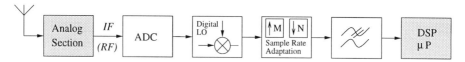

Figure 10.5   Generic Receiver for Software Radio Terminals

the 'generic' receiver is the bandwidth B which is related to the sample rate f_S and the dynamic range [1]. These parameters have either to be adapted to the different mobile communications services, or the receiver has to fulfill the strongest constraints. However, parameters such as the sample rate may vary between the different services and must be adapted for achieving the final symbol rate and performing standard specific processing.

10.2.3   Critical Functionalities

In order to determine which functionalities have to cope with the strongest constraints and which are to be parameterizable or programmable, two basic design criteria that are sensible when trying to maximize simplicity are given:

- analog components should be minimized and be fixed (non-parameterizable)

- the system clock should be fixed, i.e. the ADC and the digital signal processing hardware is clocked with a (generally) constant rate.

---

[1]*Dynamic range* is the maximum ratio of signal-power to noise-power. Since only *maximum* signal-to-noise ratios are regarded in this chapter the terms dynamic range and signal-to-noise ratio are equivalent. Since quantization noise will play a major role in the theoretical analysis of analog-to-digital converter performance, the acronym SNR is used for signal-to-quantization-noise ratio.

Table 10.1 Functionalities of Generic Receiver

Functionality	Parameters	Variability of Parameters
Antenna	bandwidth B, gain	parameterizable or fixed with widest bandwidth and highest gain
LNA	Noise Figure, gain	parameterizable or fixed with highest gain and best noise figure
System Filter	bandwidth B, loss	parameterizable or fixed with highest gain and best noise figure
Mixer (Down Conversion)	bandwidth B, image rejection	fixed with widest bandwidth and strongest image rejection
ADC	Resolution, bandwidth, sample rate	parameterizable or fixed with highest resolution, widest bandwidth
Sample Rate Adaptation	bandwidth B, SNR	parameterizable/ programmable
Channelization/ De-Spreading	bandwidth B, sample rate f_S	programmable
Channel Estimation, Equalization	algorithm	programmable
Decoding	algorithm	programmable

On the other hand the aim is to be as flexible as possible, i.e. to have as many adaptive functionalities in the receiver as possible, leading to the idea to shift most of the functionality of the receiver from the analog to the digital domain, i.e. to perform IF- or even RF-sampling. This leaves the antenna, the LNA, the system-filter, and in case of IF-sampling a down-converter to the analog domain. With regard to current technologies the effort in the analog domain has thus been minimized. The analog components have to be designed to fulfill the constraints of all services to be supported in the software radio receiver.

Still, the high dynamic range of wide-band sampled mobile communications signals (see section 10.2.1) makes highest demands on the analog components. High dynamic range in connection with relatively wide bandwidth makes the ADC the very key as well as critical component of every software radio receiver. If the signal cannot be transferred into a DSP there is no use for high-speed DSPs or sophisticated software. Thus the ADC is regarded as one of the so-called 'critical functionalities'.

264   CDMA TECHNIQUES FOR THIRD GENERATION MOBILE SYSTEMS

Very strongly related to the ADC is the task of sample rate conversion. Since the different services are generally based on different standards that are based on usually incommensurate clock- and symbol-rates the digital signal at the output of the ADC has to be adapted in sample rate. Finally the signal, still at a relatively high sample rate, has to be de-spread or channelized (meaning the channel filtering and adjacent channel interferer suppression in FDMA systems). The latter two functionalities are to be performed in the digital domain at clock rates that are well above the rates digital signal processors can cope with. Equalization, channel estimation, decoding etc. are typically standard specific tasks with high computational effort at relatively low sample rates (symbol rate). Summarizing, and not counting the critical functionalities in the analog RF-section due to the high dynamic range of the signal being passed down all the way to the ADC, the critical functionalities are

1. Analog-to-Digital Conversion

2. Digital Down-Conversion

3. Sample Rate Adaptation and Decimation

4. Channelization and Interferer Cancellation / De-spreading.

10.3   INVESTIGATION OF CRITICAL FUNCTIONALITIES

10.3.1   Analog-to-Digital Conversion

Digitization Bandwidth.   As explained above the Analog-to-Digital Conversion task is to be performed either at RF or at IF. Having in mind the purpose of the software radio receiver, i.e. receiving and processing a variety of different services, and taking into account the signal characteristics, the digitization can be performed as Full-Band Digitization and Partial-Band Digitization (Figure 10.6). While in case of Full-Band Digitization the whole bandwidth comprising all services to be supported is digitized, in Partial-Band Digitization just a part of the whole bandwidth (e.g. an equivalent to the largest channel bandwidth of all services to be supported) is digitized. Since the whole bandwidth to be supported can easily extend to some 100 MHz while the dynamic range in cellular mobile communications standards may be well above 100 dB (see GSM interferer characteristics in section 10.2.1 as example for a bandwidth of approximately 4 MHz) Full-Band Digitization does not seem to be feasible, not even in the near future. Therefore Partial-Band Digitization is the most promising candidate for software radio terminals.

Flash ADCs.   Since the signal at the input of the ADC does not comprise any static carrier (see section 10.2) the white quantization noise model can be used as a good approximation in the following analysis. With Δ being the linear quantizer step size, V_{pp} being the maximum amplitude range (peak-to-peak voltage) not saturating the quantizer, and b being the number of bits of the quantizer,

$$\Delta = \frac{V_{pp}}{2^b} \qquad (10.2)$$

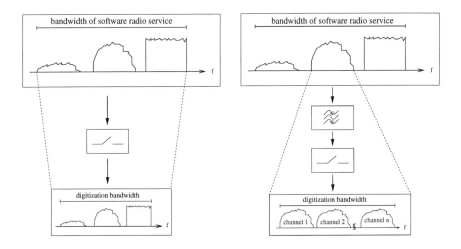

Figure 10.6 Full-Band and Partial-Band Digitization Principle

the noise density $S_{N_Q N_Q}$ of the white quantization noise is

$$S_{N_Q N_Q}(f) = \frac{\Delta^2}{12 f_S}, \quad -\frac{f_S}{2} \leq f < \frac{f_S}{2} \qquad (10.3)$$

with f_S being the sample rate of the ADC. Having a signal with Normal distribution of amplitude, with zero mean and variance σ_x^2, at the quantizer's input and an oversampling ratio U, the signal-to-quantization-noise ratio within the frequency-band-of-interest (Uth fraction of f_S) is (impedance level of 1 Ω) [8]:

$$SNR_U|_{dB} = 10.8 + 6.02b + 10 \log U + 20 \log \frac{\sigma_x}{V_{pp}} \qquad (10.4)$$

The last term in equation 10.4 is a measure of the ratio between the standard deviation of the input signal and the maximum peak-to-peak voltage of the quantizer, which is important when selecting the appropriate ADC to an application. The main points to be observed from equation 10.4 are:

1. The SNR rises with 6.02 dB per additional bit of quantizer resolution.

2. The SNR rises with 3.01 dB per doubling the oversampling ratio, which is equivalent to 1/2 bit quantizer resolution.

Thus oversampling by a factor U leads to a gain of quantizer resolution of

$$b = \frac{1}{2} \log_2 U \text{ bit}. \qquad (10.5)$$

Sigma-Delta ADCs. In contrast to Flash ADCs, whose quantization noise is theoretically uniformly distributed in frequency, Sigma-Delta ADCs are noise-shaping converters. Sigma-Delta ADCs are usually split into the Sigma-Delta Modulator (SDM), being the analog-to-digital interface, and the decimator. The latter is sometimes referred to as a serial-to-parallel converter, which is slightly confusing because it does not describe its functionality comprehensively. Generally seen, it is rather a filter followed by a down-sampler. Since the actual analog-to-digital conversion is done by the SDM we separate the modulator from the decimator. Everything that follows the modulator is digital signal processing and should therefore be looked at in conjunction with the other digital signal processing functionalities.

Sigma-Delta modulation can be regarded as a combination of Delta modulation and demodulation. Swapping the order of Delta modulation and demodulation and combining both into one system leads to the generic Sigma-Delta modulator shown in Figure 10.7. Since the Delta modulator is not an LTI system, the swapping of the order of modulator and demodulator results in a system with different characteristics. $A(z)$ is the transfer function of the former Delta demodulator, while $B(z)$ is the transfer function of the loop-filter of the former Delta modulator. Usually, it applies $A(z) = B(z)$ for Delta modulation. In case of Sigma-Delta modulation, choosing $A(z) \neq B(z)$ can result in better performance of the modulator [10]. However, many designers stick to $A(z) = B(z)$, as we do in the following analysis.

In order to understand the noise-shaping property of SDMs it is useful to first look at the generic SDM. The quantization process is determined by

1. The transfer functions $A(z)$ and $B(z)$,

2. The quantizer resolution (number of bits), and

3. The oversampling ratio.

As in case of Flash ADCs we analyze the quantization behavior of SDMs using the white quantization noise model approximation [2] and thus replace the quantizer by a white noise source $N(z)$. The noise transfer function (NTF) and the signal transfer function (STF) are

$$NTF(z) = \frac{Y(z)}{N(z)} = \frac{1}{1-B(z)} \quad (10.6a)$$

$$STF(z) = \frac{Y(z)}{X(z)} = \frac{A(z)}{1-B(z)} \quad (10.6b)$$

What we observe is basically that the transfer functions for the signal and the quantization noise are different, principally enabling one to separate the noise from the signal. For reasons of simplicity we stick to one of the classic 2nd order SDMs employing double-integration [3]. The block diagram of this SDM is sketched in Figure 10.8. With respect to equations 10.6a and 10.6b and $A(z) = B(z)$ we get

$$NTF(z) = \frac{Y(z)}{N(z)} = \left(1 - z^{-1}\right)^2 \quad (10.7a)$$

$$STF(z) = \frac{Y(z)}{X(z)} = z^{-1} \quad (10.7b)$$

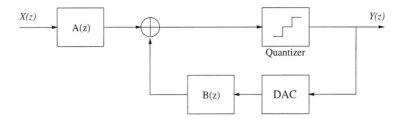

Figure 10.7 Generic Sigma-Delta Modulator

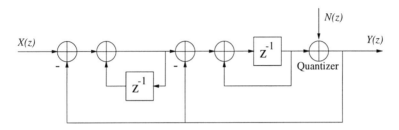

Figure 10.8 2nd order Sigma-Delta Modulator employing double integration

The input signal $X(z)$ is filtered by an all pass filter, and the quantization noise $N(z)$ by a high pass filter. For further analysis the noise transfer function is evaluated on the unit circle, and then the SNR is calculated within the Uth fraction of the total frequency band. With P_x being the signal power and N_Q the quantization noise power we receive

$$SNR_U|_{dB} = 10\log \frac{P_x}{\frac{\Delta^2}{12}} + 10\log \frac{\pi}{\int_0^{\frac{\pi}{U}} 16\sin^4 \frac{\omega'}{2} d\omega'} \qquad (10.8)$$

While the first term in equation 10.8 stands for the signal-to-quantization-noise ratio of the quantizer without noise shaping within the entire frequency band, the second term gives the enhancement of the SNR within the Uth fraction of the total frequency band caused by the noise shaping process of the oversampled Sigma-Delta modulation. Applying the 3rd order Taylor series expansion to the sine function in equation 10.8 and approximating the integral for $U \geq 4$ we get

$$SNR_U|_{dB} \approx 20\log \frac{\sigma_x}{V_{pp}} + 6.02b + 50\log U - 2.1, \ U \geq 4 \qquad (10.9)$$

Generalizing for SDMs of Lth order with sinusoidal noise shaping (employing integrators as loop filters) the SNR is [2]

$$SNR_U|_{dB} \approx 20\log \frac{\sigma_x}{V_{pp}} + 6.02b + (2L+1)10\log U$$
$$+ 10.8 + 10\log \frac{2L+1}{\pi^{2L}}, \ U \geq 4 \qquad (10.10)$$

268   CDMA TECHNIQUES FOR THIRD GENERATION MOBILE SYSTEMS

From equation 10.9 we can see that

1. The SNR rises with 6.02 dB per additional bit of quantizer resolution.

2. The SNR rises with $6(L + 1/2)$ dB per doubling the oversampling ratio, equivalent to $L + 1/2$ bit quantizer resolution.

Oversampling by a factor U with an Lth order SDM leads to a resolution gain equivalent to

$$b = \left(L + \frac{1}{2}\right) \log_2 U \text{ bit}. \qquad (10.11)$$

The analysis has been made for SDMs with an SNR maximum at DC. These 'Low-pass SDMs' can be transformed to 'Band-pass SDMs' having equivalent noise shaping characteristics, however, with the SNR maximum at any desired center frequency [2].

Flash or SDM.   Before answering this question an example will be given. Assuming the application of a combined terminal for the standards GSM, DECT, and GPS, the main parameters of Flash ADCs and SDMs fulfilling the constraints are derived. In Table 10.2 some basic parameters of the three standards are summarized.

Table 10.2   Parameters of Mobile Communications Standards

	GSM	DECT	GPS
Mobile Receive Band	935-960 MHz	1880-1900 MHz	1575 MHz
Channel Separation / Channel Bandwidth	200 kHz	1728 kHz	2046 kHz
minimum SNR at receiver ADC (hardware dependent)	9-12 dB	10 dB	6-12 dB (theoretically 1 Bit resolution)

With the parameters of Table 10.2 and the interference characteristics of GSM (Figure 10.2), it can clearly be seen that GSM poses the toughest requirements among the three standards . The dynamic range requirements of an ADC digitizing those signals at a bandwidth of 2 MHz can be derived from Table 10.3.

Requiring a dynamic range of 80 dB (in order to be on the safe side) within a band of 200 kHz, using equation 10.4 and setting

$$\frac{V_{pp}}{2} = 3\sigma_x \qquad (10.12)$$

SOFTWARE RADIO RECEIVERS   269

Table 10.3 Dynamic Range of Mobile Communications Signals at Wideband Digitization

Digitization Bandwidth = 2 MHz	GSM	DECT	GPS
Dynamic Range	75 dB	10 dB	6 dB
Channel Bandwidth	200 kHz	≈1000 kHz	2000 kHz

being a value used in practice [2] for signals having Normal distributed amplitudes, the oversampling ratio for a Flash ADC with respect to the signal bandwidth of 200 kHz can be calculated,

$$80 = 10.8 + 6.02b + 10\log U + 20\log \frac{1}{6} \qquad (10.13)$$

leading to the approximation

$$U \approx 10^{8.5} \cdot 10^{-0.6b} . \qquad (10.14)$$

For 2nd order SDMs employing double integration the oversampling ratio can be calculated with equations 10.9 and 10.12

$$80 = 20\log \frac{1}{6} + 6.02b + 50\log U - 2.1 \qquad (10.15)$$

leading to the approximation

$$U \approx 10^2 \cdot 10^{-\frac{b}{8}} . \qquad (10.16)$$

Some figures resulting from equations 10.14 and 10.16 are given in Table 10.4. One of the few available Flash ADCs that could fulfill the requirements at a power consumption of 735 mW, which is unacceptable for usage in mobile terminals, is the AD9042, sampling at 41 MS/s with a resolution of 12 bit. However, a 2nd order SDM already fulfills the requirements sampling at 15 MS/s with a 1-bit quantizer.

From Table 10.4 it can be observed that Flash ADCs designed for wideband digitization of GSM signals are highly over-designed for digitizing DECT or GPS signals at the same sample rate. With SDMs the difference between the required SNR (Table 10.3) and the achieved SNR (Table 10.4) is much smaller. Thus SDMs seem to be a better fit to the digitization task of those signals. This is principally sketched in Figure 10.9, where the power spectral density of the quantization noise of an SDM is plotted (dotted) along with the GSM interferer mask (dashed) and the minimum received signal (solid) with 12 dB SNR (see Table 10.2). The zero-dB line represents the white quantization noise floor produced by Flash ADCs.

---

[2] following the "3σ-rule", 99.7% of all signal values having Normal distributed amplitudes have magnitudes less than $3\sigma_x$, hence equation 10.12 means, that 99.7% of those signal values do not saturate the quantizer, which is sufficient to ensure reliable operation in practice

Table 10.4 Relations between sample rate, oversampling ratio and resolution of ADCs for digitizing GSM, DECT, and GPS signals at a bandwidth of 2 MHz

ADC resolution [bit]	resulting oversampling ratio	minimum sample rate for GSM ($B = 200$ kHz)	SNR at DECT bandwidth	SNR at GPS bandwidth
1 (SDM)	75	15 MHz	45 dB	30 dB
8 (Flash)	5011	1 GHz	73 dB	70 dB
10 (Flash)	316	63 MHz	73 dB	70 dB

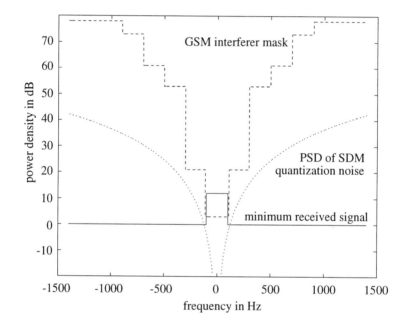

Figure 10.9 Noise shaping wide-band digitization of narrow-band signals

Based upon the analysis made in the two previous sections the appropriate type of ADC for Partial-Band Digitization (see section 10.3.1) is selected. Doing this two important aspects have to be considered:

1. Given the application in mobile communications, power consumption is a most important design criterion.

2. Aliasing caused by the digitization process has to be avoided.

Generally, higher sample rates lead to relaxed aliasing constraints and thus to simpler anti-aliasing filters at the cost of higher power consumption. Since higher sample rates yield a higher oversampling ratio the resolution gained by oversampling can be offset against the increase of power consumption. Comparing equations 10.5 and 10.11 shows that with SDMs a higher resolution gain is achieved, leading to more efficient and power-saving implementations. Moreover, the strong increase of SNR of SDMs compared to Flash ADCs is very much desirable in the field of mobile communications. This results from the typical interferer characteristics of mobile communications signals as explained in section 10.2.1.

SDMs seem to be a perfect fit to these signal characteristics and are thus ADCs tailored to the task of wideband digitization of mobile communications signals. Since the signals to be digitized are band-pass signals band-pass SDMs have to be employed. Principally, the center frequency of a band-pass SDM (i.e. the frequency where the maximum SNR can be found) can be any frequency. However, some special frequencies, e.g. a quarter or an eighth of the sample rate, have been proven to enable a relatively simple design of band-pass SDMs [15, 14].

Finally, a draw-back of the conventional type of SDMs presented above should be mentioned: Since the high SNR is only available at the center frequency of the SDM noise shaping, digitization of multiple narrow band channels is not possible. Only the center channel can be used for further processing. This is not a restriction for mobile terminals where only one channel is to be received at a time. However, for base-station applications with the need for multiple channels to be processed in parallel, the function $B(z)$ in equation 10.6a can be designed to realize a multinotch noise-transfer-function. By making $B(z)$ parameterizable the SDM could be adapted to any channel spacing. For noise-transfer-function design the reader is referred to [11].

10.3.2 Digital Signal Processing

Main Tasks. The advantages of Software Radio are reached by the application of Digital Signal Processing techniques. Its main tasks in a Software Radio receiver can be taken from Table 10.1. In section 10.2.3 the critical functionalities have been derived. On the digital signal processing side these are

1. Digital Down-Conversion

2. Sample Rate Adaptation and Decimation, and

3. Channelization and Interferer Cancellation / De-spreading.

Digital Down-Conversion. In section 10.3.1 Partial-Band Digitization of a signal at IF has been selected as an appropriate means of digitization for software radio terminals. Since signal processing is almost always simpler and thus more efficiently performed at base-band the digitized signal has to be down-converted first. Digital down-conversion after analog-to-digital conversion has one main advantage compared to analog down-conversion before AD conversion: Perfect I-Q matching and thus image rejection can be realized.

The two parameters that influence the effort of implementing digital down-conversion are the IF and the sample rate. If the ratio of IF and sample rate obeys the rule

$$f_{IF} = \frac{n}{4} f_S, \; n = 1, 3, 5, \ldots \quad (10.17)$$

i.e. the IF is an odd multiple of a quarter of the sample rate, digital down-conversion can be performed by multiplying the signal with the sequences [0 1 0 -1] and [1 0 -1 0], representing the digital sine- and cosine-signal, respectively, at a quarter of the sample rate. In case $n = 1$ the digitized version of the 'original' signal is down-converted, while in case $n > 1$ an image of the 'original' signal is digitized and then down-converted. The latter procedure is referred to as 'sub-sampling'. However, this does not influence the described method of down-conversion. The implementation of this version of digital down-conversion can be realized with the least effort.

Given the constraint of equation 10.17, digitization and down-conversion of band-pass signals can be combined. Basically the task of band-pass AD conversion is followed by an I-Q down-conversion. This is sketched in Figure 10.10. Swapping the components ADC and mixer is generally not feasible, for the multiplications necessary in the mixer are not realizable in the analog domain as in the digital domain. Obviously, there are analog mixers, however, we want to avoid the use of these (see discussion in section 10.2.2). In case of fulfilling equation 10.17 the components can easily be swapped without any restrictions concerning increasing effort of analog components or infeasibility of exact digital down-conversion. The idea is to put the down-converter between a sample-and-hold circuit and the ADC, shown in Figure 10.11. Now, the time-discrete not yet quantized signal is multiplied with the sequences [0 1 0 -1] and [1 0 -1 0], which can easily be implemented in switched-capacitor-circuitry (SC), and is eventually digitized by two ADCs in base-band. This swapping of down-conversion and AD conversion is especially useful if SDMs are employed, since low-pass SDMs are even more easily implemented than band-pass SDMs. Moreover, the power consumption can be minimized with this I-Q AD conversion, for the area effort of a band-pass SDM is similar to the area effort of two equivalent low-pass SDMs. However, when leaving out the zeros after down-conversion and interleaved clocking the I- and Q-branch the overall sample rate is halved, thus nearly halving the power consumption. The inherent mismatch between the two low-pass SDMs in the two paths can be minimized to a degree that is sufficient in mobile communications [12].

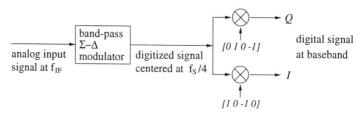

Figure 10.10 Band-pass Sigma-Delta AD conversion followed by digital down-conversion

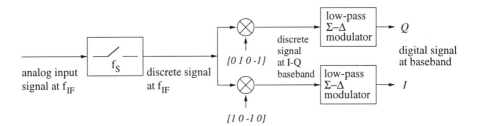

Figure 10.11 Time-discrete down-conversion followed by I-Q low-pass Sigma-Delta AD conversion

Sample Rate Adaptation and Decimation. Since a Software Radio should be able to cope with signals of different mobile communications standards it has to process signals at different sample or symbol rates. In Table 10.1 the sample rate is one of the parameters of the ADC. Basically, the sample rate can either be made adaptive to the different standards or fixed followed by a digital sample rate adaptation process. Above, oversampling Sigma-Delta ADCs have been shown to be a tailored solution to the problem of analog-to-digital conversion in mobile communications receivers, however only the SDM (i.e. the analog-to-digital interface) has been investigated. Being part of the digital signal processing, the sample rate decimation of the Sigma-Delta modulated signal can be combined with the needed sample rate adaptation. With these assumptions four approaches to sample rate adaptation can be named:

1. By keeping the ratio between IF and sample rate according to equation 10.17 both, IF and sample rate, can be made parameterizable, so that each signal can be digitized with the clock rate of the standard of current operation, keeping the opportunity for simple down-conversion.

2. By digitizing with a fixed clock rate at a fixed IF digital sample rate adaptation can be performed by means of mathematical interpolation, which can be implemented in different ways

    (a) The most obvious way of doing sample rate adaptation is combined interpolation and decimation by integer factors leading to sample rate adaptation by a rational factor. Applying this approach to signals with high dynamic range, as it is the case if no sharp channel filtering precedes the adaptation process, the image- and alias-rejection of the interpolation and decimation process have to be very strong, eventually leading to expensive implementations.

    (b) Based upon the previous method the image- and alias-rejection properties can be highly relaxed if the dynamic range of the signal to be adapted in sample rate has been reduced by means of sharp cut-off channel selection filters.

    (c) Assuming a block-wise processing, a certain number of samples per block can be dropped in order to reach the final sample rate. The thus introduced error has either to be cancelled after the dropping or the signal has to be 'predistorted' in a way that the dropping leads to a distortion-free signal. This approach will be referred to as 'asynchronous decimation'.

The last approach seems to be the simplest one, avoiding parameterization of analog components as in the first approach, and in case the error cancellation can be performed efficiently the dropping process is most elegant and straight-forward. It has to be pointed out that the error cancellation with the 'asynchronous decimation' approach is basically an interpolation process as are the approaches (2a) and (2b). The only difference is a possibly simpler implementation.

Channelization and De-Spreading. Principally, this is the first stage of signal processing where not only parameters (such as bandwidth and sample rate) have to be adapted and changed but where different algorithms have to be performed, e.g. different software runs on the hardware. Thus this is the first software programmable and not just parameterizable section of the Software Radio receiver.

Channelization is the functionality where in FDMA systems the tasks of channel filtering (channel selection) and interferer cancellation are performed. This is dependent upon the previously described task of Sample Rate Adaptation and Decimation. Since decimation filters are low-pass filters they work as coarse channel selection and interference cancellation filters. Only fine, sharp cut-off filtering has to be realized 'alone'. This 'additional' filtering (additional to decimation filtering) can principally be performed at the higher sample rate before decimation, or at the lower sample rate after decimation, while the latter is more power efficient. Any matched-filter task can be combined with the channelization task.

De-Spreading is the functionality where in spread-spectrum systems the task of de-correlation and sample rate decimation to symbol rate is realized. More generally it can be regarded as a kind of matched-filtering at symbol level. However, before de-spreading the matched-filtering at chip level has to be performed at a higher sample rate than the chip rate.

What we observe from investigating the tasks of de-spreading and channelization is that the signal is subject to two basic operations, that any filtering as well as correlation operation comprise:

- multiplication by known coefficients, and
- integration.

At the given stage in the signal processing chain in a mobile communications receiver a third task is

- sample rate decimation by an integer number.

Where these three operations can be found in the signal processing of a spread-spectrum receiver and an FDMA receiver, employing the functionalities of de-spreading and channelization, respectively, is shown in Figures 10.12 and 10.13. The FDMA system in Figure 10.12 comprises an anti-aliasing filter $h(k)$, a sample-rate decimator, and a channel filter $g(k)$; the spread-spectrum system in Figure 10.13 comprises a chip matched-filter $rc(k)$ (e.g. a raised-cosine filter), a sample rate decimator, a correlator with the code $c(k)$ followed by an integrate-and-dump circuit (realized by an integrator and a sample-rate decimator).

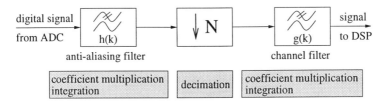

Figure 10.12 Critical Functionalities of Digital Signal Processing in FDMA Receivers

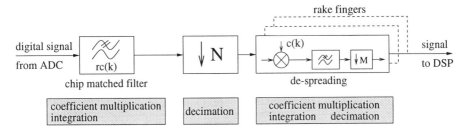

Figure 10.13 Critical Functionalities of Digital Signal Processing in Spread-Spectrum Receivers

Since the functionalities 'Channelization' and 'De-Spreading' cannot be realized on digital signal processors efficiently with today's technologies, for they have to be performed at high sample-rates as mentioned in section 10.2.3, dedicated hardware has to be designed. However, in order to be in correspondence with the software-radio approach, this hardware has to be 'general' and parameterizable to a degree where all intended standards can fit in. Anyway, from today's point-of-view there are only two basic functionalities to be realized with this hardware, namely 'Channelization' and 'De-Spreading', which both comprise of the tasks coefficient-multiplication, integration, and sample-rate decimation. Thus it seems to be sensible to design a hardware platform that can efficiently perform these tasks in different order to eventually realize the 'Channelization' and 'De-Spreading'.

Implementation Issues. It is well-known that a convolution of a signal $x(\tau)$ with any signal $h(\tau)$ can be implemented by means of a correlation of the signal $x(\tau)$ with a reverse version of $h(\tau)$, namely $h(-\tau)$. This can easily seen from the definitions of the correlation and the convolution.

$$y_{corr}(t) = \int_{-\infty}^{+\infty} x(\tau) h(t+\tau) d\tau \qquad (10.18)$$

$$y_{conv}(t) = \int_{-\infty}^{+\infty} x(\tau)h(t-\tau)d\tau \tag{10.19}$$

In order to map any general filter task on a typical de-spreading architecture, which would enable the use of a common architecture for channelization and de-spreading, the critical functionalities of digital signal processing in FDMA receivers (Figure 10.12) and in spread-spectrum receivers (Figure 10.13) are generalized.

First we realize that the de-correlation of any signal with the code $c(k)$ can be regarded as filtering the signal with an FIR filter having the impulse response $c(k)$ followed by a down-sampler. In general, if the spreading-factor is smaller than the the code-length, the filter's impulse response is time-varying, taking the contents of a sliding window shifted over the code $c(k)$. However, for showing the commonalities of channelization and de-correlation it is sufficient to simply regard $c(k)$ as an impulse response. Now, the task of down-sampling by N is swapped with the filtering by $g(k)$ in FDMA systems or $c(k)$ in spread-spectrum systems, respectively, by exploiting a general signal flow identity [7]. Since now the filtering is to be realized at the higher sample rate, the impulse response has to be interpolated, i.e. zero-padded. The interpolated impulse responses are denoted as $g'(k)$ and $c'(k)$, respectively. Finally, the anti-aliasing filter $h(k)$ or the matched-filter $rc(k)$, respectively, can be combined with the filters $g'(k)$ and $c'(k)$, respectively. The result is a generalized platform for Channelization and De-Spreading. It is shown in Figure 10.14.

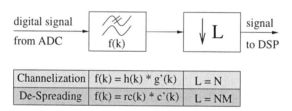

Channelization	f(k) = h(k) * g'(k)	L = N
De-Spreading	f(k) = rc(k) * c'(k)	L = NM

Figure 10.14 Generalized Platform for Channelization and De-Spreading

The architecture in Figure 10.14 is very nice and most general for realizing any filtering or correlation task, however, due to the high number of coefficients such an FIR filter cannot be implemented efficiently. Still, an efficient implementation is possible when referring to the usual implementation of a de-correlator. Its main characteristic is that the number of filter coefficients R (i.e. the spreading factor) and the decimation factor L are identical. Thus only one output-sample is to be calculated out of R input-samples, enabling the typical implementation realized by a multiplier and an integrate-and-dump circuit, which we would like to use also for the generalized platform. However, it does not seem to be applicable, since the number of coefficients R will usually extend the decimation factor L. This is due to the combination of chip-level matched-filtering with correlation in the case of 'De-Spreading' and in the 'Channelization' case due to the large number of coefficients needed to realize a sharp cut-off.

If the conventional de-correlation platform employing a multiplier and an integrate-and-dump circuit is to be used, the number of filter coefficients must be limited to the decimation factor L. This can be realized by splitting the overall impulse response of length R into sub impulse-responses of length $R_{sub} \leq L$, that are implemented in parallel as shown in Figure 10.15 for the case $R \leq 2L$.

Another approach to reduce the number of coefficients R to less than the decimation factor L is to increase the decimation factor, which again leads to an implementation of several sub-filters in parallel. The case $R \leq 2L$ is sketched in Figure 10.16.

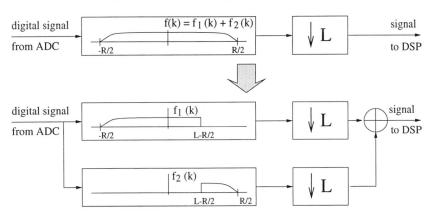

Figure 10.15 Splitting of generalized filter into two sub-filters

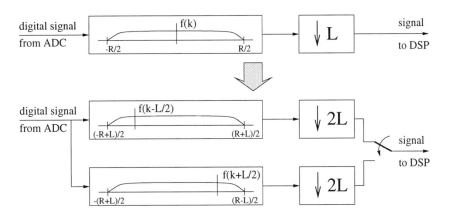

Figure 10.16 Overlapping of two Sub-Filters of the generalized Filter with an Output Commutator

These two implementations look similar to polyphase implementations of LTI systems [7]. However, the 'classic' polyphase approach decomposes a signal into different subsampled phases, whose combination yields the original signal. The architecture in Figure 10.15 could rather be called a 'polyblock implementation', while the architecture in Figure 10.16 is a 'polybranch implementation'. The main characteristic of polyphase implementations in terms of cost is that clock-rate is exchanged with hardware effort, while the 'polyblock' and 'polybranch implementations' are just parallelizations (Figure 10.17) without direct impact on the clock-rate of the system.

Obviously, there is an increasing hardware effort when implementing impulse-responses of an FIR filter in parallel. However, these parallel sections are functionally identical. This e.g. enables a reuse of existing RAKE fingers for channelization tasks. Thus the typical RAKE receiver structure employing several identical parallel de-correlators is a well-suited candidate for the generalized platform for channelization and de-spreading.

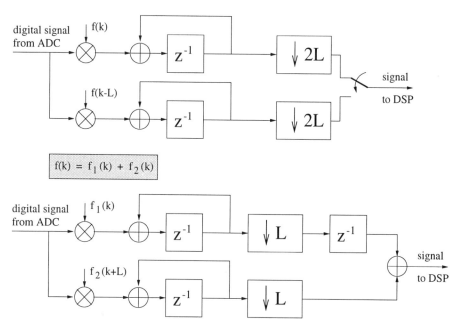

Figure 10.17 Implementation of Generalized Filter, Overlapping (top) and Splitting (bottom) Architecture

Finally, the multiplication task in the two branches is to be investigated. The signal at the input of the multiplier is multiplied by a shifted version of $f(k)$ in case of the overlapping implementation or a part of $f(k)$ in case of the splitted implementation of the generalized filter. $f(k)$ could be held in memory. However, when remembering that $f(k)$ is the result of a convolution of several filters in case of 'Channelization', or a convolution of a de-spreading code with a filter response in case of 'De-Spreading', this multiplication can be realized as shown in Figure 10.18, where the code $c(k)$ can be as well any impulse response $h(k)$ of an FIR-filter. Thus the memory to store the code $c(k)$ or the impulse response $h(k)$ would be smaller compared to the size needed to store $f(k)$, or in case of spread-spectrum systems $c(k)$ could be generated by a feed-back register-chain.

In order to evaluate the cost of the suggested implementation of the generalized filter, it is compared with the conventional implementation according to Figure 10.13.

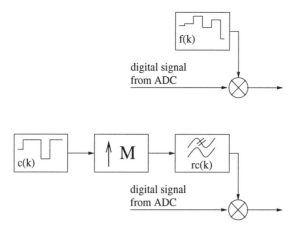

Figure 10.18   Implementation of Multiplication in Generalized Filter (top) as Filtered Code (bottom)

For reasons of simplicity we suppose a b bit input signal, and an FIR-filter ($h(k)$ in Figure 10.12 or $rc(k)$ in Figure 10.13) with n coefficients which are b bit wide. The effort is given for b-bit operations for easier comparison; the spreading code is assumed binary. Supposing $R \leq 2L$ there are two branches in the 'Filtered Code' and 'Generalized Filter Implementation'.

Table 10.5   Cost Comparison between Conventional Implementation of spread-spectrum receiver functionalities and 'Generalized Filter Implementation'

	Conventional Implementation	*Filtered Code Implementation*	*Generalized Filter Implementation*
FIR-Filter $rc(k)$	$n/2$ multiply, $n/2$ add	n add	(negligible) (initialization only)
Code Multiplication	(negligible)	2 multiply	2 multiply
Hardware Effort	$n/2$ coefficient multipliers, and adders	n adders; 2 multipliers	2 multipliers
	bn registers	$(b+1)n$ registers	$bn \times$ 'code-length' registers (memory)

Especially in case of reconfiguration, i.e. if the coefficients of the FIR-filter are to be made parameterizable and thus the coefficient multipliers are as complex as a 'normal' multiplier, the 'Filtered Code Implementation' is not only very *soft* but also is the hardware cost and the computational effort less than in the 'Conventional Implementation'.

The least effort is achievable with the 'Generalized Filter Implementation' if memory cost is negligible. However, in case of very long codes this is not applicable today, since there is an upper limit of memory size that could be implemented. Moreover, it has to be mentioned that in the case of RAKE-receivers the chip-matched filter could be shared among all RAKE-fingers in the conventional implementation, leading to lower cost, which cannot be achieved in the Filtered Code implementation, while again the Generalized Filter Implementation is the most 'Software-Radio-friendly' one. The different shifted codes can be taken from one memory only.

Anyway, the current design is always a trade-off, and a standard-optimized implementation is always more efficient than a general implementation (as is the conventional implementation in case of a RAKE-receiver). The freedom of a general implementation has to be paid for. However, it has been shown that the contradictory-looking tasks of de-spreading and channelization can be implemented on the same hardware with acceptable cost.

10.4 SUMMARY

Software Radio is feasible today at relatively low cost and low power consumption even for mobile terminals! Stating this and providing possible solution scenarios was the main intention of this chapter.

'Full Band Digitization' has been stated as being not feasible today. However, this statement has been made with respect to the application of the Software Radio principle in mobile communications receivers in cellular networks. For applications e.g. in the ISM-band, such as baby-phone, remote-control, or wireless headphones, where the typical interference characteristics of cellular communications standards do not apply and where temporary 'notches' in the quality-of-service are acceptable, 'Full Band Digitization' using simple Flash ADCs may be realizable very soon.

The proposed 'Partial Band Digitization' is a solution for 'high-end mobile terminals', where power consumption is a most restrictive constraint and strong interferers are present. Based upon signal characteristics of mobile communications signals, ADCs have been analyzed and Sigma-Delta-Modulators found to be a nearly perfect fit for the task of digitizing signals at IF. This approach limits the number of analog components and applies digital signal processing as much as possible with todays technologies. The suggested method of digital down-conversion of a signal at a quarter of the sample-rate is certainly a restriction and a trade-off in order to minimize cost. However, as long as 'Full Band Digitization' is not realizable the signal has to be converted to an IF by appropriate mixing with a variable synthesizer before analog-to-digital conversion; and in this case it is not really a restriction to fix the IF with respect to the sample rate.

The first stages of digital signal processing, namely the digital down-conversion, the sample-rate adaptation, and the de-spreading and channelization, have been investigated as 'critical functionalities', since they have to be performed at relatively high sample-rates. Sample-rate adaptation is suggested to be realized asynchronously, i.e. by dropping a certain number of samples non-equidistantly. The basic operation necessary to suppress the thus introduced error is an interpolation problem whose computational effort can be minimized by realizing it with a dedicated architecture. For the tasks of de-spreading in case of Spread-Spectrum Receivers and channelization in case of FDMA Receivers a common architecture, based on a generalization of filtering and correlation, has been presented.

The availability of common architectures for all standards to be supported is a key to the realizability of Software Radio. High performance digital signal processors enable software programmable base-band processing. We have shown that the digitization task and the IF processing can also be realized on common hardware, thus enabling the implementation and realization of software radio with technologies available today. The suggested receiver based on the generic receiver in Figure 10.5 is sketched in Figure 10.19. The task of AD conversion and digital down-conversion could as well be implemented as shown in Figure 10.11.

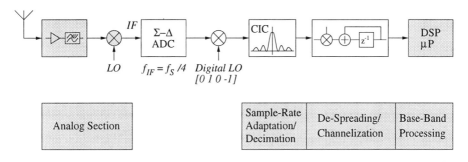

Figure 10.19 Suggested Receiver Architecture based on Generic Receiver Model

Glossary

AD	Analog-to-Digital
ADC	Analog-to-Digital Converter
CDMA	Code Division Multiple Access
DC	Direct Current
DECT	Digital European Cordless Telephone
DSP	Digital Signal Processor
FDMA	Frequency Division Multiple Access
FIR	Finite-Duration Impulse Response
GPS	Global Positioning System
GSM	Global System for Mobile Communications
IF	Intermediate Frequency
I-Q	In-phase and Quadrature-Phase
ISM	Industry, Science, Medicine: frequency band with few restrictions for general purpose use
LNA	Low Noise Amplifier
LTI	Linear Time-Invariant
PSD	Power Spectral Density
RF	Radio Frequency
Rx	Receiver

SDM   Sigma-Delta-Modulator
SNR   Signal-to-Noise Ratio, in this chapter always used as
          Signal-to-Quantization-Noise-Ratio
TDMA Time Division Multiple Access
Tx   Transmitter

Acknowledgments

This work has partially been supported by the Daimler Benz AG, Corporate Research Center Ulm, Germany and by the European Commission, ACTS Project 'Software Radio Technology'. The authors like to thank Dr. Luy of Daimler Benz AG for the support throughout the work.

References

[1] Abidi, Asad A. (1995). *Direct-Conversion Radio Transceivers for Digital Communications*. IEEE Journal of Solid-State Circuits, vol.30, no.12, Dec. 1995.

[2] Azizi, P. M., Sorensen, H. V., and van der Spiegel, J. (1996). *An Overview of Sigma-Delta Converters*. IEEE Signal Processing Magazine, Jan. 1996, pp. 61-84.

[3] Candy, J. (1985). *A Use of Double Integration in Sigma Delta Modulation*. IEEE COM, vol. 33, Mar. 1985, pp. 249-258.

[4] Candy, J. (1986). *Decimation for Sigma-Delta-Modulation*. IEEE COM, vol. 34, Mar. 1986, pp. 72-76.

[5] European digital cellular telecommunications system (Phase 2). *Radio Transmission and Reception (GSM 05.05)*. European Telecommunication Standard ETS 300 577, Sep. 1994.

[6] Fernández-Durán, A., et al. (1995). *Zero-IF Receiver Architecture for Multistandard Compatible Radio Systems, GIRAFFE Project*. IEEE, 1996.

[7] Fliege, N. J. (1994). *Multirate Digital Signal Processing*. John Wiley & Sons, Chichester, New York, Brisbane, Toronto, Singapore, 1994.

[8] Frerking, M. E. (1994). *Digital Signal Processing in Communications Systems*. Van Nostrand Reinhold, New York, 1994.

[9] Hogenauer, E. B. (1981). *An Economical Class of Digital Filters for Decimation and Interpolation*. IEEE Trans. ASSP, vol. 29, Apr. 1981, pp. 155-162.

[10] Jantzi, Stephen A., Snelgrove, W. Martin, and Ferguson Jr., Paul F. (1993). *A Fourth-Order Bandpass Sigma-Delta Modulator*. IEEE Journal of Solid State Circuits, vol. 28, Mar. 1993, pp. 282-291.

[11] Norsworthy, S. R., Schreier, R., and Temes, G. C. (1997). *Delta-Sigma Data Converters. Theory, Design, and Simulation*. IEEE Press, New York, 1997.

[12] Ong, Adrian K., and Wooley, Bruce A. (1997). *A Two-Path Bandpass $\Sigma\Delta$ Modulator for Digital IF Extraction at 20 MHz*. Proceedings IEEE ISSCC97.

[13] Páez-Borallo, José M., and Casajús Quirós, Francisco J. (1997). *Self Adjusting Digital Image Rejection Receiver for Mobile Communications*. Proceedings IEEE VTC97, vol. 2, pp. 686-690.

[14] Schreier, R., and Snelgrove, M. (1989). *Bandpass Sigma-Delta Modulation*. Electronic Letters, vol. 25, Nov. 1989, pp. 1560-1561.

[15] Thurston, A. M., Pearce, T. H., Higman, M. D., and Hawksford, M. J. (1992). *Bandpass Sigma Delta A-D Conversion*. Proc. Adv. Analog Circuit Design, April 1992.

11 BLIND SPACE-TIME RECEIVERS FOR CDMA COMMUNICATIONS [*]

David Gesbert, Boon Chong Ng[†] and Arogyaswami J. Paulraj

Information Systems Laboratory,
Durand 102, Stanford University,
Stanford, CA 94305-4055, USA.
[†] DSO National Lab 20 Science Park Drive Singapore 118230, Singapore

gesbert,paulraj@rascals.stanford.edu; nbooncho@starnet.gov.sg

Abstract: This chapter reviews space-time processing methods for CDMA mobile radio applications with emphasis on blind signal detection. We begin with a motivation for the use of blind space-time processing in CDMA. Next, we develop channel and signal models useful for blind processing. We follow this by considering first space-time single user receivers (ST-RAKE) and then review some basic theory of blind ST-RAKE algorithms. The important problem of multi-user detection (MUD) is considered next which leads to a novel technique allowing the estimation of the minimum-mean-square error linear MUD.

11.1 INTRODUCTION

Direct sequence code-division-multiple-access (DS-CDMA) is a spread spectrum multiple access communication method that is expected to gain a significant share of the cellular market. CDMA is currently deployed in certain areas across the world under the IS-95 standard and has been selected as the main multiple access technology for third generation wireless systems. DS-CDMA has several attractive properties for personal wireless communications: its efficient use of bandwidth, its resistance to interferences and its flexibility for multimedia links.

---

[*] This research was supported in part by ARPA. Support is provided by a DSO fellowship for B. C. Ng.

In CDMA the users operate in the same frequency and time channel whereas in TDMA and FDMA the users are separated in time and frequency respectively. In DS-CDMA each user has a *unique* spreading code. This code operates at a chip rate P times greater than the information data rate. The DS-CDMA link therefore needs a large bandwidth channel which can be shared by multiple users. The spreading code can be viewed as a complex symbol waveform with a large time-bandwidth product (approximately P), whereas in TDMA the time-bandwidth product of the symbol waveform is small (approximately 1). A good general reference on DS-CDMA can be found in [1]. Other books on DS-CDMA include [2], [3] and [4]. The user codes can be designed to be orthogonal or quasi-orthogonal. With orthogonal codes, and in the absence of channel delay spread, the users do not interfere with each other and signal detection is noise limited. Single-user detection schemes (RAKE receiver) can then be used successfully. If non-orthogonal signaling is used or caused by the propagation environment (multipath, asynchronism amongst the users, etc..) the users interfere with each other and detection often becomes interference limited. In this case multi-user detection is favored.

As in other multiple access systems, the use of antenna arrays in CDMA base stations can improve system capacity, quality and coverage, making space-time signal processing for CDMA a promising technology. Space-time processing can be incorporated in single user receivers (space-time (ST) RAKE) and multi-user receivers.

Space-time CDMA receivers use knowledge of the received signature waveform (channel) across time and space for the user(s) of interest. In the coherent RAKE receiver, the channel information (which in its broadest description, includes the effect of the spreading sequence, multipath propagation and user asynchronism) is normally acquired through the use of training (pilot) symbols known both by the transmitter and receiver. Blind estimation methods have been recently proposed which circumvent the use of a periodic training and allow the receiver to recover the channel information from only the observation of the received signals and knowledge of the spreading sequence for the desired user. These methods can increase the spectral efficiency, facilitate system management and allow self-startup operation of the network when necessary. The incorporation of multiple antennas is especially beneficial in blind CDMA receivers because of the additional spatial structure conferred to the signal and the increased dimensionality of the observations. As a result blind space-time CDMA currently motivates much research, as illustrated by the increasing number of publications devoted to it.

The purpose of this chapter is not to provide an extensive literature review of blind CDMA receiver design, but rather to develop certain aspects of the topic. We first review the concept of blind processing in the context of single-user receivers. More specifically, blind processing can be incorporated to estimate the spatial (or space-time) beamformer used in the ST RAKE, as was first proposed in [5]. Next we investigate the use of blind algorithms in multiuser receivers. A novel algorithm to estimate a linear minimum mean square error (MMSE) MUD is presented. The interesting features of this latter algorithm include (i) its simplicity, (ii) its ability to bypass the channel estimation step, and (iii) its generality, as it requires only knowledge of the spreading sequence for the user of interest in order to develop the multiuser receiver. We begin with a description of the space-time signal model in CDMA networks.

11.2 SPACE-TIME SIGNAL MODELS

We assume that $M > 1$ antenna elements are used only at the base station and that the mobile has a single omni antenna. The mobile transmits a possibly channel coded and modulated signal which does not incorporate any spatial (or indeed any special temporal) processing. The received continuous-time $M \times 1$ signal vector in a multiple antenna CDMA system has the following form

$$\mathbf{x}(t) = \sum_{q=1}^{Q} \mathbf{x}^q(t) + \mathbf{n}(t) \qquad (11.1)$$

where Q is the number of users sharing the same channel, $\mathbf{n}(t)$ is the additive noise vector with element-wise variance σ^2, $\mathbf{x}^q(t)$ is the vector signal contribution from a single user and is given by

$$\mathbf{x}^q(t) = \sum_{n=-\infty}^{\infty} d^q(n) \mathbf{g}^q(t - nT_c) \qquad (11.2)$$

where $\{d^q\}$ is the modulated chip stream, T_c is the chip period, and $\mathbf{g}^q(t)$ is the vector multipath channel

$$\mathbf{g}^q(t) = \sum_{l=0}^{D-1} \alpha_l^q \mathbf{a}_l^q a(t - \tau_l^q) \qquad (11.3)$$

where $a(t)$ is the chip pulse, D is the maximum number of paths for any of the users, each with complex amplitude and delay α_l^q and τ_l^q, and $\mathbf{a}_l^q$ corresponds to the array response vector for the l-th path.

We consider here a fairly general situation in which arbitrary delays may result in large inter-chip-interference (ICI) and even possibly significant inter-symbol-interference (ISI). The lack of synchronism between the cell users is also accounted for in the channel model. The modulated chip stream is given by

$$d^q(n) = s^q(k) c^q(k, n - kP) \text{ with } k = \lfloor \frac{n}{P} \rfloor \qquad (11.4)$$

where $s^q(k) \in \{-1; +1\}$ is the underlying bit[1] (or "symbol") stream for the q-th user, P is the spreading factor (length of symbol in chip durations) and $(c^q(k,0), ..., c^q(k, P -$

---

[1] To differentiate from the binary code sequence, $s^q(k)$ is referred to as "symbol" in this chapter.

1) is the spreading code used for the k-th symbol. The code may or may not be repeated across the symbols. Note that this fact will determine our ability to develop convenient multi-user receivers.

In the above model (11.3) we have assumed that the inverse signal bandwidth is large compared to the travel time across the array. Therefore the complex envelopes of the signals from a given path received by different antennas are identical except for phase and amplitude differences that depend on the path angle-of-arrival, array geometry and the element pattern. Thus, the use of multiple antennas at the receiver has merely converted a scalar channel $g^q(t)$ to a vector channel $\mathbf{g}^q(t)$.

Although blind CDMA can be defined in various ways, a common concern of it is the problem of estimating the symbols $s^q(k)$ transmitted by a single or multiple users given a minimal amount of signal/channel information available at the receiver. Existing approaches to blind CDMA differ mostly on

1. the degree of a priori signal/channel information and the generality of the channel model: τ_l^q known or unknown, synchronized or unsynchronized users, delay spread smaller or larger than a chipt period, a symbol period;

2. the receiver's structure: single-user or multiuser, linear or non-linear;

3. the estimation method of the receiver: use of finite alphabet information, second-order statistics or higher-order statistics.

Throughout this paper, the channel responses $\{\mathbf{g}^q(t)\}$ are assumed to be unknown and completely arbitrary (with the exception of the pulse shape $a(t)$ in the blind RAKE receiver) and only knowledge of the spreading code for the user of interest is available. We focus on linear (single user or multi-user) receivers. Finally, we restrict our study to blind methods exploiting solely the second order statistics of $\mathbf{x}(t)$.

MIMO equivalent model. Another interesting way to represent the CDMA multiple access model in (11.1) is to consider the stream of user's symbols $\{s^q(k)\}_{q=1..Q}$ as the signals driving a $Q \times M$ multiple inputs multiple outputs (MIMO) system, with vector output $\mathbf{x}(t)$ given by

$$\mathbf{x}(t) = \sum_{q=1}^{Q} \sum_{k=-\infty}^{\infty} s^q(k) \mathbf{h}^q(k, t - kT_s) + \mathbf{n}(t) \quad (11.5)$$

where $T_s = PT_c$ denotes the symbol period and $\mathbf{h}^q(k,t) = [h_1^q(k,t), ..., h_m^q(k,t)]^T$ is the $M \times 1$ impulse response of the channel from user q to the receive antenna array. Note that each symbol experiences a different channel in the case of aperiodic codes, hence the symbol index k in $\mathbf{h}^q(k,t)$. Using (11.2)-(11.4), the time-varying MIMO channel response is given by

$$\mathbf{h}^q(k,t) = \sum_{j=0}^{P-1} c^q(k,j) \mathbf{g}^q(t - jT_c) \quad q=1,...,Q \quad (11.6)$$

Time invariant MIMO. In the particular case of periodic codes, the MIMO channel becomes *time invariant*

$$\mathbf{x}(t) = \sum_{q=1}^{Q} \sum_{k=-\infty}^{\infty} s^q(k) \mathbf{h}^q(t - kT_s) + \mathbf{n}(t) \qquad (11.7)$$

$$\mathbf{h}^q(t) = \sum_{j=0}^{P-1} c^q(j) \mathbf{g}^q(t - jT_c) \ \forall k \ \ q = 1, ..., Q \qquad (11.8)$$

and constitutes a useful model to develop blind multi-user receivers, as we will see in section 11.4.

11.3 SINGLE-USER RECEIVERS

11.3.1 Space-time RAKE

When the channel exhibits significant multipath with delay spread larger than one chip period, a receiver can be designed to match this channel. Note that in DS-CDMA multipath has both a positive and a negative effect. On one hand, the independently fading paths can be a valuable source of diversity. On the other hand, the multipath introduces inter-path interference similar to intersymbol interference in TDMA, and since exact orthogonalilty among the codes cannot be maintained at all lags, it also introduces multiple access interference (MAI).

A popular non-blind single-user receiver in the presence of multipath is the coherent RAKE combiner first proposed by Price and Green in 1958 [6]. The RAKE receiver uses multiple correlators, one for each path, and the outputs of the correlators (called fingers) are then combined into a single output to maximize the signal to noise ratio using a matched filter approach.

When the receiver is equipped with multiple antennas, additional spatial matched filtering can be performed, giving rise to the space-time RAKE, first proposed in [5]. We define the transmitted signature waveform of the q-th user by

$$p^q(k,t) = \sum_{j=0}^{P-1} c^q(k,j) a^q(t - jT_c)$$

Therefore, the output of the coherent ST-RAKE can be expressed as follows[2]:

$$\hat{s}^q(k) = \text{sign}\{ \sum_l \alpha_l^{q*} \mathbf{a}_l^{q*} \int_{kT_s+\tau_l}^{(k+1)T_s+\tau_l} \mathbf{x}(t) p^q(k, t - kT_s - \tau_l) dt \} \qquad (11.9)$$

---

[2] In (11.9) $\mathbf{a}_l^{q*}$ plays the role of a spatial beamformer extracting the l-th path. More general beamforming solutions can also be used, e.g. max SINR beamformer.

11.3.2 Blind ST-RAKE - Principal Component Method

The implementation of the ST-RAKE requires knowledge of the delay, amplitude, and array response for each incoming path with its absolute phase information. Let us assume first that the path delays are known for the desired user. Recently, a principal component analysis technique was proposed to acquire blind estimates of $\alpha_l^q \mathbf{a}_l^q$ up to the path phase information[3], that can be exploited to implement a non coherent blind ST RAKE [7, 8].

The technique is based on the second-order statistics of both the pre-correlation data $\mathbf{x}(t)$ and post-correlation data $\mathbf{y}_l^q(k)$ where $\mathbf{y}_l^q(k)$ is defined for the l-th path from q-th user by

$$\mathbf{y}_l^q(k) = \int_{kT_s+\tau_l}^{(k+1)T_s+\tau_l} \mathbf{x}(t) p^q(k, t - kT_s - \tau_l) dt \qquad (11.10)$$

Let $\mathbf{R}_x = E(\mathbf{x}(t)\mathbf{x}(t)^*)$ and $\mathbf{R}_l^q = E(\mathbf{y}_l^q(k)\mathbf{y}_l^q(k)^*)$. These correlation matrices can be expressed as follows:

$$\begin{aligned} \mathbf{R}_x &= \mathcal{E}_l^q |\alpha_l^q|^2 \mathbf{a}_l^q \mathbf{a}_l^{q*} + \mathbf{R}_{u,l}^q \\ \mathbf{R}_l^q &= P\mathcal{E}_l^q |\alpha_l^q|^2 \mathbf{a}_l^q \mathbf{a}_l^{q*} + \mathbf{R}_{u,l}^q \end{aligned}$$

where $\mathcal{E}_l^q$ denotes the signal power in the desired path and $\mathbf{R}_{u,l}^q$ denotes the correlation matrix of all other paths and users' signals, developed in [8]. The trick here is that the covariance of the interfering paths $\mathbf{R}_{u,l}^q$ remains the same before and after despreading, whereas the processing gain P appears as a factor in the term related to the desired path after despreading. The path response $\alpha_l^q \mathbf{a}_l^q$ can be found up to a phase component as the dominant eigenvector of $\mathbf{R}_l^q - \mathbf{R}_x$.

A problem with this approach lies in the lost phase information for each path. More problematic is perhaps the estimation of the path delays τ_l^q. In fact these can be found by solving the problem above for a range of delays τ and selecting those offering the largest dominant eigenvalues of $\mathbf{R}_l^q - \mathbf{R}_x$. However this approach may be computationally intensive and is not necessarily robust given the possibly large number of both paths and users. A solution to this problem is to construct a receiver that does not use any constraint on the channel model in terms of the path parameters. This is shown below.

11.3.3 Space-time Equalizer

An equivalent way to view the ST-RAKE shown in 11.3.1 is as a receiver forming a beam in the direction of each desired path and combining the beamformed signals coherently to generate an estimate of the chip sequence $d^q(n)$. The chip sequence is then despreaded with the user's code to produce symbols estimates. However, a more general linear single-user receiver consists of using an arbitrary space-time *chip equalizer* instead of a filter matched to the channel. This equalizer can generate a better

---
[3] Phase information can be acquired from a one bit training.

estimate of the chip sequence, before despreading. As another advantage, a ST equalizer does not pose any constraint on the multipath structure. Let $(f_m^q(0), .., f_m^q(N-1))$ be the chip rate N-long impulse response of the equalizer on the m-th antenna. The equalizer's output and subsequent symbol estimates are given by:

$$\hat{d}^q(n) = \sum_{m=1}^{M} \sum_{i=0}^{N-1} f_m^q(i)^* x_m((n-i)T_c)$$

$$\hat{s}^q(k) = \sum_{j=0}^{P-1} c^q(k,j) \hat{d}^q(kP+j)$$

The MMSE chip equalizer, defined simply by[4] $\{f_m^q\} = \mathrm{argmin} E|d^q(n) - \hat{d}^q|^2$ can be obtained from the well known Wiener equation

$$\mathbf{R}_{N,x} \mathbf{f} = \mathbf{r}_N$$

where $\mathbf{R}_{N,x}$ is the ST correlation matrix of $\{\mathbf{x}(nT_c), ..., \mathbf{x}((n-N+1)T_c)\}$, $\mathbf{f}$ is the unknown ST equalizer vector and $\mathbf{r}_M$ is the unknown crosscorrelation (channel) vector. In [9] it is shown that the principal component analysis of 11.3.2 can be generalized simply to allow the blind estimation of $\mathbf{r}_N$, up to a *single* phase ambiguity. Further improvements on this technique can be found in [10]. The reader is referred to these papers for more details due to lack of space.

11.4 MULTI-USER RECEIVERS

In practical wireless environments the orthogonality of the signaling scheme cannot be maintained due to multipath propagation and user asynchronism. The lack of orthogonality between the different users' received signature waveforms causes crosscorrelation terms to arise at the output of the RAKE receiver. These terms cause significant MAI when the interfering users are received with higher power than the user of interest (near-far problem). In such cases, single user receivers are far from optimal and multi-user detectors (MUD) can be used which offer superior performance. Multi-user detectors are designed to eliminate MAI completely (in the absence of additive noise) despite the lack of orthogonality between the user's signature waveforms. MUD can be used to detect the signal of one single user of reference or possibly the signals of several cell sharing users (joint detection).

The optimum joint multiuser receiver in an AWGN channel is shown to consist of a bank of matched filters followed by a Viterbi algorithm [11]. The computational complexity of this receiver grows exponentially with the number of users. The informational complexity is also high, as the optimum MUD requires knowledge of all the users' channels. This makes the development of blind optimum MUD receivers a difficult task, although approximate and iterative maximum likelihood techniques are possible [12].

---

[4] A delay can also be introduced in the equalizer

11.4.1 Linear Multiuser Receivers

As an alternative to optimum receivers, linear multiuser (LMU) receivers for synchronous and asynchronous DS-CDMA have been proposed in [13] and [14] that exhibit the same degree of near-far resistance as the optimum multiuser receiver. They also have error rate performances comparable to the optimum multiuser receiver. When a repeated code scheme is used, LMU receivers can be designed to cancel ICI, ISI and MAI completely as we show below.

Notations and Assumptions. The spatial vector signal (11.5) is first sampled at the chip rate[5]. A total of MP samples are collected per symbol period and stacked in a signal vector[6] defined by

$$\mathbf{x}(k) = [\mathbf{x}^T(kT_s), \mathbf{x}^T(kT_s + T_c), \ldots, \mathbf{x}^T(kT_s + (P-1)T_c)]^T \quad (11.11)$$

Since periodic codes are used, the time invariant MIMO model developed in (11.7) can be exploited. We define $(L+1)T_s$ to be the unrestricted maximum length of the channels for all users. Hence the degree of the vector channels is at most L. In fact, $L = 1$ is a typical value when considering practical CDMA data rates. A severe asynchronism among the users can result in $L = 2$.

In view of (11.7) we can write

$$\mathbf{x}(k) = \sum_{q=1}^{Q} \mathbf{H}^q \mathbf{s}^q(k) + \mathbf{n}(k) \quad (11.12)$$

where the $MP \times (L+1)$-dimensional matrix $\mathbf{H}^q$ is the channel matrix for user q and its $(pM + m)$-th row (with $1 \le m \le M$ and $0 \le p \le P - 1$) is given by

$$\mathbf{H}^q(pM + m, :) = (h_m^q(pT_c), h_m^q(pT_c + T_s), \ldots, h_m^q(pT_c + LT_s)) \quad (11.13)$$

In (11.12), $\mathbf{s}^q(k) \triangleq (s^q(k), s^q(k-1), \ldots, s^q(k-L))^T$ is the current symbol vector. Let N denote the length of the LMU filter. The following MNP-dimensional output vector is introduced:

$$\mathbf{X}(k) = (\mathbf{x}^T(k), \mathbf{x}^T(k-1), \ldots, \mathbf{x}^T(k-N+1))^T \quad (11.14)$$

The vector $\mathbf{X}(k)$ can, similarly to $\mathbf{x}(k)$, be written in matrix form as

$$\mathbf{X}(k) = \sum_{q=1}^{Q} \begin{pmatrix} \boxed{\mathbf{H}^q} & & 0 \\ & \boxed{\mathbf{H}^q} & \\ & & \ddots \\ 0 & & \boxed{\mathbf{H}^q} \end{pmatrix} \begin{pmatrix} s^q(k) \\ s^q(k\text{-}1) \\ \vdots \end{pmatrix} + \begin{pmatrix} \mathbf{n}(k) \\ \mathbf{n}(k\text{-}1) \\ \vdots \end{pmatrix} \quad (11.15)$$

$$\triangleq \sum_{q=1}^{Q} \mathcal{H}^q \mathbf{S}^q(k) + \mathbf{N}(k) \quad (11.16)$$

---
[5] Oversampling can be also accommodated in this model
[6] A noise vector $\mathbf{n}(k)$ is defined similarly.

The global multiuser channel matrix $\mathcal{H} \triangleq (\mathcal{H}^1, \ldots, \mathcal{H}^Q)$ has dimension $MNP \times Q(L+N)$. At this point it is important to mention that we will need

$$MP \geq Q$$

so that $\mathcal{H}$ can be made tall (i.e. with more rows than columns) for a sufficiently large filter length N. Clearly, the use of multiple antennas ($M > 1$) allows us to support more users for a given spreading factor P. The tallness property of $\mathcal{H}$ ensures the existence of exact *zero-forcing* LMU receivers. We further assume that the conditions under which $\mathcal{H}$ has full column rank are satisfied [15]. These conditions are mild and are satisfied with probability one. In practice, the *condition number* of the matrix $\mathcal{H}$ will determine the achievable performance of the blind estimation algorithm.

The problem of linear multi-user detection in the noise free case can be stated ideally as follows: given the channel matrices $\mathcal{H}^u$ for all users $u = 1, ., Q$, find a vector $\mathbf{w}$ with MNP entries that satisfies:

$$\mathbf{w}^* \mathcal{H}^q = (0, \ldots, 0, 1, 0, \ldots, 0), \quad (11.17)$$
$$\mathbf{w}^* \mathcal{H}^u = (0, \ldots, 0), \quad u \neq q \quad (11.18)$$

where q denotes the index of the desired user and where "*" denotes complex conjugate transpose. Note that while (11.17) corresponds to the elimination of ICI and ISI, (11.18) guarantees the suppression of MAI. The position of the "1" element in (11.17) is a free parameter that indicates the reconstruction delay of the receiver, to be taken between 0 and $L + N - 1$.

11.4.2 Blind Receiver Estimation

The estimation of a blind receiver as defined above can be closely related to the research over the last few years in blind MIMO estimation, see for instance [15, 16] and the references therein. In fact, the model shown in (11.15) is not CDMA dependent. For instance it is well known that the spatial separation of co-channel TDMA users leads to a similar model. However a significant advantage of CDMA in this context over the arbitrary MIMO problem is the availability of a rich temporal (code) structure embedded into the channel which can be exploited in addition to the spatial structure to separate the users.

To a great extent the existing literature on blind CDMA has focused on the problem of estimating the users' channels as a first step to obtaining the CDMA receiver. In the so-called subspace-based techniques, the signal subspace span($\mathcal{H}$) is first obtained from the dominant eigenvectors of $\mathbf{R} = E(\mathbf{X}(k)\mathbf{X}^*(k))$. Next, the coefficients of $\mathcal{H}^q$, $q = 1, .., Q$ are adjusted to match the signal subspace, by invoking a variant of the subspace identification theorem [17]. The knowledge of the spreading code is used to differentiate amomng the users. Instances of this approach can be found in [18, 19, 20, 21].

The problem of directly estimating a multi-user detector for CDMA signals in unknown channels has been also addressed. Often a MMSE receiver is derived in the restrictive case of no delay spread [22, 23]. Motivated by the work in [22], a minimum output energy technique is introduced in [24] that results in a blind linear receiver that has a performance close to the MMSE receiver at high SNR in the presence of delay spread.

However, it is possible to estimate blindly the MMSE receiver itself in the most general case as we show below. The technique presented here is based on exploitation of a set of linear conditions that take advantage of: (1) the subspace structure associated with the code of the desired user, and (2) the estimated signal subspace of the covariance matrix for the observed signals. The main contribution shows that these conditions, when brought together, fully characterize the MMSE receiver. This technique was first presented in [25]. We now review this algorithm which is based on a two-fold projection.

11.5 MMSE RECEIVER ESTIMATION

In the presence of noise, the best (in the MMSE sense) linear receiver with delay δ, for user q, is given by:

$$\mathbf{w} = \arg\min E|\mathbf{w}^*\mathbf{X}(k) - s^q(k-\delta)|^2 \qquad (11.19)$$

The vector $\mathbf{w}$ found from (11.19) satisfies the classical Wiener equation

$$\mathbf{R}\mathbf{w} = \mathbf{r}^q \qquad (11.20)$$

where $\mathbf{R} = E(\mathbf{X}(k)\mathbf{X}^*(k))$ denotes the ST received covariance matrix and $\mathbf{r}^q = E(\mathbf{X}(k)s^{q*}(k-\delta))$ denotes the cross-correlation vector. The goal here is to identify $\mathbf{w}$ blindly without knowledge of $\mathbf{r}^q$.

11.5.1 Code Subspace Properties

Assuming i.i.d. symbols, it can be seen easily from (11.15) that $\mathbf{r}^q$ is given by

$$\mathbf{r}^q = \begin{pmatrix} \boxed{\mathbf{H}^q} & & 0 \\ & \boxed{\mathbf{H}^q} & \\ & \ddots & \\ 0 & & \boxed{\mathbf{H}^q} \end{pmatrix} \begin{pmatrix} 0 \\ \vdots \\ 1 \\ \vdots \\ 0 \end{pmatrix} \qquad (11.21)$$

Thus $\mathbf{r}^q$ coincides with the $(\delta+1)$th column of $\mathcal{H}^q$. If the detector delay δ is chosen appropriately (namely $L \leq \delta \leq N-1$ assuming $L \neq 1$), the vector $\mathbf{r}^q$ contains all the $(L+1)MP$ channel coefficients for user q, as well as $(N-L-1)MP$ zeros[7]. The point here is that for each antenna m there exists a simple selection-permutation matrix $\mathbf{T}_m$ that selects the $(L+1)P$ channel coefficients in $\mathbf{r}^q$ associated with this antenna,

---

[7]This is due to the block Toeplitz structure of the matrix $\mathcal{H}^q$.

together with $(N - L - 1)P$ zeros, and puts the selected entries in a chronologically ordered vector so that

$$\mathbf{T}_m \mathbf{r}^q = (h_m^q(0), \ldots, h_m^q((LP+P-1)T_c), 0, \ldots, 0)^T \tag{11.22}$$

For notational convenience, in what follows we will assume $\delta = L$, but the final result holds true for any admissible value of the delay. Using the relation shown in (11.8) for the m-th antenna, we can relate (11.22) to the known spreading sequence $\{c^q(n)\}$.

$$\mathbf{T}_m \mathbf{r}^q = \begin{pmatrix} c^q(0) & & 0 \\ \vdots & \ddots & \\ c^q(P-1) & & c^q(0) \\ & \ddots & \vdots \\ 0 & & c^q(P-1) \\ \vdots & & \\ 0 & & 0 \end{pmatrix} \begin{pmatrix} g_m^q(0) \\ g_m^q(T_c) \\ \vdots \\ g_m^q(LPT_c) \end{pmatrix}$$

$$\triangleq \mathbf{C}^q \mathbf{g}_m^q \tag{11.23}$$

The code matrix $\mathbf{C}^q$ of dimension $PN \times LP + 1$ can be made tall by selecting $N > L$ and is known to the receiver. In contrast, $\mathbf{g}_m^q$ is unknown. Also, $\mathbf{g}_m^q$ contains a certain number of zero entries. Note however that the knowledge of this number is not necessary for the proposed method. Therefore the knowledge of the *actual channel length* in chip durations is not required here.

Introduce $\mathbf{U}^q$, an orthonormal basis for the orthogonal complement of $\mathbf{C}^q$. Hence, $\mathbf{U}^q$ is of size $PN \times (P(N-L) - 1)$. Then, from (11.20) and (11.23), we obtain

$$\mathbf{U}^{q*} \mathbf{T}_m \mathbf{R} \mathbf{w} = \underbrace{\mathbf{U}^{q*} \mathbf{C}^q}_{0} \mathbf{g}_m^q = 0 \quad m = 1,..,M \tag{11.24}$$

Of course, (11.24) is valid for $m = 1, \ldots, M$. Thus (11.24) gives us a first set of $M(P(N-L) - 1)$ equations.

11.5.2 Signal Subspace Properties

Let us assume that the additive noise is white[8] with variance σ^2. In view of (11.16), the correlation matrix of the received signal is then given by

$$\mathbf{R} = \mathcal{H} \mathcal{H}^* + \sigma^2 \mathbf{I} \tag{11.25}$$

where $\mathbf{I}$ is the identity matrix of size MNP. The signal/noise subspace decomposition of $\mathbf{R}$ goes as follows:

$$\mathbf{R} = \mathbf{E}_s \Sigma_s \mathbf{E}_s^* + \sigma^2 \mathbf{E}_n \mathbf{E}_n^* \tag{11.26}$$

---

[8]This whiteness condition can be relaxed provided the noise covariance is known.

where the columns of $\mathbf{E}_s$ (resp. $\mathbf{E}_n$) are given by the $Q(N+L)$ dominant eigenvectors (resp. the $MNP - Q(N+L)$ least dominant eigenvectors) of $\mathbf{R}$.

A very useful result is that the desired MMSE detector always lies in the signal subspace defined by $\text{span}(\mathcal{H}) = \text{span}(\mathbf{E}_s)$, as shown below. From (11.20) and (11.25) we have

$$(\mathcal{H}\mathcal{H}^* + \sigma^2 \mathbf{I})\mathbf{w} = \mathbf{r}^q$$

which we can rewrite as

$$\mathbf{w} = (\mathbf{r}^q - \mathcal{H}\mathcal{H}^*\mathbf{w})/\sigma^2 \tag{11.27}$$

In view of (11.21), $\mathbf{r}^q$ is by construction in $\text{span}(\mathcal{H}^q)$, hence in $\text{span}(\mathcal{H})$. From (11.27) $\mathbf{w}$ also belongs to $\text{span}(\mathcal{H})$ and it follows that

$$\mathbf{E}_n^* \mathbf{w} = 0 \tag{11.28}$$

Equation (11.28) gives us an additional $MNP - Q(N+L)$ equations toward the determination of $\mathbf{w}$.

11.5.3 Algorithm

Based on the projection equations in (11.24) and (11.28), $\mathbf{w}$ can be written as the solution to

$$\mathcal{A}\mathbf{w} = \begin{pmatrix} \mathcal{A}_1 \\ \mathcal{A}_2 \end{pmatrix} \mathbf{w} = 0 \tag{11.29}$$

where $\mathcal{A}_2 = \mathbf{E}_n^*$ and $\mathcal{A}_1 = \mathcal{U}^{q*}\mathbf{TR}$, with $\mathbf{T} = (\mathbf{T}_1^*, \cdots, \mathbf{T}_M^*)^*$ and

$$\mathcal{U}^{q*} = \begin{pmatrix} \mathbf{U}^q & 0 & & 0 \\ 0 & \mathbf{U}^q & & \\ & & \ddots & \\ 0 & & & \mathbf{U}^q \end{pmatrix}$$

In order to determine $\mathbf{w}$ in a noisy situation, the projection equations (11.29) suggest optimizing the following simple quadratic cost function

$$J_{MMSE}(\mathbf{w}) = \mathbf{w}^* \{\mathcal{A}_1^* \mathcal{A}_1 + \alpha \mathcal{A}_2^* \mathcal{A}_2\} \mathbf{w} \tag{11.30}$$

where α is a tunable weight. Some additional constraint must be added in order to avoid trivial solutions. In the particular case of a unit-norm constraint, $\mathbf{w}$ is found as the minimum eigenvector of $\mathcal{A}_1^* \mathcal{A}_1 + \alpha \mathcal{A}_2^* \mathcal{A}_2$. Note that while $\mathbf{R}$ is estimated using sample averaging, $\mathbf{T}$ and $\mathcal{U}^q$ can be pre-computed.

Analysis. We now show that (11.29) completely determines the MMSE equalizer. Let us first introduce some useful notation. Let $\mathcal{C}^q$ be the tall modifed code matrix of size $MNP \times M(LP+1)$ defined as

$$\mathcal{C}^q = \begin{pmatrix} \mathbf{C}^q & 0 & & 0 \\ 0 & \mathbf{C}^q & & \\ & & \ddots & \\ 0 & & & \mathbf{C}^q \end{pmatrix}$$

Let $\bar{\mathcal{H}}$ be defined as

$$\bar{\mathcal{H}} = \mathbf{T}\tilde{\mathcal{H}}$$

where $\tilde{\mathcal{H}}$ denotes the original channel matrix $\mathcal{H}$ from which the $(\delta+1)$th column of $\mathcal{H}^q$ has been removed. The rows of $\tilde{\mathcal{H}}$ are then organized antenna-wise using $\mathbf{T}$ to form $\bar{\mathcal{H}}$. The main result can be summarized as follows:

Proposition 1:
Assuming that (i) $Q(L+N) \leq M(P(N-L)-1)$, and (ii) the matrix $(\bar{\mathcal{H}}, \mathcal{C}^q)$ has full column rank, then the matrix $\mathcal{A}$ has rank $MNP - 1$; hence the only vector in null($\mathcal{A}$) determines the MMSE equalizer completely.

Proof:
Using (11.26), $\mathcal{A}_1$ admits the following orthogonal decomposition:

$$\mathcal{A}_1 = \mathcal{A}_{11} + \mathcal{A}_{12}$$

where

$$\mathcal{A}_{11} = \mathcal{U}^{q*}\mathbf{T}\mathbf{E}_s\Sigma_s\mathbf{E}_s^*$$
$$\mathcal{A}_{12} = \sigma^2 \mathcal{U}^{q*}\mathbf{T}\mathbf{E}_n\mathbf{E}_n^*$$

Note that the row span of $\mathcal{A}_{12}$ is included in that of $\mathcal{A}_2$ and therefore that the rank of $\mathcal{A}$ is also given, due to the orthogonality property between $\mathcal{A}_{11}$ and $\mathcal{A}_2$, by:

$$\mathrm{rank}\begin{pmatrix} \mathcal{A}_{11} \\ \mathcal{A}_2 \end{pmatrix} = \mathrm{rank}(\mathcal{A}_{11}) + \mathrm{rank}(\mathcal{A}_2)$$

The rank of $\mathcal{A}_2$ is clearly $MNP - Q(L+N)$. In order to obtain the total rank of $\mathcal{A}$, we shall show that $\mathrm{rank}(\mathcal{A}_{11}) = Q(L+N) - 1$. It is clear that there exists a square invertible matrix, denoted $\mathbf{F}$, such that $\mathbf{E}_s = \mathcal{H}\mathbf{F}$. Thus $\mathcal{A}_{11}$ is given by:

$$\mathcal{A}_{11} = \mathcal{U}^{q*}\mathbf{T}\mathcal{H}\mathbf{F}\Sigma_s\mathbf{E}_s^*$$

Hence $\mathrm{rank}(\mathcal{A}_{11}) = \mathrm{rank}(\mathcal{U}^{q*}\mathbf{T}\mathcal{H})$. But, by construction, $\mathbf{T}\mathcal{H} = [\bar{\mathcal{H}}, \mathrm{Tr}^q]\mathbf{J}$ where $\mathbf{J}$ is a simple permutation matrix that shifts the last column to the $(\delta+1)$th position. Thus $\mathcal{U}^{q*}\mathbf{T}\mathcal{H} = [\mathcal{U}^{q*}\bar{\mathcal{H}}, 0]$ and it follows that $\mathrm{rank}(\mathcal{A}_{11}) = \mathrm{rank}(\mathcal{U}^{q*}\bar{\mathcal{H}})$. Making use of the fact that $\mathcal{U}^q$ is the orthogonal complement of $\mathcal{C}^q$, we easily check that $\mathcal{U}^{q*}\bar{\mathcal{H}}$

has full column rank $Q(L+N)-1$ under (i) and (ii). It thus follows that, the total rank of $\mathcal{A}$ is $Q(L+N) - 1 + MNP - Q(L+N) = MNP - 1$. ∎

Note that in the noise free case **w** given by (11.29) is not unique since there exist several admissible solutions to (11.20) as **R** has a non-empty null space. However, all solutions generate a desired MUD.

11.5.4 Numerical Examples

We validate the proposed algorithm by a number of numerical experiments. In the simulations we use arbitrary (random) channels and code sequences for all the users, illustrating that orthogonality among the codes is not necessary in MUD. The channel order is taken to be $L = 1$ corresponding to an ISI of two symbols and we chose the memory of the equalizer to be $N = 2 > L$. In all of the experiments we use the weight $\alpha = 1$. All the channels for the users are normalized so we define the transmitted and received signal-to-noise ratio as

$$\text{SNR} = 10 \log 1/\sigma^2 \qquad (11.31)$$

Since the goal is the blind estimation of the MMSE receiver, we can quantify the performance of our algorithm in terms of the mean-squared error (MSE) at the output of the receiver: $\text{MSE} = E|\hat{s}^q(k) - s^q(k)|^2$.

In Figure 11.1 we compare the proposed algorithm with the minimum output energy (MOE) method presented in [24]. MOE is formulated for the single antenna case only, so we first choose $M = 1$. As in [24], we assume a scenario with $Q = 10$ users and codes of length $P = 31$. We use 256 and 512 symbols in the estimation and assume perfect power control. We see that the proposed algorithm shows a significant performance improvement when compared to the MOE method, especially for low to medium-high SNRs. As the SNR increases, the performances of the two methods become similar, a result that is consistent with the fact that the MOE receiver approaches the MMSE solution for high SNR [26].

In Figure 11.2 we study the robustness of the proposed method to the lack of power control (the near-far problem). We assume a scenario where the power (in dB) decreases linearly with user index, and the power of the strongest and weakest of the 10 users differs by 10 dB (in this case the SNR is defined with respect to the strongest user). We plot the MSE as a function of user index, and choose $M = 4$ antennas and a spreading factor $P = 9$ to have a value of the loading ratio MP/Q comparable to the one in the previous example (512 symbols are used in the estimation). We see a small increase in the MSE as the power of the users decreases, indicating the desired near-far robustness of the proposed receiver.

In Figure 11.3 we verify that the performance of the proposed method approaches that of the asymptotic MMSE solution as the number of data points increases, hence supporting the claims of Proposition 1. We use the same multiple antenna scenario as above with perfect power control and plot the MSE of one of the users as a function of SNR for different numbers of symbols together with the asymptotic curve.

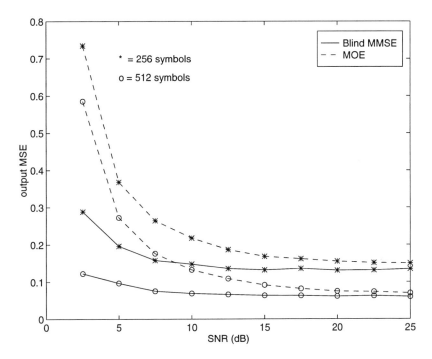

Figure 11.1 MSE as a function of SNR, for the proposed blind MMSE and the minimum output energy (MOE) methods ($Q = 10$, $P = 31$, $M = 1$). Perfect power control, 256 ("*") and 512 ('o') symbols are used in the estimation.

11.6 SUMMARY

In this chapter, we first introduced the basic concepts of blind estimation in single user space-time CDMA receivers. Space-time RAKE and its generalization were briefly reviewed. Linear multiuser detectors based on the MIMO model of CDMA were then developed. Finally, a technique that exploits the code subspace structure of CDMA signals is used to obtain a blind MMSE multiuser receiver in the context of very general channels. The new blind MMSE receiver is shown to converge to the optimal MMSE solution asymptotically in numerical experiments as well as analytically, regardless of SNR.

Further research to extend space-time blind multiuser algorithms to CDMA embedded with long codes is currently under investigation.

Acknowledgments

We gratefully acknowledge fruitful discussions with Dr. Joakim Sorelius (Uppsala University), especially regarding the proof of Proposition 1.

References

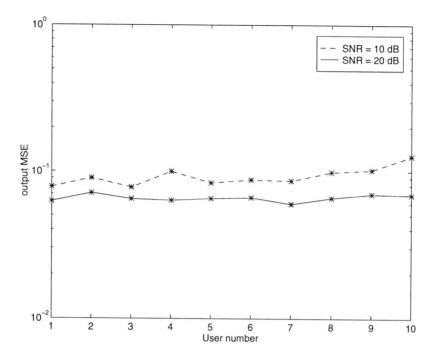

Figure 11.2 MSE as a function of user index for imperfect power control (user no. 1 has power 10dB above user no. 10). The SNR refers to the strongest user and 512 symbols are used in the estimation ($Q = 10$, $P = 9$, $M = 4$).

[1] M. K. Simon, J. K. Omura, R. A. Scholtz, and B. K. Levitt, *Spread Spectrum Communications Handbook*. New York: McGraw-Hill, revised ed., 1994.

[2] A. J. Viterbi, *CDMA: Principles of Spread Spectrum Communication*. Reading, MA: Addison-Wesley, 1995.

[3] R. Prasad, *CDMA for Wireless Personal Communications*. Boston, MA: Artech House, 1996.

[4] R. C. Dixon, *Spread Spectrum Systems*. New York: John Wiley, third ed., 1995.

[5] A. F. Naguib and A. Paulraj, "Performance of cdma cellular networks with base-station antenna arrays," in *International Zurich seminar on digital communications*, (Zurich, CH), pp. 87–100, March 1994.

[6] R. Price and P. E. Green, Jr., "A communication technique for multipath channels," *Proc. IRE*, vol. 46, pp. 555–570, Mar. 1958.

[7] A. Naguib, *Adaptive antennas for CDMA wireless networks*. PhD thesis, Stanford University, Stanford, CA, USA, October 1995.

[8] A. F. Naguib and A. Paulraj, "A base station antenna array receiver for cellular ds/cdma with m-ary modulation," in *28th Asilomar Conference on Signals, Syst. ans Comp.*, (Pacific Grove, CA), pp. 858–862, Nov. 1994.

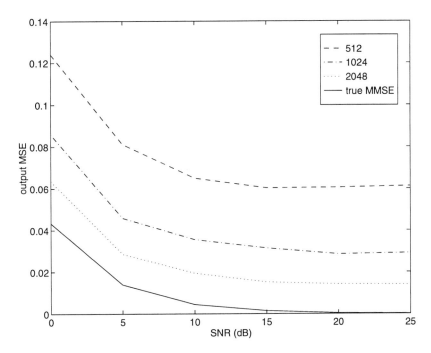

Figure 11.3   MSE as a function of SNR for different number of symbols (512, 1024 and 2048) used in the estimation, together with the MSE obtained using the true (asymptotic) covariance matrix. Perfect power control ($Q = 10$, $P = 9$, $M = 4$).

[9] H. Liu and M. D. Zoltowski, "Blind equalization in antenna array cdma systems," *IEEE Trans. on Signal Processing, Special Issue on Signal Processing for Advanced Communications*, vol. 45, pp. 161–172, Jan. 1997.

[10] J. Ramos and M. Zoltowski, "Blind space-time processor for sinr maximization in cdma cellular systems," in *First Signal Processing Workshop on Signal Processing Advances in Wireless Communications (SPAWC'97), Paris, France*, 1997.

[11] S. Verdú, "Minimum probability of error for asynchronous gaussian multiple-access channels," *IEEE Trans. on Information Theory*, vol. IT-32, pp. 85–96, Jan. 1986.

[12] J. R. Fonollosa, J. A. R. Fonollosa, Z. Zvonar, and J. Vidal, "Blind multiuser identification and detection in CDMA systems," in *Proc. ICASSP-95*, (Detroit, MI), pp. 1876–1879, 1995.

[13] R. Lupas and S. Verdu, "Linear multiuser detectors for synchronous code-division multiple-access channels," *IEEE Trans. on Info. Th.*, vol. IT-35, pp. 123–136, Jan. 1989.

[14] R. Lupas and S. Verdu, "Near-far resistance of multiuser detectors in asynchronous channels," *IEEE Trans. Commun.*, vol. 38, pp. 496–508, Apr. 1990.

[15] K. A. Meraim, P. Loubaton, and E. Moulines, "A subspace algorithm for certain blind identification problems," *IEEE Trans. on Information Theory*, vol. 43, pp. 499–511, March 1997.

[16] A. van der Veen, S. Talwar, and A. Paulraj, "Blind estimation of multiple digital signals transmitted over FIR channels," *IEEE Signal Processing Letters*, vol. (2)-5, pp. 99–102, May 1995.

[17] E. Moulines, P. Duhamel, J. F. Cardoso, and S. Mayrargue, "Subspace methods for the blind identification of multichannel fir filters," *IEEE Trans. on Signal Processing*, 1995.

[18] S. Bensley and B. Aazhang, "Subspace-based channel estimation for code division multiple access communication systems," *IEEE Trans. on Communications*, vol. 44, pp. 1009–1020, Aug. 1996.

[19] M. Torlak and G. Xu, "Blind multiuser channel estimation in asynchronous CDMA systems," *IEEE Trans. on Signal Processing*, vol. 45, pp. 137–147, January 1997.

[20] H. Liu and G. Xu, "A subspace method for signature waveform estimation in synchronous CDMA systems," *IEEE Trans. om Communications*, vol. 44, pp. 1346–1354, Oct. 1996.

[21] X. Wang and H. V. Poor, "Blind equalization and multiuser detection in dispersive cdma channels," *IEEE Transactions on Communications*, vol. 46, pp. 91–103, January 1998.

[22] M. Honig, U. Madhow, and S. Verdú, "Blind adaptive multiuser detection," *IEEE Trans. on Information Theory*, vol. 41, pp. 944–960, July 1995.

[23] X. Wang and H. V. Poor, "Blind multiuser detection: a subspace approach," *IEEE Transactions on Information Theory*, vol. 44, pp. 677–690, March 1998.

[24] M. Tsatsanis and Z. Xu, "On minimum output energy cdma receivers in the presence of multipath," in *Conf. on Info. Sciences and Systems (CISS'97)*, (Baltimore, MD), pp. 377–381, March 1997.

[25] D. Gesbert, J. Sorelius, and A. Paulraj, "Blind multi-user MMSE detection of CDMA signals," in *Proc. of Int. Conf. on Acoust. Speech and Signal Processing*, 1998.

[26] M. Tsatsanis and Z. Xu, "Performance analysis of minimum variance cdma receivers." Submitted to the IEEE Transactions on Signal Processing, 1997.

Index

A

ACTS, 16
Algorithm
   space-time RAKE, 285
   burst format and sync, 191
   constant modulus, 156
   Griffiths', 155
Algorithms
   convergence, 122
   stability, 122
Angle of arrival, 99
Antenna array processing, 59
Antenna arrays
   Algorithm comparisons, 71
   Algorithm convergence, 73
   Angle spread, 66
   Antenna array, 45
   Capacity improvement, 76
   Channel modelling, 63
   Downlink techniques, 76
   Element spacing, 66
   Motivations, 62
   Multi-user detection, 74
   Receiver algorithms, 68
   Receiver model, 66
   Uplink simulations, 71
Antenna diagram spreading, 13
Antenna diversity, 48
Antenna spacing
   optimum, 95
Asynchronous CDMA, 36

B

Baseband model, 24
Beam forming, 48
Bearing estimation, 68
BER performance
   multipath, 103
   number of antenna elements, 105
   power control error, 105
Bistatic, 10
Block fading, 218

C

Capacity, 20
Capacity average, 218
CDMA systems and MUI, 190
CDMA, 8, 16, 23
   Bandwidth, 60
   Direct-sequence, 60
   Modulation scheme, 61
   Multipath diversity, 61
   Multiple access interference, 62
   Power control, 62
   MC-CDMA, 185
Cellular spectral efficiency, 85
Channel estimation, 24, 69
Channel identification, 1, 9
Channel model
   Gaussian multiple access, 218
Channel models for antenna arrays, 59
Channel sounders, 14
Channel
   block fading, 217
   equivalent discrete-time, 221
   fading, 220
   frequency selective fading, 145
Channels
   fading
      LMMSE receivers in, 148

Circular array, 62
Code concatenation, 241
Code-division multiple-access, 23
Coding and spreading tradeoff, 225
Coding vs. spreading, 217
Coding and interleaving, 220
Combining
  equal gain (EGC), 186, 199
  maximal ratio (MRC), 186, 200
  minimum mean square error (MMSEC), 186, 200
  orthogonality restoring (ORC), 186, 199

D

Delay estimator
  conventional, 160
  minimum variance, 158
  subspace based, 159
Delay tracking
  performance, 171
Detector
  adaptive bootstrap multiuser, 117
  adaptive canceler, 130
  adaptive multistage, 128
  adaptive multiuser, 112
  decorrelating, 112
  decorrelator, 117
  decorrelator combiner canceler combiner, 134
  multi-shot matched filter, 115, 125
  multiuser decorrelating, 117
  one-shot matched filter, 113
  one-shot with singular PCC matrix, 126
  two stage multiuser, 112
Distance distribution
  mobile users, 92
Diversity, 199
Doppler effects, 65
Doppler spectrum, 34

E

Encoding, 52

F

Fading distribution, 65
Fading, 23
  block, 218
FDMA, 8
FEC
  forward error correction, 6
  turbo coding, 239
Fixed beam receiver, 68
FRAMES, 16, 166
  WCDMA, 166
Frequency diversity, 6

G

Global Navigation Satellite System (GLONASS), 15
Global Positioning System (GPS), 15

H

High Sensitivity Reception (HSR), 83, 87

I

Impulsive interference, 5
IMT-2000, 16, xxi
Information transmission systems, 6
Information transmission, 1
Interference structure, 49
Interference suppression, 70
International Mobile Telecommunications by the year 2000, xxi
Intersymbol interference (ISI), 19, 116, 125
Iterative detection techniques, 52

J

Joint detection, 17–18

L

Linear algebra, 23
Linear array, 62
Linear model, 23
LMMSE, 120
  postcombining, 145
  precombining, 145
Lognormal fading, 17
Low probability of interception, 5

INDEX 305

M

MAP decoder, 53
MC-CDMA synchronisation, 190
MC-CDMA system, 187
MC-CDMA
    cancellation schemes, 185
    detection startegies, 185
    detection strategies, 195
Mobile location
    angular, 97
    clustering, 103
    distribution, 97
Model
    DS-CDMA system, 147
    MIMO, 288
Monostatic, 10
Multi-user detection, 70
Multipath channel, 89
Multipath fading, 64
Multipath models, 40
Multipath profile, 65
Multipath propagation, 40
Multiple access interference, 17, 94, 111
Multiple antenna, 45
Multiple-antenna, 23
Multistatic, 10
Multiuser signal separation, 111
MUSIC, 69, 159
Mutual information, 224
Mutual information average, 218

N

Nyquist waveforms, 225

O

OFDM, 186
Outage probability
    vs. system capacity, 222

P

Parallel interference canceler, 128
Partial cross correlation, 114
Pathloss, 24, 64
Performance calculation

BER, 88
Power control, 24, 35, 112
Power line communications, 20
Processing gain, 28

Q

Quality of service, 19

R

Radar, 14
Raised cosine pulse, x, 29, 37
Rake combining, 26
Rake reception, 42
RAKE, 69
Range meter, 13
Rayleigh distributed envelope, 34, 40
Rayleigh fading, 24, 89
RCPC codes, 240
Re-use cluster, 86
Receiver
    Adaptive MOE, 155
    architecture, 260
    blind adaptive
        comparison, 168
    blind least squares(LS), 156
    blind space-time, 285
    communication, 260
    LMMSE postcombining, 146
    LMMSE precombining, 146, 150
    LMMSE, 145
    LMMSE-PIC hybrid, 163
    LMMSE-RAKE, 150
        adaptive, 152
    MMSE receiver estimation algorithm, 296
    MMSE receiver estimation, 294
    multi user
        blind receiver estimation, 293
        linear, 292
    multi-user, 291
    PIC, 162
        residual blind interference suppression, 162
    precombining LMMSE
        adaptive implementation, 152
        BER, 150
        delay acquisition in, 157
        delay acquisition performance, 170
        delay tracking in, 160
    RAKE, 146
    single user
        blind ST-RAKE, 290

space-time equalizer, 290
space-time RAKE, 289
software radio, 257
  critical functionalities, 262,
  critical functionalties, 264,
Root-Nyquist filtering, 26
RSC codes, 239, 243

HSR/CDMA, 107
  vs. outage probability, 222
System model
  discrete-time baseband uplink, 27
  CDMA uplink, 25
  multiple-antenna independent-multipath, 45
  multiuser CDMA, 113

S

Satellite navigation systems, 15
Sectorisation, 60
Shadowing effects, 24
Smart antennas, 85
Software radio, 20, 257
Space division multiple access (SDMA), 84, 88
Space-time RAKE, 285
Space-time receiver, 66
Space-time signal model, 287
Spatial Filtering for Interference Reduction (SFIR), 87
Spectral power density, 5
Spectral spreading, 4–5
Spectrum efficiency, 19
Spread spectrum applications
  antenna diagram spreading, 13
  CDMA for UMTS and IMT-2000, 16
  channel sounder, 14
  geodesic range meter, 13
  impulse compression radar, 14
  satellite navigation, 15
  spread spectrum, 12
  thermal flow meter, 12
Spread spectrum multiple access, 8
Spread spectrum, xxi
Spreading
  benefits, 4
SSMA, 8
Subtractive interference canceler, 128
Successive interference cancelers, 128
Synchronous CDMA, 27
System capacity

T

TD-CDMA, 16, 240
TDMA, 8
Temporal spreading, 3, 5
Thermal flow meter, 12
Time and frequency estimation, 1
Time and frequency estimation, 11
Time-dispersive, 23
Time-Division CDMA, xxi
Tracking
  error analysis, 161
Transmission power control, 44
Transmit diversity, 77
Turbo code, 239
  designing RCPTC, 246
  interleaver, 244
  iterative decoding, 244
  structure, 243

U

UMTS, xxi, 16, 239
Uniform linear array, 46, 64
UTRA, 239

W

W-CDMA, xxi, 16, 240